MEDICAL
INTELLIGENCE
UNIT

Pharmacogenomics in Admixed Populations

Guilherme Suarez-Kurtz, M.D., Ph.D.
Instituto Nacional de Câncer
Rio de Janeiro, RJ, Brazil

CRC Press
Taylor & Francis Group
Boca Raton London New York

CRC Press is an imprint of the
Taylor & Francis Group, an **informa** business

Pharmacogenomics in Admixed Populations

Medical Intelligence Unit

CRC Press
Taylor & Francis Group
6000 Broken Sound Parkway NW, Suite 300
Boca Raton, FL 33487-2742

First issued in hardback 2019

ISBN 13: 978-1-58706-311-4 (hbk)

Visit the Taylor & Francis Web site at
http://www.taylorandfrancis.com

and the CRC Press Web site at
http://www.crcpress.com

Library of Congress Cataloging-in-Publication Data

A C.I.P. Catalog record for this book is available from the Library of Congress.

About the Editor...

GUILHERME SUAREZ-KURTZ is Head of Pharmacology at the Brazilian National Cancer Institute—INCA—Coordinator of the Brazilian National Pharmacogenomics Network—REFARGEN—and a member of the Committee on Pharmacogenetics of the International Union of Pharmacology—IUPHAR. A pioneer of pharmacogenetic studies in the Brazilian population, his research explores the impact of genetic admixture on the conceptual development and the praxis of pharmacogenetics/-genomics. He is a Full Member of the Brazilian Academy of Sciences, a Senior Investigator of the Brazilian National Research Council—CNPq—and a Professor of Clinical and Basic Pharmacology at Universidade do Brasil in Rio de Janeiro, where he received his M.D. and Ph.D. degrees. He did postgraduate work at Faculté de Médecine de Paris, Columbia University, New York and University College, London. Parallel to his scientific career, he studied music and received vocal training, performing as tenor soloist in orchestral concerts and recitals in Brazil.

Dedicated to Carol, who provides loving support for my academic career and sings duets with me.

CONTENTS

EDITOR

Guilherme Suarez-Kurtz
Instituto Nacional de Câncer
Rio de Janeiro, RJ, Brazil
Email: kurtz@inca.gov.br
Chapters 5,6,14

CONTRIBUTORS

C. Adithan
Department of Pharmacology
JIPMER
Pondicherry, India
Email: adithan@vsnl.com
Chapter 8

Eleni Aklillu
Department of Laboratory Medicine
Division of Clinical Pharmacology
Karolinska Institutet
Karolinska University
Hospital-Huddinge
Stockholm, Sweden
Email: eleni.aklillu@ki.se
Chapter 7

Leif Bertilsson
Department of Laboratory Medicine
Division of Clinical Pharmacology
Karolinska Institutet
Karolinska University
Hospital-Huddinge
Stockholm, Sweden
Email: leif.bertilsson@ki.se
Chapter 7

Geoffrey K. Chambers
Victoria University of Wellington
New Zealand
Email: geoff.chambers@vuw.ac.nz
Chapter 11

Su Pin Choo
Medical Oncology
National Cancer Center
Singapore
Email: choosupin@nccs.com.sg
Chapter 9

Balram Chowbay
Clinical Pharmacology Lab
Division of Medical Sciences
Humphrey Oei Research Institute
National Cancer Centre
Singapore
Email: ctebal@nccs.com.sg
Chapter 9

Abdallah S. Daar
McLaughlin-Rotman Centre
for Global Health
Program on Life Sciences, Ethics
and Policy
University Health Network/McLaughlin
Centre for Molecular Medicine
University of Toronto
Toronto, Ontario, Canada
Email: a.daar@utoronto.ca
Chapter 13

Collet Dandara
MRC/UCT Oesophageal Cancer
Research Group
Institute of Infectious Disease
and Molecular Medicine
University of Cape Town
Rondebosch, South Africa
Email: cdandara@curie.uct.ac.za
Chapter 7

Pedro Dorado
Centro de Investigación Clínica
CICAB, SES
Hospital Universitario Infanta Cristina
Universidad de Extremadura
Badajoz, Spain
Email: pdorado@unex.es
Chapter 5

Samina Essajee
McLaughlin-Rotman Centre
for Global Health
Program on Life Sciences, Ethics
and Policy
University Health Network/McLaughlin
Centre for Molecular Medicine
University of Toronto
Toronto, Ontario, Canada
Email: s.essajee@utoronto.ca
Chapter 13

Silvia Fuselli
Dipartimento di Biologia
Università di Ferrara, Italia
Email: fss@unife.it
Chapter 2

Gerardo Jimenez-Sanchez
National Institute of Genomic Medicine
Mexico City, Mexico
Email: gjimenez@inmegen.gob.mx
Chapter 13

M. Ravindra Kumar
Department of Pharmacology
JIPMER
Pondicherry, India
Email: ravinmk@yahoo.com
Chapter 8

Suman Lal
Clinical Pharmacology Lab
Division of Medical Sciences
National Cancer Center
Singapore
Email: suman.lal@nccs.com.sg
Chapter 9

Rod A. Lea
Victoria University of Wellington
New Zealand
Email: rod.lea@esr.cri.nz
Chapter 11

Adrián LLerena
Universidad de Extremadura
Hospital Universitario Infanta Cristina
Badajoz, Spain
Email: allerena@unex.es
Chapter 5

Collen Masimirembwa
African Institute of Biomedical Science
and Technology
Harare, Zimbabwe
Email: collenmasimirembwa@yahoo.com
Chapter 7

Howard L. McLeod
The UNC Institute
for Pharmacogenomics
and Individualized Therapy
Schools of Pharmacy and Medicine
University of North Carolina
Chapel Hill, North Carolina, U.S.A.
Email: hmcleod@unc.edu
Chapter 4

Esteban J. Parra
Department of Anthropology
University of Toronto at Mississauga
Mississauga, Ontario, Canada
Email: eparra@utm.utoronto.ca
Chapter 3

Sergio D.J. Pena
Departamento de Bioquímica
e Imunologia
Universidade Federal de Minas Gerais
Belo Horizonte, Brazil
Email: spena@dcc.ufmg.br
Chapters 1,6

Sara Raimondi
Unità di Epidemiologia Molecolare e Genetica
Fondazione Policlinico Mangiagalli e Regina Elena
and
Divisione di Epidemiologia e Biostatistica
Istituto Europeo di Oncologia
Milano, Italia
Email: sara.raimondi@ieo.it
Chapter 2

Béatrice Séguin
McLaughlin-Rotman Centre for Global Health
Program on Life Sciences, Ethics and Policy
University Health Network/McLaughlin Centre for Molecular Medicine
University of Toronto
Toronto, Ontario, Canada
Email: beatrice.seguin@utoronto.ca
Chapter 13

Rashmi R. Shah
Former Senior Clinical Assessor
Medicines and Healthcare products Regulatory Agency
London, U.K.
Email: clin.safety@lineone.net
Chapter 12

Peter A. Singer
McLaughlin-Rotman Centre for Global Health
Program on Life Sciences, Ethics and Policy
University Health Network/McLaughlin Centre for Molecular Medicine
University of Toronto
Toronto, Ontario, Canada
Email: peter.singer@utoronto.ca
Chapter 13

Eduardo Tarazona-Santos
Departamento de Biologia Geral
Instituto de Ciências Biológicas
Universidade Federal de Minas Gerais
Minas Gerais, Brasil
Email: edutars@icb.ufmg.br
Chapter 2

Wei Zhang
Pharmacogenetics Research Institute
Institute of Clinical Pharmacology
Central South University
Changsha, China
Chapter 10

Hong-Hao Zhou
Pharmacogenetics Research Institute
Institute of Clinical Pharmacology
Central South University
Changsha, China
Email: hhzhou2003@163.com
Chapter 10

PREFACE

*"Da miscigenação nasce uma raça de tanto talento e resistência, tão poderosa, que supera a miséria e o desespero na criação quotidiana da beleza e da vida."**

—Jorge Amado, Tenda dos Milagres

Ethnic specificity has become an integral part of research in the overlapping sciences of pharmacogenetics and pharmacogenomics. At the writing of this preface, the PubMed database lists over 500 entries, including 120 reviews, for a query combining the terms "pharmacogen* and ethnicity". By contrast, only three reviews are listed for "pharmacogen* and admixed populations", although the realization that racial and ethnic groups are highly admixed is clearly voiced in all of these reviews. Indeed, "mixture of dissimilar individuals" was recognized as a factor of individual "varieties" by Georges-Louis Leclerc (1707-1788), one of the first naturalists who discussed human heterogeneity.

Pharmacogenomics in Admixed Populations was conceived to compile pharmacogenetic/-genomic (PGx) data from peoples of four continents: Africa, America, Asia, and Oceania, where admixture and population stratification occur in distinct patterns. The organization of the book was informed by a population PGx perspective: an initial review of the evolution of human genetic diversity leads into a series of chapters dealing with the population structure and the PGx profiles of various peoples selected on the basis of continents, countries or particular sub-groups within a continent or country. A chapter describes approaches to control for the confounding effect of admixture and population stratification on PGx association studies, and two chapters revisit the challenges and opportunities associated with the implementation of PGx on a global scale. In the final chapter, I offer a synopsis of the book content and my views on the potential role of PGx to reduce the health disparities between developing and developed nations. I suggest that this goal is unlikely to be achieved by relinquishing the notion of personalized drug therapy tailored to individual genetic characteristics—the original promise of pharmacogenetics—in favor of models of population-based drug development and prescription, with all their potential pitfalls, especially when extended to admixed populations in developing or developed nations.

As the editor of *Pharmacogenomics in Admixed Population*, I was lucky to have expert scholars accepting my invitations to contribute their excellent chapters to our book and... to meet the agreed deadlines! It is my pleasure to acknowledge my gratitude to all these friends and colleagues for lending their talent, knowledge and valuable time to bring this book to fruition.

Guilherme Suarez-Kurtz
Rio de Janeiro, March 2007

*From admixture, a race is born of so much talent and resilience, so powerful, that it overcomes misery and despair in the daily creation of beauty and life.

CHAPTER 1

The Evolution and Structure of Human Genetic Diversity

Sergio D.J. Pena*

Abstract

The conceptual development and the praxis of pharmacogenetics and pharmacogenomics will depend on a solid understanding of the evolution and structure of human genomic diversity. In this review three historically sequential views of human variability are discussed. The first, typological and essentialist, was based in the partition of humanity into races. The second involved a division into populations rather than races. The third, a new genealogical paradigm, emerged on the bases of three recent scientific developments: (1) the demonstration of absolute genome individuality in humans; (2) the genetic and paleontological demonstration of a recent and unique origin for modern man in Africa; and (3) the discovery that the human genome is structured in haplotype blocks. The new paradigm is solidly founded on human evolutionary history and stresses individuality rather than membership in populations. According to it we can envisage the human genome as composed of hundreds of thousands of small genomic blocks of high linkage disequilibrium, each one with its own pattern of variation and genealogical origin. Under this model, ideas such as that of human races or "race-targeted drugs" become meaningless.

Ex Africa semper aliquid novi
—Pliny the Elder (23-79 AD)[a]

Introduction

Pharmacogenetics and pharmacogenomics deal with variation of drug response due to genetic factors. For pharmacological agents with well characterized metabolism it is possible to assess genetic variation in pertinent loci and to use patient genotypes to direct medical treatment. This underpins the concept of personalized therapy. Unfortunately the loci relevant for the pharmacokinetics and pharmacodynamics of most drugs are not known. To try to deal with this situation some have proposed a model of population-based drug development and prescription, leading to the development of "race-targeted drugs", as exemplified by the case of BiDil® for treatment of heart failure in African Americans.[1] The theoretical foundation of such strategy is the idea that the "race" or the ethnic affiliation of a given patient may serve to replace the germane genotyping at critical pharmacogenetic loci. To be able to evaluate critically the

[a] "Out of Africa there is always something new".

*Sergio D.J. Pena—Departamento de Bioquímica e Imunologia, Universidade Federal de Minas Gerais, Belo Horizonte, MG 31270-901, Brazil. Email: spena@dcc.ufmg.br

Pharmacogenomics in Admixed Populations, edited by Guilherme Suarez-Kurtz.

appropriateness of this approach we need to understand the extent of worldwide human diversity and its degree of "racial", ethnic or geographical structure.

In this article I review human genomic variability using a historical perspective, from the recent origin of modern humans in Africa to their spread to colonize the whole planet. It is shown that the worldwide distribution of human diversity reflects such evolutionary history. In other words, the genetic relatedness of human populations can be better predicted by geography than by ethnic labels.[2] This suggests that ethnic labels will not prove to be an adequate replacement for the appropriate genotyping of patients.

Moreover, I propose that rather than thinking about populations, ethnicities or races, we should focus on the unique genome of the particular individual, which is structured as a mosaic of polymorphic haplotypes with diverse genealogical histories (Paabo, 2003). This shifts the emphasis from populations to persons. We should strive to see each individual as having a singular genome and a unique life history, rather than try to impose on him/her characteristics of a group or population. Only then we will be able to fulfill the promise of personalized therapy.

The Origin and Dispersion of Anatomically Modern Humans

Anatomically modern *Homo sapiens sapiens* is a very young species on our planet. Several lines of evidence suggest its single and recent origin, 150,000-195,000 years ago, in Africa. The first is the observation of a genetic diversity in Africa larger than in any other continent, as shown by the innumerous studies listed in Table 1. The interpretation of this finding is that a more ancient population, such as Africa, would have more time to accumulate genetic variability. Genetic trees furnish the second line of evidence. Beginning with the seminal work of Cann et al,[4] essentially all studies based on human mitochondrial DNA have produced a tree in which the first bifurcation separates African populations from those of other

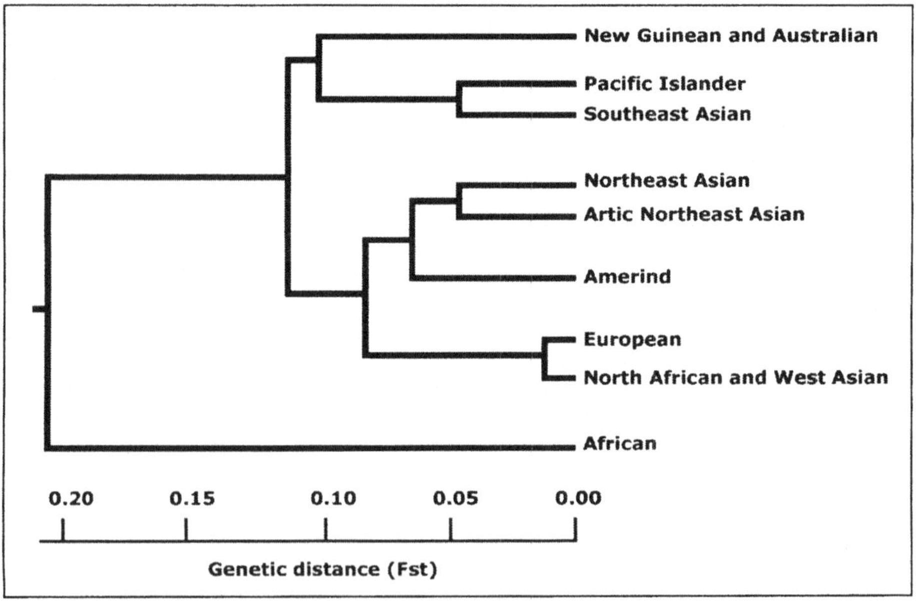

Figure 1. Population tree based on data from 120 classical markers from 1915 human populations. The genetic distance metric was F_{st} and the construction method was UPGMA. The first split separates Africa from all other regions. (Redrawn from Cavalli-Sforza and Feldman, ref. 64.)

***Table 1. Pioneer studies that provided molecular evidence for a recent African origin of modern* Homo sapiens**

Genomic Compartment	Genetic System	Reference
Mitochondrial DNA	RFLPs - coding portion	Cann et al 1987[4]
Mitochondrial DNA	D-loop sequencing	Vigilant et al 1991[53]
Mitochondrial DNA	Complete sequencing	Ingman et al 2000[54]
Autosomes	Alu insertions	Batzer et al 1994[55]
Autosomes	Microsatellites	Bowcock et al 1994[56]
Autosomes	Minisatellites	Armour et al 1996[57]
Autosomes	CD4 haplotypes	Tishkoff et al 1996[58]
Autosomes	Indels	Weber et al 2002[59]
Y chromosome	Microsatellites	Seielstad et al 1999[60]
Y chromosome	SNPs	Underhill et al 2000[61]
X chromosome	Dystrophin haplotypes Xp21	Zietkiewicz et al 1997[62]
X chromosome	Sequencing of Xq13.3 region	Kaessmann et al 1999[63]

continents. Likewise, trees built from autosomal markers (Fig. 1), X chromosome markers and Y chromosome markers present similar topology. A third compelling line of evidence for a recent African origin of modern humankind is the observation that geographic distance—not genetic distance—from East Africa along likely colonization routes is highly correlated with the genetic diversity of human populations.[5] Finally, we have dating based on the molecular clock (i.e., the known regularity of neutral mutation along time) that shows a coalescence time for mitochondrial DNA lineages around 150,000-200,000 years ago.

Until recently we were missing critical fossil evidence that could back up the "out-of-Africa" hypothesis for the origin of humankind. In 2003 White et al[6] described fossilized hominid crania found in Herto, Ethiopia, that have been isotopically dated to 160,000-150,000 years before present (YBP). These hominids, who have been named *Homo sapiens idaltu* (*idaltu* means old in *afar*, the language of Ethiopia), are morphologically intermediates between ancient hominid fossils and fossils with modern morphology and thus they were proposed as candidates for being the immediate ancestor of *Homo sapiens sapiens*. Also recently, two skulls found in the Kibish in southern Ethiopia and bearing phenotypic characteristics of anatomically modern humans (AMH) have been dated as having 195,000 years.[7] The anatomy and the antiquity of these fossils provide powerful evidence that humankind emerged recently in Africa.

Sometime, probably within the last 70,000 years, AMH left Africa and colonized other continents, decimating and replacing in their trajectory Neanderthals (*Homo sapiens neandertalensis*) and other archaic populations of *Homo sapiens*. According to this scenario, all human beings living presently on earth share a recent African ancestor.

The Races of Humanity—Typological Paradigms

A simple morphological inspection of people from different regions of earth will reveal an apparent paradox: we are at the same time very similar and yet very different. Indeed, there are great similarities among humans: the corporal plan, the erect posture, the thin skin and the relative scarcity of body hair distinguish us from the other primates. On the other hand, there are significant morphological variations among individuals: height, skin pigmentation, hair texture, facial features, etc. In special, each one of us has a morphological individuality: our relatives and our friends can identify us in a crowd without any hesitation. This morphological variety can be described at two different levels. The first is at the interpersonal level, the diversity that distinguishes a person from other within a population and that is intimately connected with personal identity. The second is at the interpopulational level, i.e., the morphological diversity that characterizes different human groups, especially in different continents.

The latter diversity is very relevant, because historically has served as a basis for the typological division of humankind into "races". The most influent proposition in this sense was that of the German anthropologist Johann Friedrich Blumenbach (1752-1840). In the 1795 edition of his book *De generis humani varietate nativa* ("On the natural varieties of humanity") he divided all humans into five groups, defined both by geography and appearance: Caucasians (the light-skinned people of Europe, Middle East, Central Asia, North Africa and India), Mongolians (East Asia), Ethiopians (the dark-skinned people of Africa), Americans (Amerindians) and Malays (Oceania). The name Caucasian has a double origin: first because in the opinion of Blumenbach the perfect human type was found in the Mountains of Georgia, in Central Asia, and second because he believed that that region had been the cradle of humankind.[8] Blumenbach's classification persists to our day, is spite of the fact that we now know that it is impossible to separate humanity in biologically significant categories, independent of the criterion adopted.

In Blumenbach's classification of humankind into different races and in subsequent attempts to do so from Ernst Haeckel in 1868[9] to Carleton Coon in 1962[10] the major emphasis was placed on the "interracial" diversity and considerable less importance was given to "intraracial" variability. In a recent conference at the University of California, Richard Lewontin[11] made the relevant observation that a mark of prejudice and racism is exactly this vision of humanity only in interpopulational terms, i.e., the inability to recognize in other "racial" groups the individuality of each person. This is often verbalized as: "they seem all equal to me, but we are all different from each other". When you deny the individuality of members of other groups, you dehumanize and objectify them.

The description of the interpersonal and interpopulational morphological variabilities belongs to the sphere of appearances, the phenotypic world. Subjacent to the observable morphological individuality there is indeed an absolute genomic individuality. However, contrary to the typological paradigm, the genomic representation of the variability between the human groups of different continents—the so-called human "races"—is very small. The physical characteristics that distinguish continental groups apparently represent morphological adaptations to the physical environment, thus being the products of natural selection acting on a very small number of genes. Let us now examine the evidence for these statements, starting with the latter.

Geography and Phenotypic Appearance

Relethford[12] showed that only 11-14% of human craniometrical diversity occurs between different continents, while 86-89% occurs between individuals within regions. When the same author partitioned the variability in skin pigmentation he observed a very different picture: 88% of variation occurred between geographical regions and only 12% within regions.[13] This discrepancy can be explained because skin pigmentation appears to be a special phenotypic feature subject to natural selection. Indeed, two opposing selective factors have been proposed to influence the adaptation of skin pigmentation to prevailing levels of environmental ultraviolet radiation: lack of synthesis of vitamin D3 when UV radiation is insufficient and destruction of folate when it is excessive.[14,15] There is an excellent correlation between levels of UV radiation and levels of skin pigmentation worldwide.[15]

The degree of skin pigmentation is determined by the amount and the type of melanin in the skin, and these in turn are apparently determined by a small number of genes (4-6) of which the melanotropic hormone receptor appears to be the most important.[16,17] This is an insignificantly small number of genes amid the 20-25,000 structural genes in the human genome.[18]

Likewise, external phenotypic features such as nose format, lip thickness and hair color and texture most likely represent adaptations to environmental conditions and/or are influenced by sexual selection. Just like the pigmentation of skin these phenotypical features depend on few genes. In summary, these iconic "race" features correlate well with the continent of origin, but depend on variation in an insignificantly small portion of the human genome. We may say that in this sense, race is skin deep. Yet, human societies have constructed elaborate systems of privilege and oppression based on these insignificant genetic differences.[19]

Genetic Markers

Subjacent to the abundant human morphological individuality, there are abundant levels of metabolic, molecular and genomic variability.[20] With the explosion of knowledge derived from the "DNA revolution" our understanding of human genomic diversity has increased exponentially in the past few years.[21]

If for the moment we ignore migrations, the dynamic of variation in allele frequencies of genomic markers is governed by the interactive forces of mutation, selection and genetic drift. Although nongeneticists have clear concepts of mutation and selection, the phenomenon of genetic drift is lesser known and deserves special elaboration.

The name genetic drift is given to the purely random variation in allele frequencies along time occurring as a sampling effect. The set of alleles of a given generation is not an exact copy of the preceding generation, but is a random sample of it, and as such is subject to statistical fluctuations, like a genetic lottery. As in every random sample, there is a variance that is inversely proportional to the size of the sample. When the effective size of a population is small, especially when there are drastic populational reductions (bottlenecks) or when a small group leaves the original population and colonizes a different region (founder effect) we can observe important allele frequency variations from one generation to the next.[22] Occasionally alleles can be fixed (reach frequency 1) or removed from the population (reach frequency zero) purely as a consequence of stochastic effects.[23] Because the vast majority of DNA markers used in the study of human diversity is selectively neutral, genetic drift, along with mutation pressure is of paramount importance in shaping the distribution of human diversity.

It is relevant to note that while the mutation rate is specific for each locus, thus varying in different parts of the genome, genetic drift depends on the demography and evolutionary history of populations, thus affecting equally all neutral loci in the genome.[24]

Partition of Human Genetic Variability

In 1972, Richard Lewontin[25] tested scientifically the notion of the existence of human races as typological entities by partitioning human genetic variability into three additive components: the variability between continents (i.e., between "races"), the variability between population groups within continents and the variability between individuals within populations. To accomplish that he researched in the available literature the allele frequencies of 17 classical genetic polymorphisms. He then grouped the populations into eight "racial" continental groups: Africans, Amerindians, Australian aborigines, East Asians (Mongoloids), South-Asians, Indians, Oceanians and Caucasians. The results came as a surprise: 85.4% of the allelic diversity occurred within population groups, 8.3% among populations of the same "race" and only 6.3% among the so-called races. These data could be better understood using a thought experiment: imagine that a nuclear cataclysm destroys all people on earth with the exception of Africans. In that case, 93% of human genetic diversity would be preserved. If only one African population remained, for instance the Zulus from South Africa, we would still maintain about 85% of human genetic variability!

This work was criticized because it made use of some polymorphisms of selective value, such as the Duffy blood group that is related to resistance to malaria. Thus, there could occur a correlation between certain allele constellations. However, the pattern is maintained even when neutral DNA polymorphisms are used. For instance, very recently we undertook a study of worldwide variability using the HGDP-CEPH Diversity Panel (1064 individuals from 52 populations) with a set of 40 biallelic short insertion-deletion polymorphisms (indels).[26] These are slow-evolving markers not subject to natural selection and thus the distribution of their variability reflects only the forces of migration and genetic drift. The 52 different populations were originated from seven geographical regions: Europe, Middle East, Central Asia, East Asia, Oceania, Americas and Sub-Saharan Africa. With the 40-indel battery we observed that 85.7% of the allelic diversity occurred within population groups, 2.3% among populations of the same "race" and only 12.1% among the so-called races. These numbers are very similar to those observed by Lewontin with classical markers. Other authors have also obtained similar figures using biallelic markers.[27-30]

These studies illustrate what can be called the "population paradigm" of human genome diversity. Mayr[31] defined population thinking, and contrasted it with the previous typological, essentialist thinking that we mentioned above. According to him "population thinkers stress the uniqueness of everything in the organic world. What is important for them is the individuals, not the type. They emphasize that every individual is uniquely different form all others. There is no "typical" individual and mean values are abstractions." [...] "The differences between biological individuals are real, while the comparison of groups of individuals are man-made inferences".

Geographical Correlations of Human Diversity

Probably the largest study of human variability undertaken thus far was that of Rosenberg et al[32] who typed the same HGDP-CEPH Diversity Panel (1064 individuals from 52 populations) that we described above with 377 autosomal microsatellites. Later they enlarged this set to 993 markers with no major change in conclusions.[33] In the sample they observed a total of 4,199 alleles, 47% of which were present in all world regions studied—only 7% of the alleles were observed in a single region, which in almost every case was Africa. These results indicate that most of human genetic diversity is shared among all regions of the world and is absolutely compatible with the recent single origin of modern humankind in Africa.

When they partitioned the variability they observed that 93-95% was contained within populations, a figure considerably higher than the one observed by Lewontin[25] (1972) or us.[26] Indeed, Excoffier and Hamilton[34] observed that the level of within-population variance observed by Rosenberg et al[32] was larger than other worldwide studies and attributed this to the fact that the authors had not used a stepwise mutation model, the most appropriate for microsatellite studies. Not taking homoplasy into account can depress the among-regions variance component.[29,35] If one associates the relatively high mutation rate of microsatellites[36] with the possibility of size constraints for their growth, different populations would tend to approach a common allelic distribution for these markers.[29]

In the same study Rosenberg et al[32] decided to ascertain the capacity of these selectively neutral microsatellites to distinguish structure in human genetic diversity without assigning them a priori to any population or geographical region. For such they used a computer program called Structure which uses a Bayesian method that tries to estimate for each individual in the sample the proportion of his/her genome that originates in a given cluster, which in turn is estimated from allele frequencies.[37] The estimation procedure is done with different and growing number K of clusters (K = 2, 3, 4, etc.), which has to be input). For each value of K the program produces a posterior probability.

Rosenberg et al[32] showed that the maximum posterior probability occurred at K = 5. Then, the clusters produced by the program corresponded to five great regions, namely: (1) Sub-Saharan Africa (2) East Asia, (3) America, (4) Oceania and (5) a cluster encompassing Europe, North Africa, Middle East and Central Asia. The study did not show any advantage in invoking a sixth cluster. It is relevant to know that under the same study protocol and using the same samples we obtained virtually identical results with our set of 40 short indels.[26]

There is an apparent and superficial correspondence between the results of this study and the five human races defined in the 18th Century by Blumenbach, i.e., Ethiopian, Mongoloid, American, Malay and Caucasian respectively. Indeed, some researchers and the press (including a recent book by the well-known science journalist Nicholas Wade[38]) have claimed that

Rosenberg's study helped reestablish the notion of human races on modern scientific grounds. However, such views are erroneous and cannot withstand close scrutiny.

First of all, we should note that even though the analysis using the Structure program was undertaken without a priori population classification, the sampling strategy was clearly population-based. Moreover, the sampled groups were few and distant from each other.[39] Serre and Paabo claimed to show using simulation studies that if sampling had been done with individuals on a geographical grid rather than being population-based, the clustering effect would be much diminished. However, such claim was later contested by Rosenberg et al.[33]

Second, we should realize that if we choose any two individuals in the same cluster they will only be on average 4-5% more similar than if any of them were compared to an individual from any other cluster. This may lead to statistically significant clustering but that does not mean that it is biologically significant. In other words, individuals from the same geographical region and individuals from different geographical regions are almost equally different!

Third, every racial classification has been based on the wrong typological idea that races were very different from each other and very internally homogeneous. That is not the picture that emerges from Rosenberg's data, which on the contrary shows very heterogeneous clusters barely different from each other. Indeed, Rosenberg's data and similar studies can be used as a strong argument that human races do not exist.[41]

We have already mentioned the work of Prugnolle et al[5] that showed that geographic distance—not genetic distance—from East Africa along likely colonization routes is highly correlated with the genetic diversity of human populations. The same authors[2] later used the data of Rosenberg et al[32] to show that pair-wise geographical distances across land-masses constitute a far better predictor of neutral allele sharing than ethnicity! In other words, the distribution of neutral human diversity reflects human evolutionary history.

The observation that allele-sharing between human populations worldwide decays smoothly with increasing physical distance is most compatible with a model of colonization of the world based on serial founder effects.[42,43] Fitting of the data to such a model translates into an estimate of the initial expansion of modern humans from East Africa *circa* 56,063 ± 5,678 years ago, from an ancestral effective population source of around 1,000 individuals. Thus, it appears that modern *Homo sapiens* remained in Africa for a long time after his origin 160,000-195,000 years ago.

Populations and Individuals

As we saw above, population thinking stresses the uniqueness of individuals within populations. However, as pointed out by Caspari[44] we should realize that such theoretical population thinking may be quite distinct from population studies in practice. In fact, many contemporary anthropologists and geneticists conceptually deal with populations in the same manner as the previous generations dealt with races.[44] Contrary to Mayr's population paradigm, what is important for them is the population and not the individual! Thus, they divide humanity into populations, which can be defined on the basis of geography, culture, religion, physical appearance or whatever other criterion that is convenient. It appears that such division of humanity into populations does not constitute the most appropriate approach to deal with human variation. Treating people, for instance, of the European population and African population, as separate categories for genetic studies tends to contribute to the public perception that the primary difference between these ways of defining populations is biological.[45] This view confounds several issues and obscures the important fact that Europeans are genealogically related to Africans, having evolved as an offshoot of the latter.

The human evolutionary history is remarkably short and the worldwide geographical distribution of genetic traits is basically due to dispersal, with ensuing mutation, selection and genetic drift. In essence, the genetic diversity observable in Europe, Asia, Oceania and the Americas is a merely a subset of the variation found in Africa.[46] As pointed out by Paabo,[3] from a genomic perspective we are all Africans, either living in Africa or in quite recent exile outside of Africa.

Thus, rather than conceptualizing humans as belonging to defined populations, it might be more appropriate to think of them as 6 billion individuals who have different degrees of pairwise relatedness along genealogical lines. This has the advantage of firmly connecting our vision of human genetic diversity to our evolutionary history and, as we will see shortly, to genome structure.

This view becomes particularly clear when seen from the perspective of what we call lineage markers. These are the uniparental maternal (mitochondrial DNA—mtDNA) and paternal (nonrecombinant regions of the Y chromosome—NRY) polymorphisms, which are haploid and do not recombine. As such, blocks of genes (haplotypes) are transmitted to the next generations and remain unaltered in the matrilineages and patrilineages until a mutation supervenes. The mutations that have occurred and reached high frequencies after the dispersion of modern man from Africa can be specific to certain regions of the globe and can serve as geographical markers. The mitochondrial DNA and the NRY provide complementary information that can trace back to several generations in the past. Two observations are of extreme importance here. The first is that if two individuals have the same mitochondrial or Y chromosome haplotype, they are genealogically related along that line, independent of which population they formally belong to.

The second is that the genealogical matrilineage to which an individual belongs to is completely independent of his patrilineages. For instance, our studies have shown that while White Brazilians carry almost exclusively Y chromosomal lineages phylogeographically related to Europe, 2/3 of them have mitochondrial DNA lineages phylogeographically related to Amerindians or Africans.[47,48] In other words, most Brazilians have these two genomic compartments of different phylogeographical origin and thus are genealogical mosaics.

What about the diploid biparental nuclear genome? The same kinds of genealogical principles that apply to lineage markers also apply in theory to nuclear genes, whose multi-generation transmission routes involve both genders.[49] In terms of formal theory the major difference from uniparental markers is a four-fold adjustment required to account for the larger effective population size of autosomal alleles. This leads to corresponding four-fold longer coalescent times. A second, less important, difference is that in autosomes, besides mutations, lineages can change because of intragenic (or intrahaplotype, see below) recombination events.

In the past few years it has become evident that much of the human genome is composed of haplotypic blocks ("hapblocks") where polymorphic markers (especially single nucleotide polymorphisms—SNPs) are strongly associated over distances as large as 170Kb.[3,22,50] The discussion of the origin of these haplotype blocks is beyond the objective of this review. Suffice to say that probably the length of haplotype blocks is influenced by both demographic factors (which is certainly responsible for most of the variation of block sizes among populations) and genomic factors, especially the existence of recombination hot-spots.[51,52] The existence of such hapblocks has high significance for the feasibility of mapping disease genes by marker association studies, since each block can be defined by typing only 4-5 SNPs. Thus, the number of SNPs needed to achieve fine genomic screen might be reduced from millions to a few hundred thousand.[22]

We can then envisage the human genome as composed of hundreds of thousands of small genomic blocks of high linkage disequilibrium (like the mtDNA or Y chromosome), each one with its own pattern of variation and genealogical origin. Rather than thinking about populations, ethnicities or races, we may then consider the genome of any particular individual as a mosaic of variable haplotypes.[3] This is the Variable Mosaic Genome (VMG) paradigm, which completely shifts the focus from populations to individuals. In other words, the paradigm emphasizes human individuality rather than membership in populations. We should strive to see each person as having an individual genome, rather than try to impose on him/her characteristics of the group or population. This is ideally suited to the practice of medicine, since in the office doctors evaluate and treat individual patients and not populations or races.

Conclusions

We reviewed three models of human genetic structure. The first, preponderant on the 19th Century and the first half of the 20th Century envisaged humanity as partitioned into well defined races. This typological model erroneously visualized races as being very different from each other, but internally homogeneous. The consequences of this model were racism, prejudice and discrimination, leading to the Nazi movement and apartheid.

Beginning in 1930-1940 the typological paradigm was replaced by a model that focused on populations, which were viewed as internally heterogeneous and differed only in allele frequencies. Although theoretically correct, in practice populations were confused with races and this second model has been associated with continuing racism and prejudice and recently led to the unfortunate development of the strategy of "race-targeted drugs". This strategy involves polygenic constellations that differ little, perhaps at most 2 or 3-fold among fairly ill-defined populations (such as African-Americans). These polygenes influence (and not determine) the pharmacogenetics and pharmacodynamics of some drugs and there is ample room left for epistasis or modification by individual genotypes. Thus, such policy is equivalent to playing a game of probabilistic black boxes and erroneously calling it personalized medicine.

Three recent scientific developments have triggered a shift to a much needed new paradigm: (1) the demonstration of absolute genome individuality in humans; (2) the genetic and paleontological demonstration of a recent and unique origin for modern man in Africa; (3) the discovery that the human genome is structured in haplotype blocks. The new paradigm is genealogical in nature and based on human evolutionary history—it stresses individuality rather than membership in populations. According to it we can envisage the human genome as composed of hundreds of thousands of small genomic blocks of high linkage disequilibrium, each one with its own pattern of variation and genealogical origin. Under this model, ideas, such as that of human races or "race-targeted drugs" become meaningless.

Such paradigm resonates well with many strands of thought in social science. For instance, recently the Nobel Laureate Amartya Sen wrote a book entitled *Identity and Violence*[65] emphasizing the necessity of humans to define their identity multidimensionally and not according to single major overarching criteria, such as color, race or creed that would inevitably lead to divisiveness and conflict. The new genealogical paradigm that emphasizes individual uniqueness is the only that does not constrain the plural definition of personal identity. Moreover, such paradigm is in perfect alignment with the concept that human rights apply to the individual and not to groups. As famously expressed by US Supreme Court Justice Anthony Kennedy:[66] "At the heart of the Constitution's guarantee of equal protection lies the simple command that government must treat citizens as individuals, not as components of a racial, religious, sexual or national class."

References

1. FDA News. FDA approves BiDil heart failure drug for black patients. 2005, (http://www.fda.gov/bbs/topics/NEWS/2005/NEW01190.html).
2. Manica A, Prugnolle F, Balloux F. Geography is a better determinant of human genetic differentiation than ethnicity. Hum Genet 2005; 118:366-371.
3. Paabo S. The mosaic that is our genome. Nature 2003; 421:409-412.
4. Cann Rl, Stoneking M, Wilson AC. Mitochondrial DNA and human evolution. Nature 1987; 325:31-36.
5. Prugnolle F, Manica A, Balloux F. Geography predicts neutral genetic diversity of human populations. Curr Biol 2005; 15:R159-R160.
6. White TD, Asfaw B, Degusta D et al. Pleistocene Homo sapiens from Middle Awash, Ethiopia. Nature 2003; 423:742-747.
7. McDougall I, Brown FH, Fleagle JG. Stratigraphic placement and age of modern humans from Kibish, Ethiopia. Nature 2005; 433:733-736.
8. Gould SJ. The geometer of race. Discover 1994; 15:65-69.

9. Haeckel EHPA. Natürliche Schöpfungsgeschichte: Gemeinverständliche Wissenschaftliche Vorträge über die Entwickelungslehre im Allgemeinen und Diejenige von Darwin, Goethe und Lamarck im Besonderen. Berlin G Reimer 1868; 606.
10. Coon CS. The Origins of Races. New York: Alfred A Knopf, 1962:320.
11. Lewontin RC. The concept of race: The confusion of social and biological reality. 2004, (http://www.uctv.tv/library-science.asp?seriesnumber=17).
12. Relethford JH. Craniometric variation among modern human populations. Am J Phys Anthropol 1994; 95:53-62.
13. Relethford JH. Apportionment of global human genetic diversity based on craniometrics and skin color. Am J Phys Anthropol 2002; 118:393-398.
14. Jablonski NG, Chaplin G. The evolution of human skin coloration. J Hum Evol 2000; 39:57-106
15. Jablonski NG, Chaplin G. Skin deep. Sci Am 2002; 287(4):74-81.
16. Sturm RA, Box NF, Ramsay M. Human pigmentation genetics: The difference is only skin deep. Bioessays 1998; 20:712-721.
17. Rees JL. Genetics of hair and skin color. Annu Rev Genet 2003; 37:67-90.
18. International human genome sequencing consortium: Finishing the euchromatic sequence of the human genome. Nature 2004; 431:931-945.
19. Bamshad MJ, Olson SE. Does race exist? Sci Am 2003; 289:78-85.
20. Pena SD, Prado VF, Epplen JT. DNA diagnosis of human genetic individuality. J Mol Med 1995; 73:555-564.
21. Cavalli-Sforza LL. The DNA revolution in population genetics. Trends Genet 1998; 14:60-65.
22. Tishkoff SA, Verrelli BC. Patterns of human genetic diversity: Implications for human evolutionary history and disease. Annu Rev Genomics Hum Genet 2003; 4:293-340.
23. Kimura M. The neutral theory of molecular evolution and the world view of the neutralists. Genome 1989; 31:24-31.
24. Luikart G, England PR, Tallmon D et al. The power and promise of population genomics: From genotyping to genome typing. Nat Rev Genet 2003; 4:981-994.
25. Lewontin RC. The apportionment of human diversity. Evol Biol 1972; 6:381-398.
26. Bastos-Rodrigues L, Pimenta JR, Pena SDJ. The genetic structure of human populations studied through short insertion-deletion polymorphisms. Ann Hum Genet 2006; 70:658-665.
27. Barbujani G, Magagni A, Minch E et al. An apportionment of human DNA diversity. Proc Natl Acad Sci USA 1997; 94:4516-4519.
28. Bowcock AM, Kidd JR, Mountain JL et al. Drift, admixture, and selection in human evolution: A study with DNA polymorphisms. Proc Natl Acad Sci USA 1991; 85:839-843.
29. Romualdi C, Balding D, Nasidze IS et al. Patterns of human diversity, within and among continents, inferred from biallelic DNA polymorphisms. Genome Res 2002; 12:602-612.
30. Watkins WS, Rogers AR, Ostler CT et al. Genetic variation among world populations: Inferences from 100 Alu insertion polymorphisms. Genome Res 2003; 13:1607-1618.
31. Mayr E. The growth of biological thought. Boston: Belknap, 1982:974.
32. Rosenberg NA, Pritchard JK, Weber JL et al. Genetic structure of human populations. Science 2002; 298:2381-2385.
33. Rosenberg NA, Mahajan S, Ramachandran S et al. Clines, clusters, and the effect of study design on the inference of human population structure. PLoS Genet 2005; 1:e70.
34. Excoffier L, Hamilton G. Comment on "Genetic structure of human populations." Science 2003; 300:1877.
35. Flint J, Bond J, Rees DC et al. Minisatellite mutational processes reduce Fst estimates. Hum Genet 1999; 6:567-576.
36. Leopoldino AM, Pena SDJ. The mutational spectrum of human autosomal tetranucleotide microsatellites. Hum Mutat 2003; 21:71-79.
37. Pritchard JK, Stephens M, Donnelly P. Inference of population structure using multilocus genotype data. Genetics 2000; 155:945-959.
38. Wade N. Before the Dawn: Recovering the Lost History of Our Ancestors. NY: Penguin Press, 2006:320.
39. Kittles RA, Weiss KM. Race, ancestry, and genes: Implications for defining disease risk. Annu Rev Genomics Hum Genet 2003; 4:33-67.
40. Serre D, Paabo S. Evidence for gradients of human genetic diversity within and among continents. Genome Res 2004; 14:1679-1685.
41. Templeton AR. Human races: A genetic and evolutionary perspective. Am Anthropol 1999; 100:632-650.

42. Ramachandran S, Deshpande O, Roseman CC et al. Support from the relationship of genetic and geographic distance in human populations for a serial founder effect originating in Africa. Proc Natl Acad Sci USA 2005; 102:15942-15947.
43. Liu H, Prugnolle F, Manica A et al. A geographically explicit genetic model of worldwide human-settlement history. Am J Hum Genet 2006; 79:230-237.
44. Caspari R. From types to populations: A century of race, physical anthropology, and the American Anthropological Association. Am Anthropol 2003; 105:65-76.
45. Foster MW, Sharp RR. Beyond race: Towards a whole-genome perspective on human populations and genetic variation. Nat Rev Genet 2004; 5:790-796.
46. Yu N, Chen FC, Ota S et al. Larger genetic differences within Africans than between Africans and Eurasians. Genetics 2002; 161:269-274.
47. Alves-Silva J, da Silva Santos M, Guimaraes PE et al. The ancestry of Brazilian mtDNA lineages. Am J Hum Genet 2000; 67:444-461.
48. Carvalho-Silva DR, Santos FR, Rocha J et al. The phylogeography of Brazilian Y-chromosome lineages. Am J Hum Genet 2001; 68:281-286.
49. Avise JC. Phylogeography: The History and Formation of Species. Boston: Harvard University Press, 2000:447.
50. Wall JD, Pritchard JK. Haplotype blocks and linkage disequilibrium in the human genome. Nat Rev Genet 2003; 4:587-597.
51. Zhang K, Akey JM, Wang N et al. Randomly distributed crossovers may generate block-like patterns of linkage disequilibrium: An act of genetic drift. Hum Genet 2003; 113:51-59.
52. Greenwood TA, Rana BK, Schork NJ. Human haplotype block sizes are negatively correlated with recombination rates. Genome Res 2004; 14:1358-1361.
53. Vigilant L, Stoneking M, Harpending H et al. African populations and the evolution of human mitochondrial DNA. Science 1991; 253:1503-1507.
54. Ingman M, Kaessmann H, Paabo S et al. Mitochondrial genome variation and the origin of modern humans. Nature 2000; 408:708-713.
55. Batzer MA, Stoneking M, Alegria-Hartman M et al. African origin of human-specific polymorphic Alu insertions. Proc Natl Acad Sci USA 1994; 91:12288-12292.
56. Bowcock AM, Ruiz-Linares A, Tomfohrde J et al. High resolution of human evolutionary trees with polymorphic microsatellites. Nature 1994; 368:455-457.
57. Armour JA, Anttinen T, May CA et al. Minisatellite diversity supports a recent African origin for modern humans. Nat Genet 1996; 13:154-160.
58. Tishkoff SA, Dietzsch E, Speed W et al. Global patterns of linkage disequilibrium at the CD4 locus and modern human origins. Science 1996; 271:1380-1387.
59. Weber JL, David D, Heil J et al. Human diallelic insertion/deletion polymorphisms. Am J Hum Genet 2002; 71:854-862.
60. Seielstad M, Bekele E, Ibrahim M et al. A view of modern human origins from Y chromosome microsatellite variation. Genome Res 1999; 9:558-567.
61. Underhill PA, Shen P, Lin AA et al. Y chromosome sequence variation and the history of human populations. Nat Genet 2000; 26:358-361.
62. Zietkiewicz E, Yotova V, Jarnik M et al. Nuclear DNA diversity in worldwide distributed human populations. Gene 1997; 205:161-171.
63. Kaessmann H, Heissig F, Von Haeseler A et al. DNA sequence variation in a noncoding region of low recombination on the human X chromosome. Nat Genet 1999; 22:78-81.
64. Cavalli-Sforza LL, Feldman MW. The application of molecular genetic approaches to the study of human evolution. Nat Genet 2003; 33(Suppl):266-275.
65. Sen A. Identity and Violence: The Illusion of Destiny. New York, W.W. Norton & Company, 2006:215.
66. Kennedy A. Miller v. Johnson, 2006; 515 U.S. 900, 91.

Chapter 2

Controlling the Effects of Population Stratification by Admixture in Pharmacogenetics

Eduardo Tarazona-Santos,* Sara Raimondi and Silvia Fuselli

Abstract

Admixture is a common type of gene flow in human populations and occurs when individuals from two or more parental populations that have been isolated for several generations form a new hybrid population. Admixed populations are common in North and Latin America, Central Asia and South Africa. Population structure (or stratification), the presence in one population of subgroups that differ in allele frequencies, can affect the results of population genetics and epidemiological studies. In this review, we analyze how population stratification and admixture affect the design and results of population genetics and association studies involving pharmacogenetic loci and drug-response related traits. Specifically, we discuss how admixture and population stratification are related to the allelic architecture of complex traits and illustrate how admixture may be a confounder in population genetics studies designed to infer the action of natural selection. We also analyze how admixture and population stratification affect case-control association studies, and give some real and hypothetical examples relevant in pharmacogenetics. Finally, we briefly describe methods and software developed to control for the effect of population stratification: genomic control, regression, structured association methods and Principal Component Analysis.

Pharmacogenomics, Admixed Populations and Population Stratification

Admixture is a type of gene flow that occurs when individuals from two or more parental populations that have been isolated for several generations, form a new hybrid population.[1] Different admixture dynamics are possible. Two extreme cases occur when (Fig. 1):[2,3] (A) parental populations contribute to the hybrid one in a unique generation of admixture (i.e., an intermixture or hybrid-isolation model); and (B) there is continuous gene-flow across several generations from the parental populations to the admixed one (i.e., a continuous gene flow model). Admixture is conceptually different from amalgamation or mixture, which happens when parental populations merge, but endogamy within these populations persists.[1] In the real world, individuals in the hybrid population seldom mate randomly nor show absolute parental-population endogamy. Instead, the level of mating among individuals from the differ-

*Corresponding Author: Dr. Eduardo Tarazona Santos—Departamento de Biologia Geral, Instituto de Ciências Biológicas, Universidade Federal de Minas Gerais, Av. Antonio Carlos 6627, Pampulha, Caixa Postal 486, Belo Horizonte, MG, CEP 31270-910, Brasil. Email: edutars@icb.ufmg.br

Pharmacogenomics in Admixed Populations, edited by Guilherme Suarez-Kurtz.

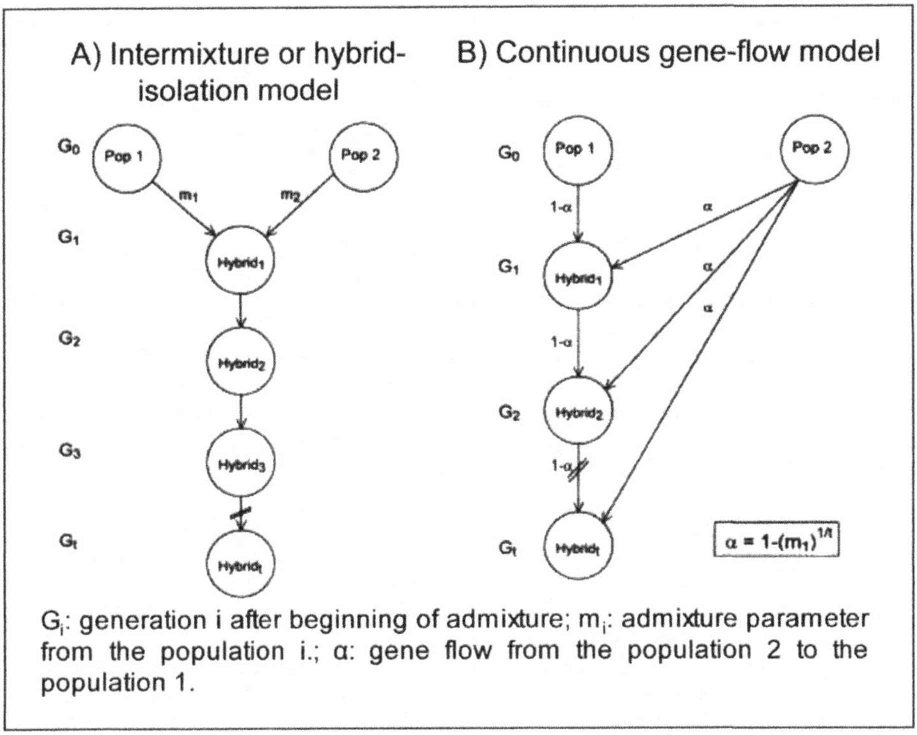

Figure 1. Models of gene flow. (Figure adapted from Long[2] and reproduced with permission from Pfaff et al.[3])

ent parental populations usually holds between these two extreme situations. Admixture events have been common during human evolutionary history, originating hybrid populations in different continents: Latin-American, Caribbean, African American and US-Hispanic in the Americas;[4-6] Central Asian populations;[7] and South African groups that have received contributions from Europeans and African autochthonous ethnic groups.[8]

Population genetics and association studies in admixed populations are sensitive to population structure or population stratification, which is the presence in one population of subgroups that differ in allele frequencies. The existence of subpopulations may be evident or more or less cryptic. Population stratification is a potential problem in population genetics and epidemiological studies. In population genetic studies, failure to recognize admixture can prevent proper characterization of the **haplotype**[a] structure for a group of populations, leading to wrong inferences about the evolutionary factors that have modeled the observed patterns of genetic diversity. In epidemiological studies, ignoring population stratification due to admixture can lead to false positive or false negative results.

In this chapter, we analyze how admixture affects the design and results of pharmacogenetic studies, which can be roughly classified as: (1) Population genetics studies, which try to infer the evolutionary factors that have shaped the pattern of diversity of genes important in pharmacogenetics and,[9-12] (2) Association studies, which test if a phenotypic trait is associated with a specific genetic variant (allele, haplotype or genotype). Phenotypic traits can be pharmacokinetic (often inherited in a Mendelian fashion) or pharmacodynamic variables; or more com-

[a] Terms in bold characters are included in a Glossary of Terms at the end of the chapter.

plex traits such as the presence of adverse drug reactions or measures of therapeutic efficacy. The type of pharmacogenetic variable to be considered in a study determines the statistical methods to be used to measure association with genetic traits and to control for confounding factors such as population stratification. Different genetic traits are used in association studies. For instance, it is possible to use individual SNPs (Single Nucleotide Polymorphisms) or haplotypes. The selection of SNPs to be studied can be guided merely on the basis of previous studies (even ignoring if they are the functionally important SNPs) or may rely on biological criteria (i.e., non-synonymous substitutions or SNPs located in the promoter or in evolutionarily conserved regions).[13] Studies based on haplotypes require criteria to select the SNPs used for haplotype definition. A possible and useful approach is to capitalize the information about **linkage disequilibrium** to select those SNPs that better and more exhaustively represent the common variants in a gene or genomic region (tag-SNPs).[14-16] In the simplest association study design, the case-control study, the statistical association among a binary pharmacogenetic trait and genetic variants can be measured using statistical tests such as the **Chi-square (χ^2) test**, or the **Fisher exact test**, or calculating a measure of association such as the **Odds Ratio (OR)**. The tests for trend, such as the **Cochran-Armitage-trend test**, verify the assumption of a linear trend in risk as exposure increases.

To recognize an admixed population is often important and may be a complex issue in planning pharmacogenetic studies. Criteria used by investigators to define populations are influenced not only by the cultural background of the studied population, but also by that of the researcher.[17] For instance, Fuselli et al have recently studied the pattern of nucleotide diversity of the N-Acetyltransferase (*NAT2*) coding region in Latin America,[12] including an urban sample from the shantytown of Las Pampas (Lima, Peru), which was considered a priori as admixed or "*mestizo*" because of the population history of Lima, a city that has received contributions from European, Native American and African gene pools. However, we verified that this sample has a *NAT2* haplotype structure similar to a native sample of farmers from the Central Andes. Using unlinked markers, we estimated that the genetic contributions of Native Americans, Europeans and Africans to the Las Pampas sample are 82%, 12% and 6% respectively, which shows that this population has a predominant Native American gene pool that is higher than some populations that are traditionally classified as Native American.[18] Categories such as admixed or "*mestizos*" have a strong cultural and socio-economic basis and do not necessarily reflect the genetic background of individuals or populations. Often, in Latin America, the same individual can be considered Native American or admixed at different ages, depending on cultural and socio-economic changes. Therefore, these categories should be used with caution in genetic studies on Latin American and other admixed populations from different parts of the world. This example also illustrates the operational problem of defining the minimum non-autochthonous contribution to consider a population as admixed. Ideally, this should be independent from where the population is settled, in a large urban centre (as the Las Pampas population) or in a small village from the Amazon Region or the Andean mountains. We suggest that the answer to this problem should be operational, depending on the scientific question we posit and the sensitivity of the used methods to the effect of admixture. In the following paragraphs we will review why and how admixture and population stratification should be taken into account in the design of population genetics and association studies in pharmacogenetics.

Population Genetics Studies, Pharmacogenetics and Admixture

Recently, a historical barrier between human population genetics and genetic epidemiology has vanished.[19] Population genetics studies designed to understand the pattern of diversity of genes important in pharmacogenetics have become popular. Scholars that for years had used population genetics models almost exclusively to study historical human evolution, have shown increasing interest for issues closely related to health problems. This trend has been motivated by factors such as (1) the completion of the human genome project,[20] (2) interest on the genetic basis of common complex diseases,[21] (3) the explosion of information about SNPs in general and polymorphisms in candidate genes for complex diseases in particular,[22,23] (d) technical possibilities of genotyping at continuously reducing costs a large number of polymorphisms and to mapping mutations responsible for complex diseases by capitalizing the pattern of linkage disequilibrium observed across the human genome.[24] Three examples illustrate why recognizing admixture is important in population genetics studies: (1) the problem of the allelic architecture of complex diseases; (2) how admixture can affect the genetic structure of human populations and (3) how admixture can interfere with inferences about the action of natural selection in human populations.

One of the biomedical problems recently addressed using a population genetics approach was the allelic architecture of complex diseases: if the genetic basis for these and other complex traits were determined by common or rare variants. Reich and Lander[25] defended the first scenario—known as the CV/CD hypothesis (CV/CD: Common Variants/Common Disease), by considering a population genetic model that incorporated **mutation-selection equilibrium** and the demographic growth of the human population. They showed that this model is compatible with the persistence in non-African populations of common susceptibility alleles that would have existed in the ancestral human population before the **Out of Africa** migration. On the other hand, since the seminal work of Ewens,[26] we know that under neutrality, a population shows more rare polymorphisms than common ones. Inspired by this result, Zwick et al[27] proposed that the class of rare variants may be responsible for common traits, and Pritchard[28] developed this idea using a stochastic model that incorporates neutrality and weak purifying selection (an evolutionary force that prevent deleterious alleles to reach high frequencies in a population). When we were turning into the 21st Century, whether the allelic architecture of common complex traits fitted better the common- or the rare-variants hypotheses was controversial. Five years later, some common diseases have shown association with common variants,[29] while other are associated with rare ones.[30 32] This seems valid also in pharmacogenetics: fast-, intermediate- and slow-metabolizer phenotypes for several Drug-Metabolizing-Enzymes (DME) genes often depends on common alleles, as in the case of *NAT2*.[11,12] On the other hand, Maitland et al,[33] by using a comparative genomic approach and a resequencing analysis, have showed that the Uridine 5'-diphosphate Glucuronosyltransferase-1A genes (*UGT1A* gene cluster), associated with irinotecan and tranilast toxicity, has low-frequency coding SNPs in highly conserved regions that are strong candidates to be considered in association studies. When performing association studies in admixed populations, the allelic architecture of genes is relevant. Common variants are usually old and because they predate the expansion of human populations across the world after the Out of Africa migration, they are shared by most human populations. Therefore, common alleles associated with complex traits in admixed populations will be the same independently of the parental populations of provenience. Conversely, rare alleles are usually younger and specific for different ethnic groups. If rare variants are responsible for a pharmacogenetic trait, parental populations contribute to the admixed population with different constellations of these rare alleles associated with complex traits.

Recognition of admixed populations is also important to understand the evolutionary forces that have shaped the genetic structure of human populations (i.e., the partition of genetic variance within- and between-populations). Admixture can affect patterns of diversity that otherwise would fit population genetic models such as **isolation-by-distance**, or would show gradients on allele or haplotype frequencies. For instance, European populations show low between-populations genetic variability if compared with other continents,[34] while Native Americans with no admixture show in general, a high inter-population differentiation.[35] In Latin America, gene flow of European ancestry individuals into Native populations has continuously occurred since the 16th Century. Because European immigrants are relatively homogeneous, gene flow into different Latin American native populations has a homogenizing effect, reducing the F_{ST} (a measure of between-population differentiation) for Native American populations. On the other hand, it is not clear if gene-flow from Africa due to slave trade has the same effect, since African populations are more structured than European ones. The effect of Post-Columbian admixture on the genetic structure of Native American populations may modify the genetic structure also for pharmacogenetic loci. For instance, we have recently reported low inter-population differentiation for *NAT2* genetic variation across Native American populations.[12] Even if we did not have data to quantify the contribution of European admixture on the observed low differentiation (F_{ST} = 0.03 across the Americas, P<0.01), Post-Columbian admixture might have played a role in determining the observed low level of population structure.

Inferences about natural selection, a topic traditionally restricted to evolutionary studies, have recently come into attention of biomedical research, because if natural selection acts, it implies that genetic variation on the target gene is functionally important. Recently, it has been show that diversity of *NAT2*[11] and *CYP3A* (Cytocrhome P450, family 3, subfamily A),[10] two important pharmacogenetic loci, have been likely shaped by natural selection. Natural selection is usually inferred assuming the null **hypothesis of neutrality** and using a variety of statistical tests.[36] In this context, admixture can act as a confounder. When parental populations are differentiated, admixture increases the variability of the hybrid population and the proportion of common variants with respect to the neutral expectation. In this case, admixture mimics the effect of balancing natural selection (that maintains different alleles at high frequencies in the population).[36] Ignoring admixture can lead in this case to spuriously reject the null neutral hypothesis. Conversely, directional natural selection usually generates an excess of rare alleles respect to neutral expectation, either because positive selection makes one allele predominant or because negative selection maintains deleterious alleles at low frequencies.[36] Testing directional natural selection in an admixed population is anticonservative, since admixture increases the proportion of common alleles, rendering more difficult to document a significant excess of rare alleles in a population. In this case, admixture may generate false negative results. Therefore, caution is required when testing the action of natural selection in admixed populations. Fortunately, today it is possible to incorporate the effect of admixture in the null hypothesis of neutrality to be tested, by using computer simulations based on the **coalescent theory**.[37] Software such as *ms*[38] or *Simcoal*[39] allow the researcher to implement more realistic null hypotheses that incorporate admixture.

Population Stratification and Admixture in Epidemiological and Pharmacogenetic Studies

Confounding and modifying variables are classical problems in epidemiology. Statistical association between a pharmacogenetic trait and a locus can have three explanations. (1) The best possible one is that variants at the locus we studied (SNP or haplotype) can be the cause of the association. Mutations defining alleles *2 and *3 of the Thiopurine S-methyltransferase (*TPMT*) polymorphism are classical examples of causal mutations that can be screened to predict toxicity of thiopurine drugs.[40] (2) The locus could not be the causal factor, but may be physically and statistically linked to the causal one, and due to their linkage disequilibrium, we capture the association. The *ABCB1* gene (ATP-binding cassette, subfamily B, member 1), also known as Multi-Drug Resistance 1 (*MDR1*), encodes the membrane-bound ATP-dependent pump P-glycoprotein (Pgp), known to be the transporter of a wide range of drugs. A silent polymorphism in exon 26 shows correlation with levels of *MDR1* expression and activity.[41] Although there is evidence that this substitution is not the causal variant, it is likely to be in linkage disequilibrium, therefore a marker of a causal variant that influences the activity of the transporter.[42] (3) The worst scenario is that association may be due to a confounding exogenous variable, therefore, a spurious association or false positive result is observed. This can be caused by population stratification in general or admixture in particular. In case-control studies, population stratification is a problem that can emerge when the genetic background differs across individuals with different outcomes. On the other hand, false negative results are also possible as a consequence of ignored modifying variables.

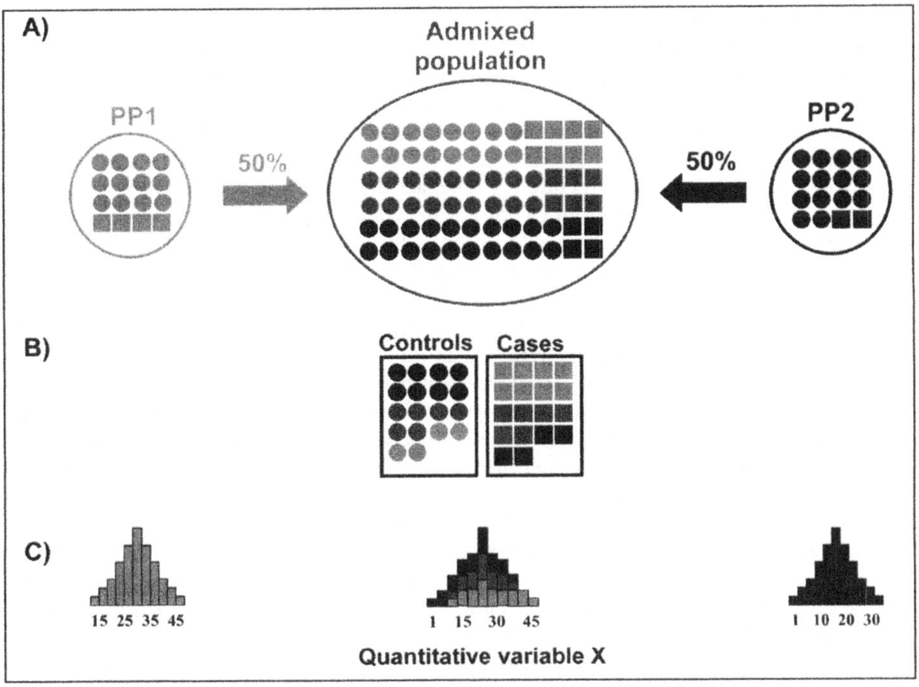

Figure 2. Representation of admixture and the effect of population stratification on genetic association studies. A) Each of the parental populations PP1 (light gray) and PP2 (black) has contributed with 50% to the gene pool of the admixed population (dark gray). Individuals from the admixed population have different levels of admixture. In the figure we have simplified this situation by representing only three levels of individual admixture: individuals whose genomes derive completely from PP1 (light gray), from PP2 (black) and admixed individuals (dark gray). In this figure, normal and disease individuals are denoted by circles and squares, respectively. Note that the incidence of the disease is higher in PP1 than in PP2. In the admixed population, the incidence of the disease is higher among the light gray individuals, intermediate among the dark gray individuals and lower among the black individuals. B) A case-control study: cases are represented by squares and controls by circles. Because the incidence of the disease is higher among the individuals with higher light gray (PP1) ancestry, these will be over-represented among the cases. As a consequence, any allele that has a higher frequency in PP1 than in PP2 may show association with the disease, independently if it actually confers a higher risk. C) Effect of population stratification on the distribution of a quantitative phenotypic variable X, distributed in a normal fashion. In this case, the light gray population (PP1) shows higher values of the variable X than the black population (PP2). As a consequence, in the admixed population, individuals showing low values tend to have more black (PP2) ancestry, and individuals showing high values tend to have more light gray (PP1) ancestry. Also in this case, any allele with higher frequency in PP1 can show correlation with higher values of the phenotypic variable X, even not being causal mutations.

The following two examples illustrate the roles of confounders and modifiers in pharmacogenetic studies. First, in admixed populations, ancestry can be a confounding variable (Fig. 2). Consider a pharmacogenetic study that measures how toxicity after ingestion of a drug depends on the presence of a slow-metabolizer allele (S) of a DME locus that also has a fast-metabolizer allele (F). The presence of adverse drug reactions may be recorded and codified by a dummy variable that classifies individuals as those who showed any toxicity (cases) and those that did not (controls). Assume that this study is performed on a hybrid population with parental populations

P1 and P2 and that the allele S is more frequent in P1. If, for any reason (genetic or not), toxicity is more frequent among individuals with more P1 ancestry, in the studied admixed population these individuals can be over-represented among the group with toxicity (cases) and thus, any allele showing higher frequency in the P1 population than in P2, can show association with toxicity, independently if they are actually responsible (or are in linkage disequilibrium with the responsible allele). The DME locus too, even not being involved in toxicity, can show a spurious association due to population structure (i.e., the fact that cases and controls do not match for ancestry). In this case, ancestry from P1 (or P2) is a confounder. Controlling for ancestry -in a way that cases and controls match, would eliminate the spurious association.[43] In the presented case, the danger of spurious association increases with the differences in incidences of toxicity and in allele frequencies of S among populations P1 and P2.

Second, there are also examples of modifier variables in pharmacogenetic studies. The polymorphism of the Apolipoprotein E (*ApoE*) is an interesting example for studies that evaluate drug efficacy in Alzheimer disease patients. ApoE has three isoforms that differ in efficacy in maintaining and repairing neurons. Among these variants, *ApoE4* is associated with neuropathies in general[44] and with Alzheimer disease in particular.[45] Pharmaceutical clinical trials are commonly classified in four phases, performed on increasing number of individuals, to assess: safety and tolerability of a new drug (Phase I), its clinical efficacy (Phase II and Phase III), and post-launch safety surveillance (Phase IV). Phase II trials are divided into IIA, aimed to defining dosing requirements, and IIB, designed to measure efficacy. If poor efficacy or toxicity is observed during Phase II, the development of a new drug fails and no Phase III will be undertaken. A Phase IIB clinical trial for Alzheimer's disease performed to test cognitive effects of rosiglitazone failed to show significant improvement in 511 patients.[46] On the basis of this negative result, the development program would have terminated. However, when the clinical trial population was genotyped for *ApoE4* and a stratified analysis was performed, each different dose of rosiglitazone showed improvement in patients without the *ApoE4* allele. *ApoE4* carriers did not show any significant improvement. This is an example where a genetic modifier (i.e., an allele responsible for a more severe form of the disease) leads to a false negative result. Identifying this factor and controlling it by a stratified analysis is useful.

Summarizing, when stratification is present and it is possible to identify a stratification variable, comparing a measure of statistical association such as the Odd Ratios (ORs) for the different strata with the overall crude OR could lead to the following situations:

a. In the first example, we have no association for any of the strata (OR = 1), but a different overall OR. In this case, the stratification variable (P1 ancestry) is a confounding variable, associated with both the exposure (alleles, haplotypes or genotypes) and the outcome of interest (higher incidence of toxicity in the P1 population than in the P2). The confounder (or confounding variable) is responsible at least partially for the statistical association between the exposure and the outcome.[47] Epidemiology textbooks suggest that an unbiased estimate of the overall OR, adjusted for the stratification variable, could be obtained through a procedure to combine contingency tables (one for each strata), such as the proposed by Mantel and Haenszel.[48] In pharmacogenetic studies, when the studied population has an internal structure (or it is stratified), there is the risk of false positive association if additionally: (a) the frequencies of genetic variants to be genotyped differ among subpopulations and (b) the frequencies of the phenotypic trait also differ among subpopulations.
b. In the second example, there are significant differences among the ORs for the different strata (alleles of *APOE-4*). Significance of these differences among ORs across strata could be assessed using the Breslow-Day test.[49] Therefore, the stratification variable (the *APOE-4* allele) modifies the effect of the drug on the drug-response variable. If this is the case, adjusting the overall OR for the stratification variable is not appropriate, since the crude OR represents the average association between the *APOE* alleles and the drug-response variable, but it does not take into account the information concerning the different effects observed by the stratification variable across the strata. When the stratification variable modifies the studied association, a stratified analysis is suggested.

c. Finally, if no significant differences among ORs for each strata and the overall OR are observed, the stratification variable does not influence the studied association, and can be excluded from further multivariate analyses.

Therefore, population stratification is important in pharmacogenetics, since different types of variables could confound or modify the studied association. When it is possible to identify the stratification variable, combining different contingency tables or stratified analysis could clarify the roles of confounders and modifiers. However, a problem could occur when several confounders or modifiers are potentially present. In this case, the most common solution is to apply multivariate analyses such as regression models, to adjust the overall OR for the independent effect of the confounding variables. Setakis et al[50] have recently shown that when ancestry is a potential confounder, the classical regression approach performs as well as more recent and sophisticated analyses developed for genetic data.

A different choice to avoid the problem of confounders in pharmacogenetic studies is to use a matching analysis. This is a particular and extreme type of stratified analysis, where each "stratum" includes only one case and the correspondent, matching, control(s). The adjusted OR could be also calculated using the procedure of Mantel and Haenszel.[48] A limitation of matching studies is over-matching, that leads to under-estimate the association. Over-matching may happen when cases and controls are matched for a variable that represents a phase of the studied phenotypic trait. For instance, if the outcome of a study is the therapeutic effect of a drug, matching for a pharmacokinetic trait such as fast- and slow-metabolizer phenotypes can under-estimate or conceal the association, if it exists, because it will reduces the observed differences among cases and controls. Overmatching also happens when cases and controls are matched for a variable associated with the pharmacogenetic variant, but not with the drug-response variable, and therefore is not a confounding variable. In general, when it is not clear if some variables can be confounders or modifiers, it is better to perform a stratified analysis rather than a matching analysis.

Methods for Controlling Population Stratification in Admixed Populations Using Genetic Data

Three approaches are possible in genetic epidemiology and pharmacogenetic studies to control for the effect of population stratification due to admixture: (1) genomic control, (2) structured association and (3) regression.

Genomic Control Approach

The statistical association between a pharmacogenetic trait and genetic variants can be measured using statistical tests such as the Chi-square (χ^2) test, the Fisher exact test, or calculating a measure of association such as the OR. The tests for trend, such as the Cochran-Armitage-trend test, verify the assumption of a linear trend in risk as exposure (i.e., the number of alleles or haplotypes carried by an individual) increases.

For instance, the χ^2 value measures the strength of the association given a number of degrees of freedom, but it is also necessary to measure the significance of the observed value, which indicates the probability of observing that value by chance if the null hypothesis (no association) is true. In addition to the polymorphism to be tested for association, the Genomic Control approach (GC) uses a set of unlinked markers to assess the level of stratification among cases and controls.[51,52] The statistical association between the polymorphism and disease in a 2x3 contingency table of genotypes vs. cases and controls, for an additive genetic model is tested by the Cochran-Armitage-trend test. However, to account for the effect of population stratification, the statistic employed is corrected by a factor λ, which is recovered from the genomic control data. Therefore, the GC approach firstly estimates the inflation factor(s) from the set of "null" loci specified by the user. The software that implemented the method uses these estimates to test for effects at either a single locus (marginal) or two loci (interaction) with relation to a specified response variable. The GC approach has been implemented in three computer programs (Table 1): (1) the *GC program* implements the genomic control models introduced by Devlin and Roeder,[51] (2) the *GCF program* implements the basic Genomic Control approach, but adjusts the p-values for uncertainty in the estimated effect of substructure.[54] Finally, (3) Devlin and Roeder[51] also implemented a Bayesian GC approach, more appropriate for data at genomic scale. This approach treats the putative associated SNP as an outlier with respect to markers used as genomic control and is implemented in the *Gcontrol* software (Table 1).

Structured Association Approach

The most popular structured association approach was developed by Pritchard et al[53-55] to control for population stratification in case-control studies, and is based on three steps:

a. A statistical test is performed to ascertain if cases and controls are sampled from the same population.[55] This test is implemented in the software *Strata* (Table 1).
b. If population stratification is detected, a set of unlinked markers is used to infer aspects of population stratification. In the case of an admixed population, the algorithm implemented in the software *Structure*[54] (Table 1) attempts to estimate the contribution of each of the K parental populations. Actually, K is imposed by the user, but when the number of parental populations is not clear, it is important to explore models with different K (numbers of parental populations) to determine the model with the highest likelihood. In general, the algorithm attempts to redistribute the individuals across K subpopulations fitting as better as possible the classical definition of populations for each of the K subgroups: **Hardy-Weinberg equilibrium** across the loci and linkage equilibrium among them. Because achieving this task is computationally demanding, the algorithm uses a Markov Chain Monte Carlo technique (MCMC). Loosely speaking, to solve the problem of defining the genetic structure of the K parental populations and the observed contribution of these to the considered samples, an ideal approach would be to consider all the possible configurations of genetic structure of parental populations, calculating the likelihood (i.e., the probability of the data given a model) for each of them. However, this task is impossible from the computational point of view because there are too many possible configurations. The MCMC approach (used by the *Structure* algorithm) overcomes this difficulty by browsing across the space of all possible solutions, but sampling a subset of configurations in a way that configurations with higher likelihood have higher chance to be sampled than those with lower likelihood. This is possible because in the mathematical space of all possible solutions, similar solutions (with similar likelihoods) are closer than very different solutions. The MCMC algorithm browses across this space, spending more time in regions where the configurations are more likely and less time in regions where configurations are less likely. If the algorithm browses the space of possible solutions long enough (the length of the browsing is defined by the length of the run specified by the user of the program), it will be able to obtain a reliable likelihood curve from which it can recover the more likely (or maximum likelihood) solutions. The *Structure* algorithm uses a Bayesian approach. Differently from a pure maximum likelihood approach, the Bayesian approach allows to incorporate information which is external with respect to the probabilistic model. This external information can be data or beliefs the researcher has and should be formalized in the form of prior probabilities that are multiplied by the likelihood. In the case of the analysis of population stratification in admixed populations, allele frequencies in at least some of the parental populations can be incorporated into the model, to improve inferences about admixture. Summarizing, the *Structure* algorithm (when an admixture model is used) will estimate for each individual, which portion of the sampled individual genome belong to which of the K parental populations. After this, a third step can be performed by applying a test for association controlling for population stratification.
c. The final step is testing for association for each locus (using the software *Strata*,[55] Table 1), conditioning on the ancestry of individuals in the sample.

Table 1. Methods and software to control for population stratification in epidemiological studies

Method	Software	Environment	Ref.	Website
Genomic control	*GC*	R programming language for Linux	51	http://wpicr.wpic.pitt.edu/WPICCompGen/software.html
Corrected Genomic control	*GCF*	R programming language for Linux	52	http://wpicr.wpic.pitt.edu/WPICCompGen/software.html
Bayesian Genomic control	*Gcontrol*	Windows and Solaris executables, C source code	51	http://wpicr.wpic.pitt.edu/WPICCompGen/software.html
Structured association	*Structure and Strata*	Windows and Linux/Unix	53-55	http://pritch.bsd.uchicago.edu/software.html
Structured association	*AdmixMap*	Window and Linux	56	http://www.ucd.ie/genepi/admixmap/index.html
Principal Component Analysis	*Eigenstrat*	Linux	62	http://genepath.med.harvard.edu/~reich/EIGENSTRAT.htm
Regression using haplotypes	*HaploStats*	R environment	60-61	http://mayoresearch.mayo.edu/mayo/research/biostat/schaid.cfm

A similar structured association method has been proposed by Hoggart et al[56] and implemented in the software *AdmixMap* (Table 1). This method also uses a MCMC approach and combines classical and Bayesian approaches to measure association controlling for ancestry.

Regression Approach

A classical regression approach can also be used to control for admixture and population stratification. Recently, Wang et al[57,59] have used logistic regression to control for population stratification in case-control association studies. Although Wang et al[57] have illustrated the regression approach using a single marker model, it is straightforward to expand this approach to a larger set of markers used to control for ethnicity. The logistic regression approach is intuitive, flexible, computationally fast and easy to implement using different statistical and genetic epidemiology packages, and has recently shown that provides good protection against the effect of substructure.[50]

The logistic regression model assumes that the probability of disease depends on a set of variables x_i in the following way:

$$P(d = 1 | x_i) = 1/\{1+\exp[-(\beta_0 + \beta_1 x_1 + \ldots + \beta_p x_p)]\}$$

In this formula, d indicates the presence (d = 1) or absence (d = 0) of the disease, x_i are a set of variables and the β_i coefficients are parameters that represent the effects of the variables x_i on the risk of the disease.[59] In genetic epidemiology studies, at least one of the x_i variables is the genotype of the candidate polymorphism which association with the disease is evaluated. The other variables can be potential confounders for which the investigator has collected data. Ethnicity or admixture is one of the potential confounders in admixed populations. Genotypes for a set of markers can be used to control for the effect of population stratification in admixed populations. For this purpose, it is accepted that markers that show large differences in frequency among ethnic groups (Ancestry Informative Markers, AIMs) are more informative and should be preferentially used.

For the seek of generality, it is natural to consider that the logistic regression approach proposed by Wang et al[57] can be accommodated in the **Generalized Linear Model (GLM)** framework proposed by Schaid et al[60] and Tate et al[61] and implemented in the software *HaploStats* (Table 1). This method is suitable for haplotypes of candidate genes and allows also the calculation of ORs and their confidence intervals for each considered haplotype. The GLM approach is adapted for different experimental designs because the test of association is not restricted to a binary trait. Instead, it allows for ordinal or quantitative variables as outcomes. Moreover, the *HaploStats* method shows some advantages: (1) it allows for different models of inheritance (dominant, recessive or additive); (2) considering that haplotypes are not observed but inferred from genotypes, it assesses the significance of association taking into account the uncertainty about haplotype inferences. Finally, (3) it infers the chromosome phase and haplotype frequencies using an expectation-maximization algorithm and conditioning on all the observed data and estimated regression parameters. Although there has not been any systematic assessment of how well *HaploStats* handles population stratification, the encouraging results of Wang et al[57] and Setakis et al[50] with logistic regression suggest that regression approaches can be successfully used to control for population stratification in admixed populations.

Recently, Price et al[62] have used **Principal Component Analysis**, a multivariate method largely used in historical population genetics, to correct for stratification in association studies. The algorithm, implemented in the software *Eigenstrat* (Table 1) seems to perform better than the genomic control approach.

Concluding Remarks

Stratification is an important factor to take into account in pharmacogenetics and pharmacogenomics, either in population genetics and association studies. In population genetics studies, recognizing and controlling admixture is important to make correct inferences about the evolutionary forces that have shaped the patterns of diversity observed in genes relevant in pharmacogenetics. In association studies in admixed populations, differences in ethnic backgrounds of people showing different outcomes can produce false positive results. Fortunately, it is possible to use genetic markers to control for the effect of population stratification. We have currently available on public databases a large set of markers from which we can select a set of informative polymorphisms to be used to measure ancestry of individuals participating in a pharmacogenetic study. In principle, using markers with large differences among the parental populations (AIMs) is convenient because they require a relatively small number of loci to be genotyped to achieve good admixture estimations. However, it is not clear yet how the ascertainment of the used AIMs can bias the results, particularly if these markers are not equally informative for the different parental populations. Studies on this direction are necessary. Moreover, we have today a set of statistical approaches to control for population stratification and admixture, which include classical methods such as regression, and recently developed structured association methods that rely on Bayesian inferences and computationally intensive use of MCMC algorithms (*Structure* and *AdmixMap*). With a conscious use of these resources by the investigators, we hope that the danger of false positive results will be reduced in pharmacogenetic studies.

Acknowledgements

Authors are grateful to Dr. Guilherme Suarez-Kurtz for the invitation to write this chapter and his suggestions, and to an anonymous reviewer who contributed to improve the first version of the manuscript. E.T-S. research is supported by the Brazilian Federal agency CNPq (Conselho Nacional de Desenvolvimento Científico e Tecnológico).

Glossary of Terms

Chi-Square (χ^2) Test: Test of association to assess the hypothesis that exposure categories (alleles, genotypes or haplotypes) are associated with outcome categories such as the presence of a disease, adverse drug reactions or response to a therapy (in general classified as cases and controls). Data are arranged in a two-dimensional table with C columns (outcome categories) and R rows (exposure categories). The Chi-square statistic involves the differences between observed and the expected frequencies, where the expected frequencies are computed under the hypothesis of independence between the exposure and the outcome. For large values of the Chi-square statistic, the null hypothesis is rejected in favor of the alternative hypothesis of association. The number of exposures categories depends on the nature of genetic data. Alleles or haplotypes can be used to test if their frequencies differ among cases and controls. Genotypes can also be used, considering codominant models with three exposure categories (AA, Aa and aa) or dominant/recessive models of inheritance. The χ^2-test does not perform well if the number of observation for a cell in the table to be tested is small (lower than 5). In this case, the Fisher exact test is more appropriate. Moreover, when there is a 2 x 2 table, the Chi-square test could be replaced with an Odds Ratio (see the term Odds Ratio in this glossary), which has a clearer interpretation.

Coalescent Theory: Theory that models the evolution of k copies of genes (or a genomic region) sampled from a population. Using stochastic processes, the theory describes the genealogy of the sampled genes backward in time across generations, until the existence of the Most Recent Common Ancestor (MRCA) of the sampled genes. During the history of this genealogy, coalescent events happen: the fact that at certain generations, two copies of a gene share a common ancestor gene in the previous generation. The theory allows predictions about patterns of genetic diversity and to easily implement computer simulations that model (backward in time) the demographic history of populations. The standard coalescent considers k copies of a gene sampled from a single population without recombination, considers only genetic drift and allows to superimpose different mutation models on the gene genealogy. However, further developments can incorporate other evolutionary factors such as recombination, population structure, gene flow and changes in population sizes.

Cochran-Armitage-Trend Test: A statistical test to verify the assumption of a linear trend in risk as exposure increases. This raises the question of how well the variation in risk over the ordinal exposure categories is captured by a linear assumption. The test is performed on a contingency table, as in the Fisher exact test or the Chi-square test. In genetic association studies, it can be used to test an additive model. For instance, given a locus with two alleles A and a, it is possible to test if the effect of the allele A is 0 for the genotype aa, e for the genotype Aa and 2e for the genotype AA. In this case, the Fisher exact test or the Chi-square test would only consider genotypes as categorical variables, without incorporating any trend. The Armitage-trend test can also be used to test the effect of the presence of a specific haplotype H. Calling any of the other observed haplotypes as X, it is possible to test the hypothesis that the distribution of genotypes XX (no copies of the H haplotype), XH (1 copy of the H haplotype) and HH (2 copies of the H haplotypes) differ among cases and controls.

Fisher Exact Test: Non-parametric test of association between two categorical variables. Its use in genetic association studies is very similar than the χ^2 test (see the term Chi-square test in this glossary), but is preferred when the observations for any of the cells of the contingency table is small. This is because the Fisher exact test performs inferences based on exact distributions rather than approximations. From the practical point of view, many statistical packages perform the Fisher exact test for 2 x 2 tables, but for larger tables the task is computationally more intensive.

***F_{ST}*:** It quantifies the proportion of the total genetic variance observed in a set of populations that is due to inter-population comparisons. F_{ST} can be used as a genetic distance between populations (i.e., a measure of how different are allele or haplotype frequencies among two populations).

Generalized Linear Model (GLM): A general set of models to assess the role of explanatory variables on dependent variables. The class of GLM is an extension of traditional linear models and is, therefore, applicable to a wider range of data analysis problems. These models include regression for continuous response variables, logistic and probit models for binary data (such as those included in case-control designs), and log-linear models for multinomial data.

Haplotype: The combination of allelic states of a set of polymorphic markers lying on the same DNA molecule (linked group), e.g., a chromosome or a region of a chromosome.

Hardy-Weinberg Equilibrium: The simplest model of equilibrium in population genetics. It implies that: (1) given a locus with two alleles A and a and frequencies p and q respectively, the expected genotype frequencies for AA, Aa and aa are p^2, 2pq and q^2 and (2) that these allele and genotypes frequencies do not change in time.

Hypothesis of Neutrality: Hypothesis that the pattern of genetic variation observed for a gene or a genomic region, can be explained solely on the basis of mutation and genetic drift (random changes in allele frequencies), without requiring the action of natural selection. Neutrality is usually used as null hypothesis to test the action of natural selection. When the observed pattern of diversity is incompatible with the action of only mutation and drift, it is accepted to infer that natural selection has acted. The null hypothesis of neutrality may be rendered more realistic by incorporating other evolutionary forces such as recombination, gene flow among subpopulations and changes in population size. These more complex null hypotheses can be tested using computer simulations.

Isolation-by-Distance: The decline of population similarity with geographical distance due to the fact that geographically close populations exchange more migrants than distant ones, so that the degree of relative isolation between localities is roughly proportional to their geographic distance.

Linkage Disequilibrium: Statistical association between two allelic states on the same chromosome. Given the loci A (with alleles A1 and A2) and B (with alleles B1 and B2), under linkage equilibrium (independence), it is expected to observe the haplotype A1-B1 with a frequency that is the product of the frequencies for each of the two alleles. If the observed haplotype frequency is significantly higher or lower, there is linkage disequilibrium. There are several measurements of linkage disequilibrium, but the most popular and robust one seems to be the square of the correlation coefficient (r^2) for the presence of two alleles across the sampled chromosomes.

Mutation-Selection Equilibrium: Mutation and selection are opposing forces determining the frequency of mutant alleles in the population. A stable level of genetic diversity (equilibrium) is reached when the rate at which new variants are introduced by mutation is balanced by their eventual elimination by selection.

Odds Ratio (OR): A measure of association for a 2 x 2 contingency table, that is useful to assess the association in case-control studies, when the exposure variable presents only two categories, such as two alleles or two classes of genotypes based on dominant or recessive models of inheritance (see also Chi-square test in this glossary). Specifically, the OR is the ratio of the odds of the exposed individuals respect to the non-exposed individuals. In absence of association, OR = 1. An OR >1 represents positive association and OR <1 represents negative association. The strength of association increases with the deviation from 1. The OR could be interpreted as the risk to develop the disease for exposed subjects, compared with non-exposed subjects. For example, an OR = 2 means that exposed subjects have a double risk to develop the disease compared with non-exposed subjects. Confidence intervals for an OR give a measure of the significance of an association: if the lowest value of the interval is >1 then there is a significant positive association, while if the upper value of the interval is <1 then there is a significant negative association.

Out of Africa: A model to explain the origin of modern humans which proposes that the transition from *Homo erectus* to *Homo sapiens* took place less than two hundred thousand years in Africa, and that these humans replaced the hominids already present on other continents. Several findings support this hypothesis: humans lack genetic diversity reflecting a recent origin, global genetic diversity is a subset of African diversity; Neanderthals lie outside the range of human diversity showing their own evolutionary history.

Principal Component Analysis (PCA): A multivariate statistical technique to reduce dimensionality on datasets problems involving several variables. In the case of PCA, the goal of dimensionality reduction is to reduce a large set of variables to a subset of new variables, the Principal Components, which are linearly dependent on the original sets of variables, eliminating redundancy and identifying sets of variables that contain independent information among them. The procedure works in a way that each Principal Component is preferentially determined by a set of original variables that are correlated among them, and therefore, contain redundant information. On the other hand, the different components are computed in a way that correlations among them are as low as possible. Principal Components are sorted (first, second, ..., k-th), and represent decreasing portions of the total variance contained in the original dataset. In this way, if the original variables contain redundant information, few principal components are usually enough to adequately represent the data. In the use of PCA to control for population stratification by admixture, the technique should be able to identify set of alleles whose allele frequencies are correlated because they derive from the same parental population.

References

1. Chakraborty R. Gene admixture in human populations: Models and predictions. Am J Phys Anthrop 1986; 29(S7):1-43.
2. Long JC. The genetic structure of admixed populations. Genetics 1991; 127:417-28.
3. Pfaff CL, Parra EJ, Bonilla C et al. Population structure in admixed populations: Effect of admixture dynamics on the pattern of inkage disequilibrium. Am J Hum Genet 2001; 68:198-207.
4. Parra EJ, Kittles RA, Argyropoulos G et al. Ancestral proportions and admixture dynamics in geographically defined African Americans living in South Carolina. Am J Phys Anthropol 2001; 114:18-29.
5. Salzano FM, Bortolini MC. The evolution and genetics of Latin American populations. Cambridge University Press, 2002.

6. Miljkovic-Gacic I, Ferrell RE, Patrick AL et al. Estimates of African, European and Native American ancestry in Afro-Caribbean men on the island of Tobago. Hum Hered 2005; 60:129-33.
7. Comas D, Calafell F, Mateu E et al. Trading genes along the silk road: mtDNA sequences and the origin of central Asian populations. Am J Hum Genet 1998; 63:1824-38.
8. Loubser O, Marais AD, Kotze MJ et al. Founder mutations in the LDL receptor gene contribute significantly to the familial hypercholesterolemia phenotype in the indigenous South African population of mixed ancestry. Clin Genet 1999; 55:340-5.
9. Fuselli S, Dupanloup I, Frigato E et al. Molecular diversity at the CYP2D6 locus in the Mediterranean region. Eur J Hum Genet 2004; 12:916-24.
10. Thompson EE, Kuttab-Boulos H, Witonsky D et al. CYP3A variation and the evolution of salt-sensitivity variants. Am J Hum Genet 2005; 75:1059-69.
11. Patin E, Barreiro LB, Sabeti PC et al. Deciphering the ancient and complex evolutionary history of human arylamine N-acetyltransferase genes. Am J Hum Genet 2006; 78:423-36.
12. Fuselli S, Gilman RH, Chanock SJ et al. Analysis of nucleotide diversity of NAT2 coding region reveals homogeneity across Native American populations and high intra-population diversity. Pharmacogenomics J 2006, [Epub ahead of print].
13. Andersson T, Flockhart DA, Goldstein DB et al. Drug-metabolizing enzymes: Evidence for clinical utility of pharmacogenomic tests. Clin Pharmacol Ther 2005; 78:559-81.
14. Need AC, Motulsky AG, Goldstein DB. Priorities and standards in pharmacogenetic research. Nat Rev Genet 2005; 37:671-81.
15. Stram DO. Tag SNP selection for association studies. Genet Epidemiol 2004; 27:365-74.
16. Carlson CS, Eberle MA, Rieder MJ et al. Selecting a maximally informative set of single-nucleotide polymorphisms for association analyses using inkage disequilibrium. Am J Hum Genet 74:106-20.
17. Foster MW, Sharp RR. Beyond race: Towards a whole-genome perspective on human populations and genetic variation. Nat Rev Genet 2004; 5:790-6.
18. Sans M. Admixture studies in Latin America: From the 20th to the 21st century. Hum Biol 2000; 72:155-77.
19. Morton NE. Genetic epidemiology. Ann Hum Genet 1997; 61:1-13.
20. Lander ES, Linton LM, Birren B et al. Initial sequencing and analysis of the human genome. Nature 2001; 409:860-921.
21. Risch N, Merikangas K. The future of genetic studies of complex human diseases. Science 1996; 273:1516-7.
22. Sachidanandam R, Weissman D, Schmidt SC et al. A map of human genome sequence variation containing 1.42 million single nucleotide polymorphisms. Nature 2001; 409:928-33.
23. Packer BR, Yeager M, Staats B et al. SNP500Cancer: A public resource for sequence validation and assay development for genetic variation in candidate genes. Nucleic Acids Res 2004; 32:D528-32.
24. Zondervan KT, Cardon LR. The complex interplay among factors that influence allelic association. Nat Rev Genet 2004; 5:89-100.
25. Reich DE, Lander ES. On the allelic spectrum of human disease. Trends Genet 2001; 17:502-10.
26. Ewens WJ. The sampling theory of selectively neutral alleles. Theor Popul Biol 1972; 3:87-112.
27. Zwick ME, Cutler DJ, Chakravarti A. Patterns of genetic variation in Mendelian and complex traits. Annu Rev Genomics Hum Genet 2000; 1:387-407.
28. Pritchard JK. Are rare variants responsible for susceptibility to complex diseases? Am J Hum Genet 2001; 69:124-37.
29. Lohmueller KE, Pearce CL, Pike M et al. Meta-analysis of genetic association studies supports a contribution of common variants to susceptibility to common disease. Nat Genet 2003; 33:177-82.
30. Mockenhaupt FP, Cramer JP, Hamann L et al. Toll-like receptor (TLR) polymorphisms in African children: Common TLR-4 variants predispose to severe malaria. Proc Natl Acad Sci USA 2006; 103:177-82.
31. Fearnhead NS, Wilding JL, Winney B et al. Multiple rare variants in different genes account for multifactorial inherited susceptibility to colorectal adenomas. Proc Natl Acad Sci USA 101:15992-7.
32. Cohen JC, Kiss RS, Pertsemlidis A et al. Multiple rare alleles contribute to low plasma levels of HDL cholesterol. Science 2004; 305:869-72.
33. Maitland ML, Grimsley C, Kuttab-Boulos H et al. Comparative genomics analysis of human sequence variation in the UGT1A gene cluster. Pharmacogenomics J 2006; 6:52-62.
34. Barbujani G, Goldstein DB. Africans and Asians abroad: Genetic diversity in Europe. Annu Rev Genomics Hum Genet 2004; 5:119-50.
35. Mulligan CJ, Hunley K, Cole S et al. Population genetics, history, and health patterns in native Americans. Annu Rev Genomics Hum Genet 2004; 5:295-315.
36. Bamshad M, Wooding SP. Signatures of natural selection in the human genome. Nat Rev Genet 2003; 4:99-111.

37. Tarazona-Santos E, Tishkoff SA. Divergent patterns of inkage disequilibrium and Haplotype structure across global populations at the interleukin-13 (IL13) locus. Genes Immun 2005; 6:53-65.
38. Hudson RR. Generating samples under a Wright-Fisher neutral model of genetic variation. Bioinformatics 2002; 18:337-8.
39. Laval G, Excoffier L. SIMCOAL 2.0: A program to simulate genomic diversity over large recombining regions in a subdivided population with a complex history. Bioinformatics 2004; 20:2485-7.
40. Schaeffeler E, Fischer C, Brockmeier D et al. Comprehensive analysis of thiopurine S-methyltransferase phenotype-genotype correlation in a large population of German-Caucasians and identification of novel TPMT variants. Pharmacogenetics 2004; 14:407-17.
41. Hoffmeyer S, Burk O, von Richter O et al. Functional polymorphisms of the human multidrug-resistance gene: Multiple sequence variations and correlation of one allele with P-glycoprotein expression and activity in vivo. Proc Natl Acad Sci USA 2000; 97:3473-8.
42. Soranzo N, Cavalleri GL, Weale ME et al. Identifying candidate causal variants responsible for altered activity of the ABCB1 multidrug resistance gene. Genome Res 2004; 14:1333-44.
43. Suarez-Kurtz G. Pharmacogenomics in admixed populations. Trends Pharmacol Sci 2005; 26:196-201.
44. Mahley RW, Weisgraber KH, Huang Y. Apolipoprotein E4: A causative factor and therapeutic target in neuropathology, including Alzheimer's disease. Proc Natl Acad Sci USA 2006; 103:5644-51.
45. Farrer LA, Cupples LA, Haines JL et al. Effects of age, sex, and ethnicity on the association between apolipoprotein E genotype and Alzheimer disease: A meta-analysis. APOE and alzheimer disease meta analysis consortium. JAMA 1997; 278:1349-56.
46. Tang MX, Stern Y, Marder K et al. The APOE-epsilon4 allele and the risk of Alzheimer disease among African Americans, whites, and Hispanics. JAMA 1998; 279:751-5.
47. Szklo M, Nieto FJ. Defining and assessing heterogeneity of effects: Interaction. In: Szklo M, Nieto FJ, eds. Epidemiology-Beyond the Basics. Aspen Publisher, 2000:Cap 6:221-253.
48. Mantel N, Haenszel W. Statistical aspects of the analysis of data from retrospective studies of disease. J Natl Cancer Inst 1959; 22:719-48.
49. Breslow NE, Day NE. Statistical methods in cancer research. Volume I - The analysis of case-control studies. IARC Sci Publ 1980; 32:5-338.
50. Setakis E, Stirnadel H, Balding DJ. Logistic regression protects against population structure in genetic association studies. Genome Res 2006; 16:290-6.
51. Devlin B, Roeder K. Genomic control for association studies. Biometrics 1999; 55:997-1004.
52. Devlin B, Bacanu SA, Roeder K. Genomic Control to the extreme. Nat Genet 2004; 36:1129-30.
53. Pritchard JK, Rosenberg NA. Use of unlinked genetic markers to detect population stratification in association studies. Am J Hum Genet 1999; 65:220-8.
54. Pritchard JK, Stephens M, Donnelly P. Inference of population structure using multilocus genotype data. Genetics 2000; 155:945-59.
55. Pritchard JK, Stephens M, Rosenberg NA et al. Association mapping in structured populations. Am J Hum Genet 67:170-81.
56. Hoggart CJ, Parra EJ, Shriver MD et al. Control of confounding of genetic associations in stratified populations. Am J Hum Genet 72:1492-1504.
57. Wang Y, Localio R, Rebbeck TR. Bias correction with a single null marker for population stratification in candidate gene association studies. Hum Hered 2005; 59:165-75.
58. Wang Y, Localio R, Rebbeck TR. Evaluating bias due to population stratification in case-control association studies of admixed populations. Genet Epidemiol 2004; 27:14-20.
59. Schlesselman JJ. Case-control studies, design, conduct, analysis. Oxford University Press, 1982.
60. Schaid DJ, Rowland CM, Tines DE et al. Score tests for association between traits and Haplotypes when linkage phase is ambiguous. Am J Hum Genet 2002; 70:425-34.
61. Lake SL, Lyon H, Tantisira K et al. Estimation and tests of Haplotype-environment interaction when linkage phase is ambiguous. Hum Hered 2003; 55:56-65.
62. Price AL, Patterson NJ, Plenge RM et al. Principal components analysis corrects for stratification in genome-wide association studies. Nat Genet 2006; 38:904-9.

Chapter 3

Admixture in North America

Esteban J. Parra*

Abstract

The history of North America has been marked by the encounter of populations from different continents. The discovery of the New World began a period defined by human migrations at a much larger scale than in previous history. This movement of people, voluntary or forced, changed profoundly the human landscape. As populations came into contact, admixture followed in varying degrees depending on the circumstances. Today, many people living in the US, Mexico, Canada and the Caribbean can trace their ancestry to more than one continent. The most important genetic contributions came from the indigenous Native American groups, Western Europeans and West Africans, although there have also been influences from other regions, such as East Asia and South Asia. We can reconstruct and interpret this history of migration and admixture using genetic markers. A very complete perspective can be obtained when analyzing autosomal markers (located on any of the chromosomes other than the sex-determining chromosomes, which are inherited from both parents), maternally transmitted mtDNA markers and Y-chromosome markers, which are transmitted from fathers to sons. In many cases, the maternal and paternal admixture histories are remarkably different, so including mtDNA and Y-chromosome markers can provide a much better picture than that offered by the autosomal markers. Thus, using genetic markers we can reconstruct history at the individual and population level, even in the absence of a historical record. In this chapter, I provide an overview of admixture in North America, with a particular emphasis on the two major admixed groups: African Americans (and African Caribbeans) and Hispanics. I also discuss the implications of the history of admixture for the distribution of the genetic variation involved in drug metabolism and drug response and the potential consequences of population stratification in candidate gene association studies in recently admixed populations.

Brief History of the Main North American Admixed Populations

African Americans and African Caribbeans

The history of African Americans can be traced back to 1619, when the first Africans arrived at the British colonies (Jamestown), although as early as 1526 the presence of African slaves was reported in Spanish expeditions to what would become the United States (South Carolina, Georgia, Florida and New Mexico). Institutional slavery began very soon after. Nevertheless, it was not until the beginning of the 18th century that the arrival of enslaved Africans reached increased rates, in parallel with the demand for workers to cultivate the tobacco, indigo, and rice plantations in the southern colonies. The highest numbers arrived in the decade from 1790-1800 and the early years of the 19th century. In 1808 slave trade became illegal but it continued at a low rate for several more decades. Estimates of the total number of enslaved

*Esteban J. Parra—Department of Anthropology, University of Toronto at Mississauga, 3359 Mississauga Rd. N., Mississauga, Ontario L5L 1C6, Canada. Email: eparra@utm.utoronto.ca

Pharmacogenomics in Admixed Populations, edited by Guilherme Suarez-Kurtz.

Africans brought into the United States differ but generally accepted numbers range between 380,000 and 570,000.

Although it is very difficult to precisely determine the ethnic origin of the enslaved Africans, information from shipping lists has provided an approximate picture of their geographic provenance. The slave trade affected a very wide area of western and western-central Africa, mainly the coastline between the present day countries of Senegal in the north and Angola in the south. The most important regions were Senegambia (Gambia and Senegal), Sierra Leone (Guinea and Sierra Leone), Windward Coast (Ivory Coast and Liberia), Gold Coast (Ghana), Bight of Benin (from the Volta river to the Benin river), Bight of Biafra (east of Benin river to Gabon), and Angola (Southwest Africa, including part of Gabon, Congo and Angola). Curtin[1] has offered, based on data on the English trade of the 18th century (the peak of the Atlantic slave trade), estimates of the proportional contribution by areas. His analysis shows that Angola and Bight of Biafra were the regions contributing the highest numbers of slaves imported into the North American mainland (around 25% each). It is important to point out, however, that there were significant differences in ethnic origin depending on the port of entry in the US, and the figures for the Colonies of Virginia and South Carolina differed considerably.

The history of African Americans has been marked not only by their forced migration from Africa, but also by their admixture with the other ethnic groups that they met when they arrived in North America, namely Europeans and Native Americans. However, few historical records address the issue of admixture. Additionally, there have been important factors that, in the time since the abolition of slavery until the present, have configured the present African-American population. Of special interest is the pattern of migrations of African Americans within the US over the past 150 years. In this sense, the redistribution of African Americans in the southern states during the 19th century and the Great Migration from the rural south to the urban areas in the north beginning after World War I are of particular relevance, and have had an enormous impact in defining the present distribution of the African-American population in the US.[2] According to the 2004 American Community Survey, there are approximately 34.8 million people who identify themselves as Black or African American in the US (12.2% of the total population). The majority of the African American population lives in the eastern states of the US and the largest percentages of African Americans with respect to the total population are found in the District of Columbia, Maryland, South Carolina, Georgia, Louisiana and Mississippi (Fig. 1).

Finally, during colonial times the situation in the islands of the Caribbean was quite different from the US mainland. In addition to the Spanish, who concentrated their colonization effort in Cuba, Hispaniola and Puerto Rico, other European powers (mainly England, France, and Holland) established colonies in the Caribbean during the 17th century.[3] The indigenous populations had been decimated by warfare, disease and forced labor, and millions of Africans were forcefully brought to the Caribbean during the slave trade in order to work in the sugarcane plantations, which were the mainstay of the economy of the islands. It has been estimated that the British and French islands received around three million enslaved Africans between 1601 and 1870.[4] In terms of absolute numbers, many more enslaved Africans were brought to the Caribbean than to the US during the slave trade. Additionally, the enslaved Africans in the islands of the Caribbean far outnumbered the white settlers. For example, in 1775 the number of enslaved Africans in Jamaica was approximately 200,000 and the number of white settlers around 12,000.[5] The conditions of slavery were particularly harsh in this region and the death rates were extremely high, surpassing the birth rates.[4] The abolition of the slave trade in 1807 and slavery in 1833 created a shortage of labor in the British West Indies, and tens of thousands of indentured servants (laborers under contract for a specified time in return for payment of travel expenses and maintenance) were brought to the region (mainly to Trinidad, British Guyana and Jamaica). Most of the indentured servants came from South Asia (India), China and the Portuguese islands of Madeira and the Azores, further defining the demographic picture of the islands of the Caribbean.[6]

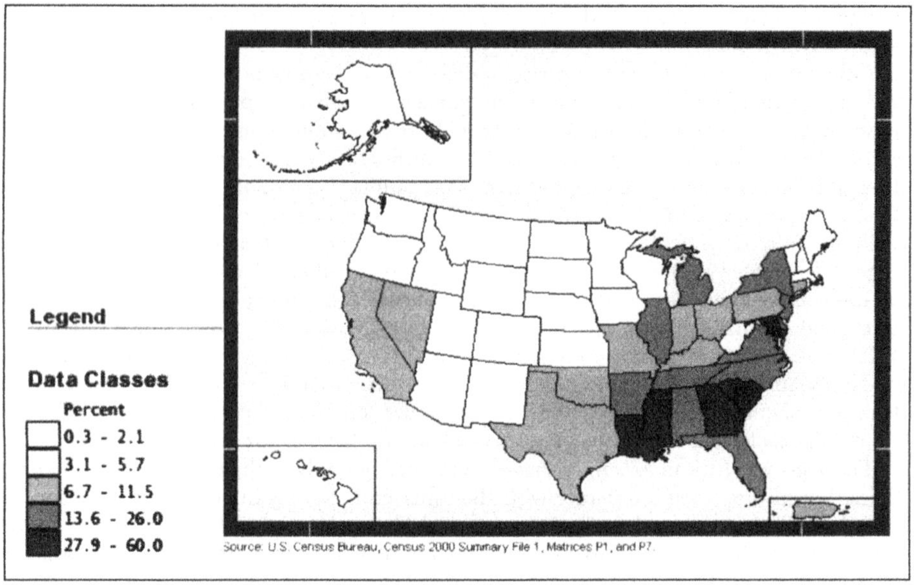

Figure 1. Percentage of the population in the US who are black or African American. Map showing data by state. Source: US Census Bureau, available at http://www.census.gov/.

Hispanics

In the US, the terms "Hispanic" or "Latino" are generally employed to identify persons of Latin American origin or descent. Although this definition lumps together people with very different historical, cultural and linguistic backgrounds, this classification is widely used. Even though Central America, the Caribbean, and South America were for centuries under the domination of Spain and Portugal, they have had quite different regional histories, both before and after the Colonial period. Populations from four continents, North and South America, Europe, and Africa have contributed to the formation of contemporary Hispanic populations. My main objective here is to discuss the anthropological background of the three main Hispanic groups currently living in North America: Mexicans (and Mexican Americans), Cubans and Puerto Ricans.

In Mexico, intermixture of Spanish men with Indigenous American women began soon after the arrival of Hernán Cortés in 1521 and continued through the three centuries of Spanish domination in "New Spain", configuring the Mexican population both biologically and culturally. Today, the majority of Mexicans (approximately 60% of the total population of more than 100 million people) are "mestizos", a term used to describe persons of mixed European/Native American ancestry. It is important to mention that there was also a substantial African presence in Mexico during the Spanish rule. Curtin[1] estimated the number of enslaved Africans brought in Mexico during the entire period of the slave trade to be around 200,000. In fact, in the early colonial period, enslaved Africans outnumbered the Spanish settlers in Mexico.[7] The West African contribution was higher around the Gulf coast (Campeche, Yucatan, Tabasco and Veracruz) and areas of Southwest Mexico (Oaxaca, Guerrero), regions where the largest Afro-Mexican communities in Mexico are located today.

In the Caribbean Colonies (Cuba and Puerto Rico), the Native American population was far smaller than in Mexico, and was decimated by slavery and disease soon after the first contact with the Europeans. Nevertheless, the rate of admixture during the initial phases of the colonization was high enough to result in an appreciable genetic contribution from the Arawaks (Taino) and Caribs, the original inhabitants of the Hispanic Caribbean. Another distinctive

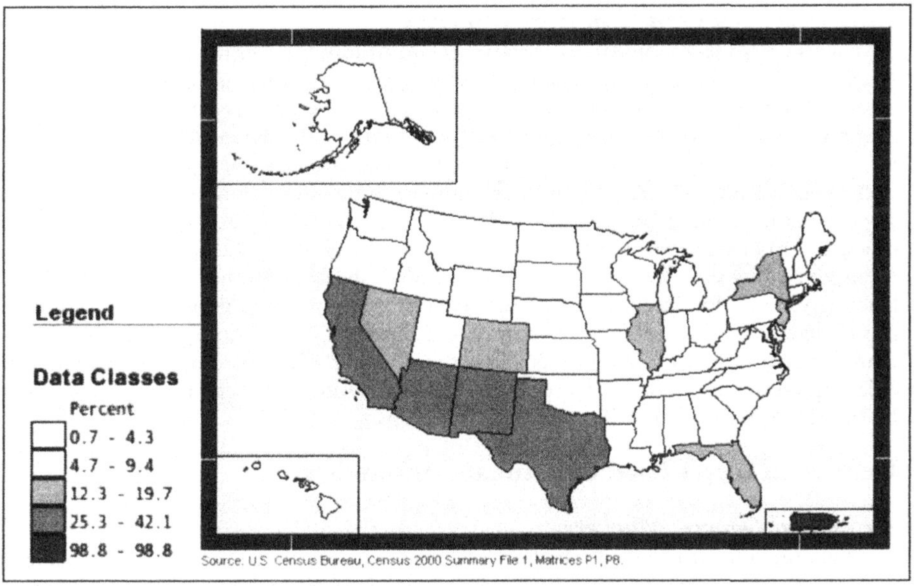

Figure 2. Percentage of the population in the US who are Hispanic. Map showing data by state. Source: US Census Bureau, available at http://www.census.gov/.

feature of this region is a significant African influence, which is also reflected in many aspects of the present societies of countries like Cuba, Puerto Rico, and the Dominican Republic. Enslaved Africans were brought to work in the sugar plantations in large numbers, even outnumbering the population of European origin.[8] Accordingly, the African contribution in contemporary Cubans and Puerto Ricans has been higher than in other Hispanic populations.

Hispanics have become the largest minority in the US. According to the American Community Survey, in 2004 there were around 40.5 million Hispanics in the US (approximately 14% of the total US population). Hispanics constitute a large proportion of the population in the southwest of the US (California, Arizona, New Mexico and Texas) and also in states such as New York, New Jersey, Illinois and Florida (Fig. 2). Mexicans are the largest Hispanic group, with more than 25 million people (64% of the total Hispanic population). The Mexican presence in what is now the US has a long history. A substantial portion of southwestern US was Mexican territory, until it was annexed by the US after the US-Mexican war (1846-1848). However, most of the Mexican migration to the US has occurred since the 20th century. By the 1900s, the total Mexican-American population was estimated to be between 380,000 and 560,000. The first major wave of migration to the U.S. occurred after the Mexican revolution, between 1910 and 1930. Mexican immigration, both legal and illegal has continued to increase since then, particularly after the 1950s. In the US, the states with the largest communities of Mexican ancestry are California, Texas, Arizona and Illinois. The second largest Hispanic group in the US is the Puerto Rican community, with approximately 3.9 million people (around 10% of the total Hispanic population). Migration from Puerto Rico (a self-governing commonwealth in association with the US) to the mainland began in the early 1940s and concentrated mainly in the northeastern states. Currently, the largest Puerto Rican communities in the US are found in New York, Florida, New Jersey, Pennsylvania and Massachusetts. Finally, there is also a large Cuban community in the US, comprising 1.4 million people (3.6% of the total Hispanic population). There were several waves of migration from Cuba to the US starting in 1959, and the majority of the Cubans have settled in the state of Florida.

Other Admixed Groups in North America

This chapter focuses on African Americans and African Caribbeans and also on the main North American Hispanic groups (Mexican and Mexican Americans, Cubans and Puerto Ricans). These are the admixed groups for which most of the genetic data have been collected. However, it is important to mention that admixture has been widespread in North America and has not been restricted to these groups. In Canada, African Canadians comprise approximately 2% of the population (662,000 people). Many of them are immigrants from the Caribbean, but some trace their roots to Black Loyalists who supported the British during the American Revolution and to fleeing slaves who arrived in Canada through the Underground Railroad.[9] The Métis, one of the three federally recognized Canadian aboriginal groups, is an admixed population resulting from the admixture of indigenous Canadians and Europeans. This admixture can be traced back to the initial colonization of Canada by the French and the British.[10] Currently, there are almost 300,000 Métis living in Canada.[9] Finally, European Americans and Asian Americans also show evidence of genetic contributions from other continental groups, but to a lesser extent than African American or Hispanic populations.[11-13]

Genetic Markers Used to Estimate Admixture

Autosomal Genetic Markers

Obtaining precise estimates of admixture proportions at the individual and global levels requires the use of autosomal genetic markers. The pioneering admixture work of Glass and Li,[14] Workman[15] and Reed[16] employed "classical" genetic markers: blood groups, red cell enzymes and plasma protein polymorphisms. The advent of the PCR technique[17] has made it possible to work directly at the DNA level and currently a wide range of genetic markers are used in admixture studies, including Single Nucleotide Polymorphisms (SNPs), Insertion/Deletion Polymorphisms (Indels), Alu insertions and Short Tandem Repeats (STRs or microsatellites). Irrespective of the type of genetic markers used to estimate admixture, it is advisable to use markers informative for ancestry rather than randomly selected markers because the former maximize the information about admixture while minimizing the genotyping effort. These markers are known as Ancestry Informative Markers (AIMs). Several criteria have been proposed to measure ancestry information content, such as the absolute allele frequency difference between parental populations (delta -δ),[18] the Fisher information content (f)[19] and the Informativeness for assignment (I_n).[20] Although a widely used criterion, delta has some limitations because it doesn't take into account all available information about allele frequencies and its application to more than two parental populations or multiallelic loci is not straightforward.[20,21] The f and I_n measures give different absolute values but they rank markers similarly with respect to ancestry information content.[22] It is important to emphasize that precise estimation of individual ancestry proportions requires many more markers than the estimation of the average ancestry proportions in a sample. While two or three dozen AIMs will provide precise estimates of ancestry proportions in a sample, many more markers (>100) would be required to estimate individual admixture proportions with similar precision. In the past, the lack of useful AIMs was an important problem in admixture studies but fortunately this is no longer a limitation and thousands of AIMs have been selected for studying admixture between continental populations.[23-27] Table 1 shows a list of 100 AIMs that are optimal for studying admixture proportions and dynamics in populations throughout the Americas. The list includes markers showing high frequency differences between Europeans and Native Americans, Europeans and West Africans, and Native Americans and West Africans. A substantial effort is currently underway to develop informative markers to study genetic structure and admixture at the continental level. Due to the more limited genetic differentiation within continental populations, the identification of this type of AIMs is far more challenging than the identification of AIMs to study admixture between the major continental population groups.

Table 1. Panel of ancestry informative markers to estimate admixture proportions in North American admixed populations

rs#	Chr	WAF/EU	Delta NA/EU	WAF/NA	rs#	Chr	WAF/EU	Delta NA/EU	WAF/NA
140864	1	0.117	0.579	0.462	2695	9	0.040	0.713	0.674
2225251	1	0.751	0.567	0.184	1980888	9	0.043	0.850	0.807
725667	1	0.741	0.075	0.816	1327805	9	0.432	0.717	0.285
963170	1	0.125	0.800	0.925	951308	10	0.899	0.350	0.549
2814778	1	0.998	0.000	0.998	1594335	10	0.816	0.091	0.725
723822	1	0.106	0.775	0.668	2207782	10	0.721	0.750	0.029
725416	1	0.818	0.025	0.793	1891760	10	0.034	0.808	0.842
6003	1	0.600	0.087	0.687	2366882	10	0.444	0.861	0.417
2065160	1	0.434	0.770	0.336	714857	11	0.733	0.197	0.536
1506069	1	0.924	0.041	0.883	1487214	11	0.708	0.075	0.783
2752	1	0.461	0.209	0.252	594689	11	0.353	0.334	0.019
1861498	2	0.808	0.139	0.947	720496	11	0.262	0.788	0.526
1063	2	0.032	0.841	0.810	1042602	11	0.480	0.456	0.024
1526028	2	0.816	0.025	0.791	1800498	11	0.495	0.554	0.058
1435090	2	0.065	0.867	0.802	1079598	11	0.072	0.438	0.509
3287	2	0.571	0.001	0.570	5443	12	0.381	0.091	0.472
1350462	2	0.599	0.830	0.231	726391	12	0.725	0.409	0.316
1344870	3	0.079	0.925	0.846	708156	12	0.649	0.833	0.184
17203	3	0.614	0.633	0.020	717091	13	0.715	0.303	0.412
768324	3	0.109	0.863	0.753	2078588	13	0.899	0.050	0.849
1465648	3	0.821	0.025	0.796	1900099	14	0.161	0.825	0.664
2317212	3	0.525	0.872	0.347	1152537	14	0.770	0.075	0.695
938431	3	0.200	0.747	0.547	1800404	15	0.521	0.176	0.345
1316579	3	0.513	0.725	0.212	2862	15	0.185	0.539	0.354
719776	4	0.824	0.100	0.724	724729	15	0.735	0.132	0.867
951784	4	0.046	0.717	0.763	1153849	15	0.731	0.764	0.034
1112828	4	0.124	0.750	0.874	4646	15	0.045	0.427	0.381
1403454	4	0.132	0.718	0.850	2351254	15	0.614	0.725	0.111
3309	5	0.101	0.407	0.306	764679	16	0.186	0.712	0.526
3317	5	0.587	0.112	0.699	292932	16	0.000	0.727	0.727
1431948	5	0.868	0.121	0.747	2228478	16	0.405	0.072	0.478
1461227	5	0.284	0.714	0.430	2891	17	0.534	0.140	0.394
3340	5	0.142	0.499	0.642	2816	17	0.517	0.458	0.059
2077681	6	0.820	0.008	0.828	1074075	17	0.720	0.150	0.870
951554	6	0.139	0.650	0.511	717962	17	0.043	0.700	0.743
1935946	6	0.620	0.725	0.105	1369290	18	0.821	0.100	0.921
1881826	6	0.175	0.750	0.575	386569	19	0.082	0.813	0.895
2763	7	0.055	0.446	0.501	4884	19	0.094	0.653	0.746
2161	7	0.135	0.350	0.215	3138520	20	0.473	0.086	0.386
2396676	7	0.697	0.025	0.672	718092	20	0.711	0.795	0.084
2341823	7	0.796	0.321	0.475	718387	21	0.708	0.175	0.533
1320892	7	0.053	0.789	0.842	878825	22	0.538	0.818	0.280
285	8	0.478	0.051	0.529	16383	22	0.519	0.675	0.156
983271	8	0.588	0.833	0.245	723337	X	0.021	0.825	0.846
3176921	8	0.579	0.085	0.665	2213602	X	0.871	0.025	0.896
1373302	8	0.090	0.813	0.722	721003	X	0.900	0.275	0.625
1808089	8	0.186	0.792	0.605	2064722	X	0.342	0.738	0.396
720966	8	0.842	0.108	0.950	1986586	X	0.577	1.000	0.423
1987956	8	0.000	0.921	0.921	2188457	X	0.816	0.100	0.716
1928415	9	0.787	0.125	0.912	1415878	X	0.083	0.892	0.975

Delta is the absolute frequency difference between the parental populations. WAF: West Africa (Mende from Sierra Leone), EU: Europe (Spanish from Valencia), NA: Native American (Nahua from Mexico).

mtDNA Polymorphisms

The unique characteristics of mtDNA make this marker a useful tool in anthropological and evolutionary studies. In particular, mtDNA doesn't recombine and shows maternal inheritance, making it possible to trace the pattern of female migrations through time. It is important to mention that a recent study has identified a case of mtDNA paternal transmission in humans.[28] However, to my knowledge, this is the only such case described in humans, and maternal inheritance is supported by thousands of maternal-offspring comparisons and is still regarded as the rule.[29] Another useful feature of mtDNA is its high mutation rate,[30,31] making this marker ideal to explore recent evolutionary events. Finally, mtDNA has an effective population size of one quarter that of autosomal DNA markers, and it is more susceptible to the effects of genetic drift. As a result, mtDNA polymorphisms show more geographic structure than autosomal markers and this is a useful attribute for admixture studies. Typically, analysis of mtDNA is carried out using PCR-RFLP to characterize polymorphisms defining common Continent-specific mtDNA haplogroups (groups of related sequences defined by shared diagnostic mutations) and/or sequence analysis of the two hypervariable sequences located on the mtDNA control region (HVR1 and HVR2). In terms of admixture studies in the Americas, it is important to mention that macrohaplogroup L (comprising haplogroups L1, L2 and L3) is restricted to Africa, haplogroups H, I, J, K, T, U, V and W to Europe and haplogroups A, B, C and D to Asia and the New World.[32]

Y-Specific Polymorphisms

The Y chromosome has been a relative newcomer in evolutionary studies. For decades, efforts to find polymorphic markers in this chromosome were unsuccessful,[33] but the situation has changed dramatically in the past few years. Tens of thousands of SNPs and hundreds of STRs have been identified in the nonpseudoautosomal region of the Y chromosome. Of these markers, at least 200 SNPs and 30 STRs have been reasonably well characterized in human populations.[34] Similarly to the mtDNA, the Y-specific region of the Y-chromosome does not recombine and has an effective population size of one quarter that of the autosomal markers. Due to the strong influence of genetic drift and the effect of patrilocality the Y-specific polymorphisms show a remarkable geographic clustering. The Y-chromosome is inherited from father to son, so studying Y-specific polymorphisms provides a unique perspective on the patterns of male migration, serving as a complement to the information on female migration provided by the mtDNA. However, it is important to realize that due to the pattern of inheritance of the mtDNA and the Y-specific region (uniparental and nonrecombinant) they behave as a single genetic locus and only provide a partial perspective on the history of admixture. The global, multigenerational picture of admixture history can only be obtained by incorporating autosomal markers.

Admixture in African American and African Caribbean Populations

The availability of AIMs has made it possible to gain new insights on admixture in African American and African Caribbean groups. The history of these populations is complex for a number of reasons: the enslaved Africans came from various regions within the African continent and there were differences in ethnic origins depending on the North American port of entry. Additionally, there have been regional differences in the history of migration and admixture within the US and the Caribbean. The research carried out during the last decade reports substantial geographic variation in admixture levels, but several trends can be identified. The European contribution tends to be lower in African Caribbean communities than in US African Americans. There have been several admixture studies in the Caribbean, and the estimates of the European contribution are typically below 15%. The lowest contribution has been described in Tobago, with an average European ancestry of 4.6%.[35] Our estimate of European ancestry in Jamaica was slightly higher (6.8%),[11] while the European contribution to African Caribbeans living in Britain was 10.2%.[12] The highest European contributions described in this region correspond to Trinidad, with an average value around 16%.[36]

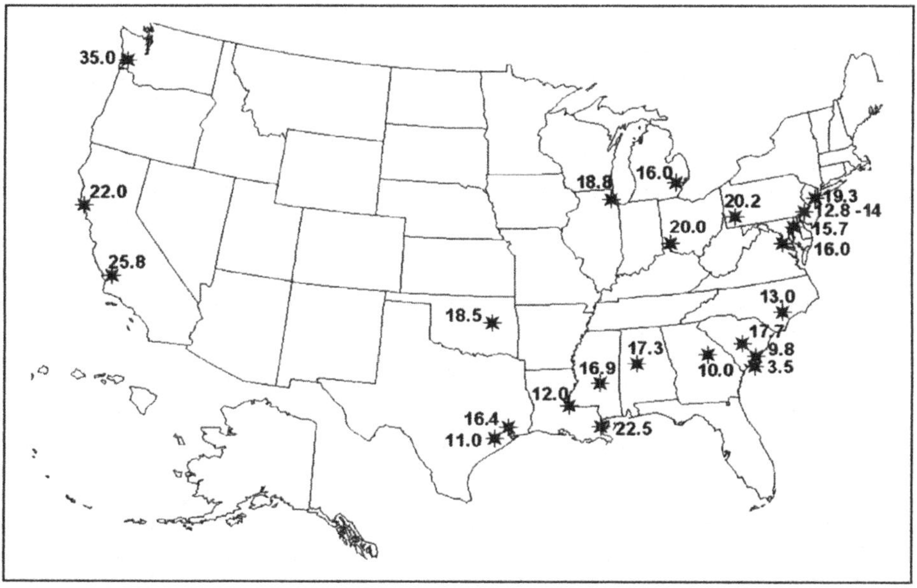

Figure 3. Map of the US showing estimates of the percentage of European contribution to several African American communities. Courtesy of Rick Kittles, University of Chicago.

In the continental US, the European contribution to African American communities also shows a substantial variation, but in general is higher than in the Caribbean. Figure 3 depicts a map showing the percentage of European ancestry observed in several African American samples throughout the US. These estimates were obtained using especially selected AIMs and are quite precise. The pioneering admixture studies that took place in the 1950s and 1960s seemed to indicate that northern populations have a higher level of European ancestry than southern populations.[14-16,37,38] However, the recent data summarized in Figure 3 show that the situation is much more complex than previously thought. Although the European admixture proportions in some southeastern African American communities are relatively low (e.g., South Carolina and Georgia), there is substantial variation in admixture levels within northern and southern populations. Additionally, the west coast communities tend to show the highest admixture estimates, although the number of samples analyzed in this area is still very limited and additional studies will be required to confirm this point.

Among the African American groups, the Gullah-speaking Sea Islanders from South Carolina show the lowest European contribution (3.5%).[39] This is in agreement with historical, cultural and anthropological data indicating that the Gullah have been relatively isolated throughout history and have retained numerous African characteristics in language, social organization, religion, magic, art, folklore and music.[40] The proportion of European ancestry is also relatively low in the "Low Country" around Charleston, SC (ranging from 9.9% and 14%) and in Georgia (10%). The European contribution tends to be higher in other regions of the US, ranging from 11.0% to 22.5% in other southern states, from 12.8% to 20.2% in the northeast and from 20% to 35% in the west coast.[41] It is clear that the differences in admixture proportions observed between African American groups cannot simply be explained in terms of geography (e.g., differences between northern and southern states). For example, the admixture proportions in New Orleans (22.5%) are higher than in most African American communities in the northeast. This could be due to the unique history of Louisiana. This area was under French rule for a substantial period, until it became part of the Spanish territory in 1763, and finally part of the US in 1803. Both the geographic origin of the enslaved Africans imported to Louisiana and their status during the French domination have been distinct from what occurred in the British Colonies, and there have been historical accounts of substantial intermarriage in the New Orleans area.[42,43] Finally, it is also important to consider the recent demographic history of the US African American populations. In the period following World War I there were significant changes in the distribution of African Americans in the US. In the largest internal migration in the history of North America, southern African Americans, constituting the immense majority of the total African-American population (around 90%), left the rural south in search of new opportunities in the urban areas of the north. It is known that big cities like Chicago, Detroit, New York, Philadelphia, Pittsburgh and Baltimore, experienced a very significant increase in the number of African-American residents, both in absolute and in relative terms.[2,44] These internal migrations have probably had a major effect in defining the current distribution of admixture proportions depicted in Figure 3.

In addition to the data on autosomal markers, further insights on the nature and dynamics of admixture can be obtained using maternally and paternally transmitted markers (mtDNA and the nonpseudoautosomal region of the Y chromosome, respectively). As described in the previous section, the effect of genetic drift is much more pronounced in mtDNA and the nonpseudoautosomal region of the Y-chromosome than in the autosomes. Consequently, in human populations the mtDNA and the Y-specific region show continent-specific polymorphisms that can provide useful insights on the admixture process.[32,34] In a study published in 1998, we estimated male and female European contributions to eight African American groups throughout the US and one African Caribbean sample (Jamaica).[11] The results of this analysis are shown in Table 2. The first column of the table depicts the estimate of maternal European ancestry based on the L macrohaplogroup and the H haplogroup. The second column depicts the estimate based on the YAP insertion. There is strong evidence of a sex-biased European contribution. In every population analyzed there is evidence of a higher European male contribution, as shown by the Y/mtDNA ratios. Therefore, even if marriages between African-American males and European-American females are presently more common than marriages between African-American females and European-American males,[43,45] it seems clear that during a substantial part of African-American history European males have made a more significant genetic contribution to the African-American gene pool than

Table 2. European contribution to 10 African American samples and an African Caribbean sample from Jamaica estimated using mtDNA (macrohaplogroup L and haplogroup H) and Y-specific markers (YAP)

	European Contribution	
Sample	**mtDNA (%)**	**Y Chromosome (%)**
Maywood, Ill.	8.31	24.32
Detroit	0.00	30.33
New York	9.11	18.58
Philadelphia 1	11.02	22.94
Philadelphia 2	2.84	23.55
Pittsburgh	9.90	23.87
Baltimore	14.94	22.79
Charleston, SC	6.46	NA
New Orleans	7.04	46.88
Houston	6.80	8.55
Jamaica	12.93	17.89

Source: Parra et al.[11]

European females. This is in accordance with the historical data regarding the period of slavery in the U.S.[42]

Another relevant issue to African American history is the extent of the Amerindian contribution. There have been accounts of substantial contact among Native Americans and people of African descent in specific periods of US history, especially in regions like the Mississippi delta and Florida.[46] Some early anthropological reports emphasized the high proportion of African-American college students claiming some Amerindian ancestry.[47,48] The same can be said of some African Caribbean populations. In fact, the Garifuna, also known as Black Caribs, living in Guatemala, Honduras and Belize trace their ancestry to the indigenous Caribbean populations (Caribs and Arawaks) and West Africans.[49] Several studies have evaluated the extent of the Native American contribution to African Americans and African Caribbeans, using a panel of AIMs that incorporate markers showing large frequency differences between West African and Native American populations and European and Native American groups. These studies indicate that the Native American contribution has been relatively low, with values generally lower than 5%. Using a panel of 38 AIMs, we reported that the average Native American ancestry in African Americans living in Washington D.C. was 2.7% and the Native American contribution to British African Caribbeans was 1.9%.[12] A recent study in three African American populations (Winston-Salem, Sacramento and Pittsburgh)[41] also reported low Native American contributions (3.9%, 4.9% and 0.9%, respectively). The Native American ancestry in African Caribbean samples from Trinidad and Tobago was estimated as 10% and 1.4% respectively.[35,36] Research using the maternally inherited mtDNA shows general agreement with the autosomal data. We tested 10 African-American samples and a sample from Jamaica for the presence of the common Amerindian-specific mtDNA haplogroups (A, B, C and D) and detected just four individuals with an Amerindian haplogroup among more than 1,000 individuals.[11] Therefore, the available data seem to indicate that the Native American contribution to African Americans and to African Caribbeans has been quite limited. However, it is important to emphasize that the information remains very incomplete, and given the diverse history of these populations, it is quite possible that the Native American ancestry will be higher in some groups. For example, the Native American contribution seems to be relatively high in the Black Caribs, a finding that is consistent with the unique history of this Caribbean population. A study in a population of Black Caribs of Guatemala using classic genetic markers (erythrocyte antigens, serum proteins and enzymes) reported 75.2% West African, 22.4% Native American and 2.4% European contributions.[50] Additionally, two mtDNA studies were carried out in Black Carib communities from Belize and Honduras. In Belize,[49] the majority of the mtDNA corresponded to African lineages, and only one out of 28 individuals showed the 9-bp deletion characteristic of Native American populations. In Honduras,[51] the maternal Native American contribution was estimated to be around 16%.

Admixture in Hispanic Populations

As mentioned in the introduction, the term "Hispanic" or "Latino" is used to refer to individuals with very diverse cultural and historical backgrounds. One of the characteristics that most of the "Hispanic" populations share is a history of recent admixture, which began soon after the arrival of Columbus to America in 1492. However, this history of admixture has been quite heterogeneous. In this section, I summarize information available for the three major North American Hispanic groups: Mexicans (and Mexican Americans), Puerto Ricans and Cubans.

The majority of the contemporary Mexican population consists of "mestizos" or admixed individuals. According to the 2000 Mexican census, 60% of the population are mestizos, 30% are Native Americans and 9% are people of European ancestry.[52] The available genetic data indicate that there is a wide dispersion of admixture proportions in mestizo populations throughout Mexico, with the Native American contribution ranging from 28% to 76%, the

European contribution from 16% to 71%, and the West African contribution from 1% to 40% (reviewed in Bonilla et al ref. 53). There are several reasons for the wide range of ancestry estimates reported in the literature: the type and number of genetic markers used in the different studies, differences in the regional histories of the Mexican states, and differences in the characteristics of the samples (e.g., socioeconomic status). Some of the early admixture studies were based on blood groups, serum or red cell enzyme polymorphisms, which are not nearly as informative for inferring admixture proportions as the AIMs used in the most recent research. It is also important to emphasize regional differences in population history within Mexico. For example, the high West African contributions that have been reported in the states located on the east coast of Mexico (e.g., Campeche, Yucatan, Tabasco and Veracruz), where West African admixture proportions range between 20% and 40%[53] are consistent with historical reports indicating a substantial West African presence around the Gulf coast and areas of southwest Mexico (Oaxaca, Guerrero), regions where the largest Afro-Mexican communities in Mexico are located today.[7] Additionally, there is evidence that socioeconomic status is strongly related to individual admixture proportions in Mexico. In a recent study in Mexico City, we observed a positive association of education and European ancestry. In a logistic model with education as a dependent variable, the odds ratio for higher educational status associated with an increase from 0 to 1 in European admixture proportions was 9.4 (95% credible interval: 3.8-22.6). These data indicate that, in agreement with previously published studies, in Mexico not everyone has the same access to education.[54] Therefore, we would expect estimates of mean admixture proportions to vary between studies that have sampled different socioeconomic groups.

There have also been several admixture studies in Mexican-American populations. The results are consistent with our earlier discussion focused on Mexican populations. Native American ancestry ranges from 18% to 52%, West African ancestry is typically lower than 10%, and the balance of the gene pool is of European origin.[23,55-61] There is also evidence of population stratification in Mexican Americans. For example, Relethford et al[59] demonstrated the effect of social class subdivision on admixture levels in Mexican Americans living in San Antonio, Texas. The wealthier socioeconomic group, residing in the transitional neighborhoods, exhibited the highest European ancestry (82%), while European ancestry levels for individuals from the low-income barrios were significantly lower (54%).

The information from the uniparentally transmitted markers (mtDNA and Y-specific markers) indicates that the admixture process has been sex-biased in Mexico, with the Native-American contribution coming mainly from the females and the European contribution from the males. In our recent study in Mexico City, we analyzed 69 autosomal AIMs and several mtDNA and Y-specific polymorphisms. The average autosomal Native American contribution was 65%. In contrast, the Native American genetic contribution for the maternally inherited mtDNA was estimated as approximately 90%, and the paternal Native American genetic contribution around 40%. The European-specific markers showed the reverse picture, with a European maternal contribution of around 7% and a paternal contribution of 60% (the average autosomal European contribution was 30%). This sex-biased contribution has already been described in many other admixture studies throughout the Americas[62-68] and is consistent with historical reports indicating that during colonial times Spanish men embarking on the conquest of America commonly practiced unions with Native American women.[69]

The other major Hispanic groups, Puerto Ricans and Cubans, have not been studied as extensively as Mexican Americans. Hanis et al[57] were the first authors to report admixture estimates for these two groups. Their estimates of the European, Native American and West African genetic contributions in a sample of Puerto Ricans were 45%, 18% and 37%, respectively. For the Cuban sample, the estimates were 62%, 18% and 20%, respectively. Recent studies in Puerto Ricans show general agreement with Hanis' estimates. Using a panel of 35 AIMs, an analysis of a sample of Puerto Rican women living in New York reported ancestry proportions of 53% European, 18% Native American and 29% West African.[70] Choudhry et al[61] recently studied a group of Puerto Rican asthmatics and controls with 44 AIMs and observed European contributions of 65.5% (cases) and 59.7% (controls), Native American contributions of 18.3% (cases) and 19.1% (controls) and West African contributions of 16.2% (cases) and 21.3% (controls). Therefore, the available data indicate that the proportion of West African ancestry in Hispanic Caribbean populations is, on average, higher than in Mexican American groups. In contrast, Mexican Americans typically have higher Native American proportions than Puerto Ricans and Cubans. This is not surprising given the historical evidence described above, although it is important to remember that within Mexico there are substantial regional differences in admixture. The known disparities in the history of admixture between Mexico and the Caribbean, combined with the differential geographic distribution of the major Hispanic groups in the US, explain the east/west differences in admixture proportions observed in the Hispanic population as a whole. As Bertoni et al[71] reported in a recent systematic study of admixture in Hispanics, in the west of the US, where Hispanics are predominantly of Mexican origin, the Native American contribution tends to be higher than in the east, where the majority of Hispanics are of Puerto Rican or Cuban origin. Conversely, the West African contribution tends to be higher in the east of the US.

Finally, I discuss the issue of sex-biased gene flow in the Caribbean Spanish colonies. In this regard, there have been several mtDNA studies in Puerto Rico that can shed some light on this topic. Bonilla et al. directly compared autosomal and mtDNA ancestral contributions to a sample of Puerto Rican women from New York (ref. 72 and unpublished data). The ancestry proportions estimated using the autosomal markers were 53.3% European, 29.1% West African and 17.6% Native American. The ancestry proportions estimated using the mtDNA haplogroups were approximately 9.4% European, 26.4% West African and 64.2% Native American. Similarly to what has been reported in many other Latin American populations, there is strong evidence of sex-biased gene flow. The majority of the mtDNA contribution is of Native American origin, while the autosomes show a major European component. It is important to mention that the estimates of maternal contributions obtained by Bonilla et al. show remarkable agreement with a systematic mtDNA analysis recently carried out by Martinez-Cruzado et al. in Puerto Rico.[72] Their estimates of maternal contributions were 61.3% Native American, 27.2% sub-Saharan African and 11.5% West Eurasian. Therefore, although historical reports indicate that the Taino disappeared from Puerto Rico late in the 16th century, the majority of present-day Puerto Ricans have Native American ancestry in their maternal lineages. The information compiled in the last decade using autosomal and uniparental genetic markers clearly indicates that the current Mexican and Puerto Rican populations have been defined by the admixture process that took place between European males and Native American females. In Mexico, approximately 90% of the maternal lineages are of Native American ancestry, implying that there has been very little European female contribution throughout colonial and post-colonial history. In Puerto Rico the situation is slightly different. The initial admixture process that took place in a relatively short period (mainly the 16th century) between the Spanish males and the Taino women and the subsequent growth of the admixed population explain the large Native American maternal contribution observed today (~60%), but there has been a higher West African and European female gene flow than in Mexico. Unfortunately, the information for Cuba is very limited. An mtDNA study carried out by Torroni et al[73] in the Cuban province of Pinar del Rio suggested a European maternal contribution of 50%, a West African contribution of 46% and a Native American contribution of 4%. These values are very different to what has been described in Puerto Rico, and seem to indicate a very different pattern of gene flow (much less sex-biased than in Puerto Rico). However, further studies will be required to confirm this point.

Implication of Admixture for Pharmacogenomics

Admixture and Distribution of Genetic Variation Involved in Drug Metabolism and Drug Response

Anthropological and genetic evidence indicate that the human species has a recent origin, which can be traced back to Africa around 200,000 years ago.[74] Given this recent origin it is not surprising that the genetic differences between continental groups are, in general, small. Numerous studies indicate that in humans the average autosomal Fst value (the percentage of the genetic variance explained by differences between continental populations) is typically between 0.10-0.20.[75] However, Fst values show a wide dispersion around the mean (Fig. 4). For most loci, the differences between geographic regions are small, but there is a subset of loci for which the diversity between regions is high. This subset may include loci that are related to drug metabolism and drug response. In 2001, Wilson et al[76] showed that four out of six drug-metabolizing enzymes had significant frequency differences between three ethnically labeled groups. Similarly, Tate and Goldstein[77] reported that out of 42 genetic variants that have been significantly associated with drug response in at least two studies, more than two-thirds show significant differences in frequency across populations and nearly one-third have allele frequency differences higher than 0.20. Obviously, these findings have important implications in pharmacogenomics.[78] In this sense, when there are large differences in the frequency of genetic variants involved in the biological response to drug treatment between continental populations, this information can be taken into account during the therapeutic decision-making process. With respect to the North American admixed groups, the history of admixture and the relative contributions of the parental populations will determine the distribution of the relevant variants. It is important to emphasize that, although knowledge about the population distribution of susceptibility variants is helpful, what really is critical is the individual genotype

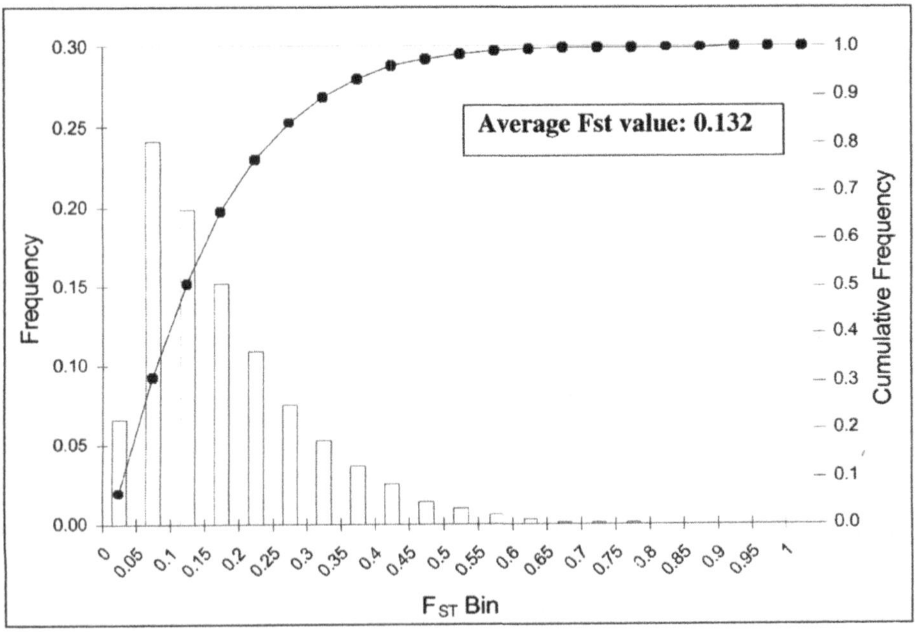

Figure 4. Histogram and cumulative distribution of Fst values (Weir and Cockerham's unbiased estimator) for 8,525 autosomal SNPs. Source: Shriver et al.[25]

(and the complex interaction with other genotypes and the environment). Due to the recent separation of human populations there will always be, to some extent, overlap in genotype frequencies between populations. The ultimate goal will be to have individualized treatment according to the genetic makeup of each person, but this goal is still beyond reach, so at this point increasing the information available at the population level is extremely important.

Population Stratification in Admixed Populations: Implications for Candidate Gene Association Studies

Candidate gene association studies have been widely used to understand the genetics of drug response.[79,80] Although genome-wide association studies are becoming increasingly feasible in terms of cost and genome coverage,[81] candidate gene studies will remain useful in the near future. However, the presence of population stratification can cause spurious associations between a phenotype and unlinked candidate loci in population-based association analysis.[82-84] This problem is particularly important in recently admixed populations because individual admixture proportions often vary significantly within the population. This variation in admixture proportions indicates a departure from random mating expectations and an increased potential for false positive results in association studies. Figure 5 shows a triangular plot depicting the variation in individual admixture proportion in three North American admixed samples: Mexicans, Puerto Ricans and African Americans from Washington DC. Note the substantial dispersion of individual admixture estimates around the average estimate for each sample. Using pigmentation as a model phenotype, we have shown that this "admixture stratification" can have a profound effect on population-based association studies.[12,85] We measured quantitatively skin pigmentation and estimated admixture proportions using 33 AIMs in two admixed samples (African Americans and African Caribbeans). Approximately one-half of the AIMs were significantly associated with skin pigmentation in these samples, even though most of the AIMs are located in genomic regions with no pigmentation candidate genes. In these samples, many markers gave a significant result not because they have a functional effect on pigmentation, but because they are informative for ancestry and pigmentation and ancestry show a strong correlation due to admixture stratification. After adjusting for individual admixture proportions, most of the significant effects disappeared and only two AIMs located within two pigmentation candidate genes (TYR and OCA2) remained significant. In admixed populations, admixture stratification is also reflected in a larger than expected proportion of significant associations between unlinked AIMs. We have detected the presence of admixture stratification in all the African American and Hispanic admixed population samples that we have analyzed but there is substantial variation in the degree of stratification.[86]

In admixed populations, two factors can be responsible for the observed admixture stratification: continuous gene flow and assortative mating. Continuous gene flow refers to an admixture model in which there has been an ongoing contribution from one or more parental populations to the admixed population over a period of time extending into the recent past. Obviously, the process of admixture in North America has been taking place for several centuries, and it is still actively configuring our present societies, so this undoubtedly explains some of the stratification present in admixed populations. In this sense, recent gene flow is of particular significance because recombination does not have enough time to break the associations between unlinked markers. Assortative mating can be defined as nonrandom mating according to phenotypic characteristics. If there is assortative mating based on any factor (e.g., socioeconomic status, education, skin pigmentation) that is correlated with ancestry, any population structure originally present in the admixed population will be maintained through the generations. There is evidence indicating that this may explain some of the stratification observed in Hispanic populations, in particular populations of Mexican ancestry. As mentioned above, Relethford described a strong association of social status with ancestry in Mexican Americans living in San Antonio, TX,[59] and we have recently observed a strong association of education status with ancestry in a large sample from Mexico City

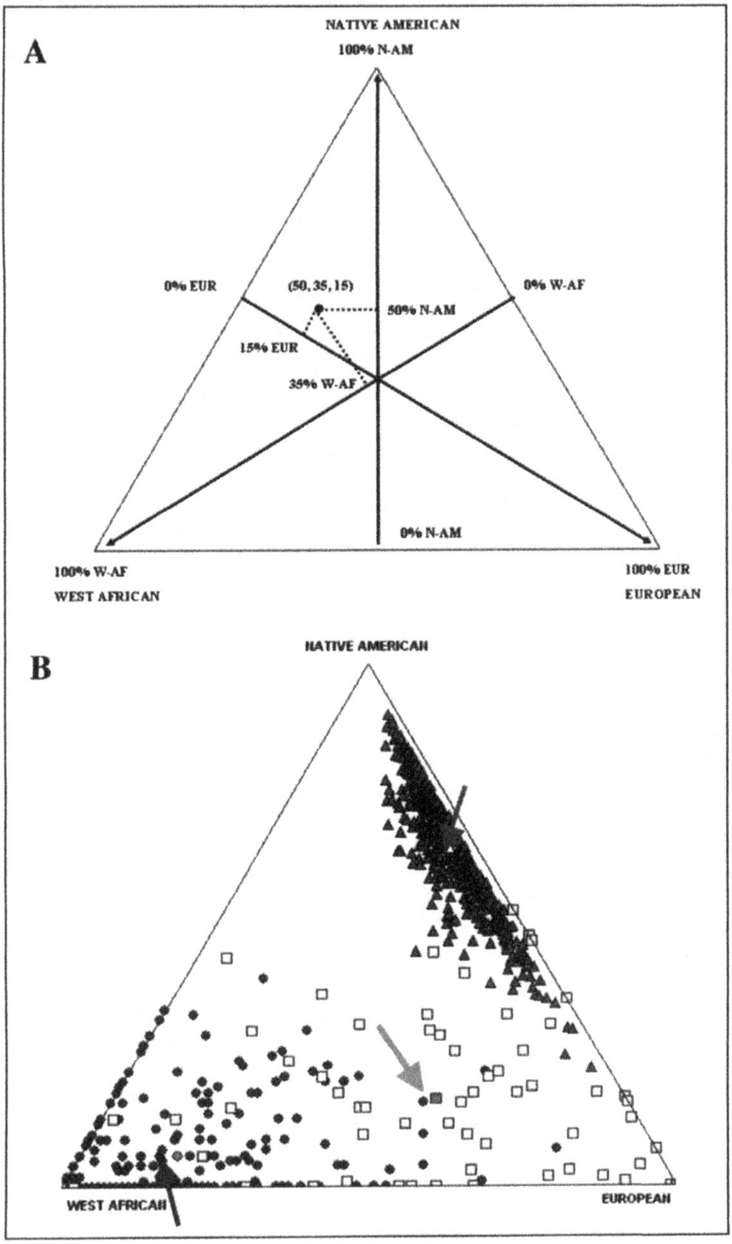

Figure 5. Triangular representation of individual ancestry estimates. A) The figure shows how to interpret triangular plots depicting the relative Native American (N-AM), European (EUR) and West African (W-AF) genetic contributions in any individual. Shown is a hypothetical example of an individual with 50% N-AM contribution, 35% W-AF contribution and 15% EUR contribution. B) Triangular plot showing individual ancestry proportions in African Americans from Washington DC (blue circles), Puerto Ricans (yellow squares) and Mexicans (red triangles). Average ancestry proportions in the samples are indicated with blue, yellow and red arrows, respectively. A color version of this figure is availabe online at www.eurekah.com.

(Parra, unpublished results). Because in most societies mating is assortative with respect to socioeconomic status, if socioeconomic status is correlated with ancestry, admixture stratification will be present in the population. The implications of these associations are profound. For example, if there are differences in socioeconomic status between a group of cases and controls, there will also be differences in ancestry proportions between both groups, and any marker showing large differences in frequency between the parental populations will be associated with the disease, even if it is not functionally relevant. In order to avoid these spurious associations, it is critical to take measures to control for potential confounding due to admixture stratification. In admixed populations, a good strategy to avoid confounding is to genotype a large panel of AIMs to estimate individual admixture proportions and include individual ancestry as a variable in the analysis.[86,87] The important issue of population stratification in admixed populations is the topic of another chapter in this book (Genomic control in admixed populations, by Eduardo Tarazona-Santos).

Conclusion

In the previous sections, I have provided a comprehensive review of admixture in the main North American admixed groups. The use of autosomal markers, in particular the powerful AIMs, in combination with mtDNA and Y-chromosome polymorphisms provides a complete picture of the history and dynamics of the admixture process. Although there are still many gaps in our knowledge, and some admixed groups and geographical areas have been poorly covered in the genetic studies, one of the clear messages from this research is that there is substantial heterogeneity in admixture proportions and population stratification within and between the main admixed groups (e.g., African Americans or Mexicans). This heterogeneity has the following implications in the field of pharmacogenomics: (1) the frequency of some alleles that may be important in drug metabolism or drug response may show variation within and between admixed groups, and (2) in candidate gene association studies there will be differences in the extent of confounding due to population stratification. For these reasons, in pharmacogenetic studies in admixed populations, it is advisable to characterize the admixed sample with a panel of AIMs in order to estimate admixture proportions and to detect and correct for admixture stratification.

Acknowledgements

I wish to express my gratitude to Mark Shriver for the population data on AIMs to study admixture in North America, to Rick Kittles for the map of European admixture in African American communities, to Carolina Bonilla for the unpublished mtDNA data from Puerto Rico and to Laura Simmonds for research assistance. Esteban J. Parra is supported by the Natural Sciences and Engineering Research Council of Canada (NSERC), the Canadian Institutes of Health Research (CIHR) and the Banting and Best Diabetes Centre of the University of Toronto.

References

1. Curtin P. The Atlantic slave trade. Madison: University of Wisconsin Press, 1969.
2. Johnson DM, Campbell RR. Black migration in America: A social demographic history. Durham: Duke University Press, 1981.
3. Wilson S. The indigenous people of the Caribbean. Gainesville: University Press of Florida, 1997.
4. Rogozinski J. A brief history of the Caribbean. New York: Facts On File, 1999.
5. Black CV. History of Jamaica. London: Collins Educational, 1958.
6. Look Lai W. Indentured labour, Caribbean sugar: Chinese and Indian migrants to the British West Indies, 1838-1918. Baltimore: The John Hopkins University Press, 1993.
7. Aguirre Beltran G. La población negra de Mexico. In: Secretaría de la, Reforma Agrarias, eds. Estudio Etnohistórico Mexico. Centro de Estudios del Agrarismo en Mexico, 1981.
8. Kanellos N, Perez C. Chronology of Hispanic-American History: From pre-columbian times to the present. New York: Gale Research, 1995.
9. http://www.statcan.ca/.

10. http://www.metisnation.ca/.
11. Parra EJ, Marcini A, Akey J et al. Estimating African American admixture proportions by use of population-specific alleles. Am J Hum Genet 1998; 63:1839-1851.
12. Shriver MD, Parra EJ, Dios S et al. Skin pigmentation, biogeographical ancestry and admixture mapping. Hum Genet 2003; 387-399.
13. Hammer MF, Chamberlain VF, Kearney VF et al. Population structure of Y chromosome SNP haplogroups in the United States and forensic implications for constructing Y chromosome STR databases. Forensic Sci Int 2005, [Epub ahead of print].
14. Glass B, Li CC. The dynamics of racial intermixture - An analysis based on the American Negro. Am J Hum Genet 1953; 5:1-19.
15. Workman PL. Gene flow and the search for natural selection in man. Hum Biol 1968; 40:260-279.
16. Reed TE. Caucasian genes in American Negroes. Science 1969; 165:762-768.
17. Mullis K, Faloona F, Scharf S et al. Specific enzymatic amplification of DNA in vitro: The polymerase chain reaction. Cold Spring Harbor Symp Quant Biol 1986; 51:263-273.
18. Shriver MD, Smith MW, Jin L et al. Ethnic-affiliation estimation by use of population-specific DNA markers. Am J Hum Genet 1997; 60:957-964.
19. McKeigue PM. Mapping genes that underlie ethnic differences in disease risk: Methods for detecting linkage in admixed populations by conditioning on parental admixture. Am J Hum Genet 1998; 64:171-186.
20. Rosenberg NA, Li LM, Ward R et al. Informativeness of genetic markers for inference of ancestry. Am J Hum Genet 2003; 73:1402-1422.
21. Pfaff CL, Barnholtz-Sloan J, Wagner JK et al. Information on ancestry from genetic markers. Genetic Epidemiology 2004; 26:305-315.
22. McKeigue PM. Prospects for admixture mapping of complex traits. Am J Hum Genet 2005; 76:1-7.
23. Collins-Schramm HE, Phillips CM, Operario DJ et al. Ethnic-difference markers for use in mapping by admixture linkage disequilibrium. Am J Hum Genet 2002; 70:737-750.
24. Collins-Schramm HE, Chima B, Morii T et al. Mexican American ancestry-informative markers: Examination of population structure and marker characteristics in European Americans, Mexican Americans, Amerindians and Asians. Hum Genet 2004; 114:263-271.
25. Shriver MD, Kennedy GC, Parra EJ et al. The genomic distribution of population substructure in four populations using 8,525 autosomal SNPs. Hum Genomics 2004; 1:274-286.
26. Shriver MD, Mei R, Parra EJ et al. Large-scale SNP analysis reveals clustered and continuous patterns of human genetic variation. Hum Genomics 2005; 2:81-89.
27. Smith MW, Patterson N, Lautenberger JA et al. A high-density admixture map for disease gene discovery in African Americans. Am J Hum Genet 2005; 74:1001-1013.
28. Schwartz M, Vissing J. Paternal inheritance of mitochondrial DNA. N Engl J Med 2002; 347:576-580.
29. Pakendorf B, Stoneking M. Mitochondrial DNA and human evolution. Annu Rev Genomics Hum Genet 2005; 6:165-183.
30. Jazin E, Soodyall H, Jalonen P et al. Mitochondrial mutation rate revisited: Hot spots and polymorphism. Nat Genet 1998; 18:109-110.
31. Ingman M, Kaessmann H, Pääbo S et al. Mitochondrial genome variation and the origins of modern humans. Nature 2000; 408:708-713.
32. http://www.mitomap.org.
33. Dorit RL, Akashi H, Gilbert W. Absence of polymorphism at the ZFY locus on the human Y chromosome. Science 1995; 268:1183-1185.
34. Jobling MA, Tyler-Smith C. The human Y chromosome: An evolutionary marker comes of age. Nat Rev Genet 2003; 4:598-612.
35. Miljkovic-Gacic I, Ferrell RE, Patrick AL et al. Estimates of African, European and Native American ancestry in Afro-Caribbean men on the island of Tobago. Hum Hered 2005; 60:129-133.
36. Molokhia M, Hoggart C, Patrick AL et al. Relation of risk of systemic lupus erythematosus to west African admixture in a Caribbean population. Hum Genet 2003; 112:310-318.
37. Glass B. On the unlikelihood of significant admixture of genes from the North American Indians in the present composition of the Negroes of the United States. Am J Hum Genet 1955; 7:368-385.
38. Adams J, Ward RH. Admixture studies and detection of selection. Science 1973; 180:1137-1143.
39. Parra EJ, Kittles RA, Argyropoulos G et al. Ancestral proportions and admixture dynamics in geographically defined African Americans living in South Carolina. Am J Phys Anthropol 2001; 114:18-29.
40. Pollitzer WS. The Gullah people and their African heritage. Athens, GA: The University of Georgia press, 1999.

41. Reiner AP, Ziv E, Lind DL et al. Population structure, admixture, and aging-related phenotypes in African American adults: The Cardiovascular Health Study. Am J Hum Genet 2005; 76:463-477.
42. Williamson J. New people: Miscegenation and mulattoes in the United States. Baton Rouge: Louisiana State University Press, 1995.
43. Piersen WD. From Africa to America. New York: Twayne Publishers, 1996.
44. Tanner HH. The settling of North America. New York: MacMillan, 1995.
45. Wilkinson DY. Black male/white female. Cambridge, Massachusetts: Schenkman Publishing Company, 1975.
46. Katz WL. Black Indians: A hidden heritage. New York: MacMillan, 1986.
47. Herskovits M. The anthropometry of the American Negro. New York: Columbia University Contributions to Anthropology, 1930.
48. Meier A. A study of the racial ancestry of the Mississippi college Negro. Am J Phys Anthropol 1949; 7:227-240.
49. Monsalve M, Hagelberg E. Mitochondrial DNA polymorphisms in Carib people of Belize. Proc R Soc Lond B Biol Sci 1997; 264:1217-1224.
50. Crawford MH, Gonzalez NL, Schanfield MS et al. The black caribs (Garifuna) of livingston, guatemala: Genetic markers and admixture estimates. Hum Biol 1981; 53:87-103.
51. Salas A, Richards M, Lareu M et al. Shipwrecks and founder effects: Divergent demographic histories reflected in Caribbean mtDNA. Am J Phys Anthropol 2005; 128:855-860.
52. http://presidencia.gob.mx/mexico/.
53. Bonilla C, Gutierrez G, Parra EJ et al. Admixture estimates of a rural population of the state of Guerrero, Mexico. Am J Phys Anthropol 2005; 128:861-869.
54. http://pnd.presidencia.gob.mx/index.php?idseccion=41.
55. Long JC, Williams RC, McAuley JE et al. Genetic variation in Arizona Mexican Americans: Estimation and interpretation of admixture proportions. Am J Phys Anthropol 1991; 84:141-157.
56. Hanis CL, Chakraborty R, Ferrell RE et al. Individual admixture estimates: Disease associations and individual risk of diabetes and gallbladder disease among Mexican-Americans in Starr County, Texas. Am J Phys Anthrop 1986; 70:433-441.
57. Hanis CL, Hewett-Emmett D, Bertin TK et al. Origins of U.S. Hispanics. Implications for Diabetes. Diabetes Care 1991; 14:618-627.
58. Cerda-Flores RM, Kshatriya GK, Bertin TK et al. Gene diversity and estimation of genetic admixture among Mexican-Americans of Starr County, Texas. Ann Hum Biol 1992; 19:347-360.
59. Relethford JH, Stern MP, Gaskill SP et al. Social class, admixture, and skin color variation in Mexican-Americans and Anglo-Americans living in San Antonio, Texas. Am J Phys Anthropol 1983; 61:97-102.
60. Bonilla C, Parra EJ, Pfaff CL et al. Admixture in the Hispanics of the San Luis Valley, Colorado, and its implications for complex trait gene mapping. Ann Hum Genet 2004; 68:139-153.
61. Choudhry S, Coyle NE, Tang H et al. Population stratification confounds genetic association studies among Latinos. Hum Genet 2006; 118:652-664.
62. Merriwether DA, Huston S, Iyengar S et al. Mitochondrial versus nuclear admixture estimates demonstrate a past history of directional mating. Am J Phys Anthropol 1997; 102:153-159.
63. Dipierri JE, Alfaro E, Martinez-Marignac VL et al. Paternal directional mating in two Amerindian subpopulations located at different altitudes in northWestern Argentina. Hum Biol 1998; 70:1001-1010.
64. Batista dos Santos SE, Rodrigues JD, Ribeiro-dos-Santos AK et al. Differential contribution of indigenous men and women to the formation of an urban population in the Amazon region as revealed by mtDNA and Y-DNA. Am J Phys Anthropol 1999; 109:175-180.
65. Sans M. Admixture studies in Latin America: From the 20th to the 21st century. Hum Biol 2000; 72:155-177.
66. Rodriguez-Delfin LA, Rubin-de-Celis VE, Zago MA. Genetic diversity in an Andean population from Peru and regional migration patterns of Amerindians in South America: Data from Y chromosome and mitochondrial DNA. Hum Hered 2001; 51:97-106.
67. Martinez Marignac VL, Bertoni B, Parra EJ et al. Characterization of admixture in an urban sample from Buenos Aires, Argentina, using uniparentally and biparentally inherited genetic markers. Hum Biol 2004; 76:543-557.
68. Bedoya G, Montoya P, Garcia J et al. Admixture dynamics in Hispanics: A shift in nuclear genetic ancestry of a South American population isolate. PNAS 2006; 103:7234-7239.
69. Grant J. Representing native peoples: Recent ethnohistories of colonial Mesoamerica and its frontiers colonial Latin American review. 1999; 8(1):145-152.

70. Bonilla C, Shriver MD, Parra EJ et al. Ancestral proportions and their association with skin pigmentation and bone mineral density in Puerto Rican women from New York City. Hum Genet 2004; 115:57-68.
71. Bertoni B, Budowle B, Sans M et al. Admixture in Hispanics: Distribution of ancestral population contributions in the continental United States. Hum Biol 2003; 75:1-11.
72. Martinez-Cruzado JC, Toro-Labrador G, Viera-Vera J et al. Reconstructing the population history of Puerto Rico by means of mtDNA phylogeographic analysis. Am J Phys Anthropol 2005; 128:131-155.
73. Torroni A, Brown MD, Lott MT et al. African, Native American, and European mitochondrial DNAs in Cubans from Pinar del Rio Province and implications for the recent epidemic neuropathy in Cuba. Cuba Neuropathy Field Investigation Team. Hum Mutat 1995; 5:310-317.
74. Foley R. The context of human genetic evolution. Genome Res 1998; 8:339-347.
75. Tishkoff SA, Verrelli BC. Patterns of human genetic diversity: Implications for human evolutionary history and disease. Annu Rev Genomics Hum Genet 2003; 4:293-340.
76. Wilson JF, Weale ME, Smith AC et al. Population genetic structure of variable drug response. Nat Genet 2001; 29:265-269.
77. Tate SK, Goldstein DB. Will tomorrow's medicines work for everyone? Nat Genet 2004; 36:S34-S42.
78. Daar AS, Singer PA. Pharmacogenetics and geographical ancestry: Implications for drug development and global health. Nat Rev Genet 2005; 6:241-246.
79. Goldstein DB, Tate SK, Sisodiya SM. Pharmacogenetics goes genomic. Nat Rev Genet 2003; 4:937-947.
80. Goldstein DB. The genetics of human drug response. Philos Trans R Soc Lond B Biol Sci 2005; 360:1571-1572.
81. Carlson CS. Agnosticism and equity in genome-wide association studies. Nat Genet 2006; 38:605-606.
82. Pritchard JK, Rosenberg NA. Use of unlinked genetic markers to detect population stratification in association studies. Am J Hum Genet 1999; 65:220-228.
83. Cardon LR, Palmer LJ. Population stratification and spurious allelic association. Lancet 2003; 361:598-604.
84. Freedman ML, Reich D, Penney KL et al. Assessing the impact of population stratification on genetic association studies. Nat Genet 2004; 36:388-393.
85. Hoggart CJ, Parra EJ, Shriver MD et al. Control of confounding of genetic associations in stratified populations. Am J Hum Genet 2003; 72:1492-1504.
86. Parra EJ, Kittles RA, Shriver MD. Implications of correlations between skin color and genetic ancestry for biomedical research. Nat Genet 2004; 36:S54-S60.
87. Pfaff CL, Kittles RA, Shriver MD. Adjusting for population structure in admixed populations. Genet Epidemiol 2002; 22:196-201.

CHAPTER 4

Pharmacogenetics in the African American Population

Howard L. McLeod*

Abstract

There is great heterogeneity in the way humans respond to medications, often requiring empirical strategies to define the appropriate drug therapy for each patient. Genetic polymorphisms in drug metabolizing enzymes, transporters, receptors, and other drug targets provide putative markers for predicting which patients will experience extreme toxicity and treatment failure. Both quantitative (allele frequency) and qualitative (specific allele) differences for polymorphic genes have been observed between different population groups. For example, the frequency of altered function variants in the membrane transporter ABCB1 is higher in African and African American populations than European-derived populations. The wide-ranging degree of admixture of African, European, and other ancestral genome structure makes the African American population of particular interest in applying genetics to allow for comprehensive strategies for using the genome to optimize therapy for patients.

Introduction

Inter-individual variability in therapeutic drug response and drug toxicity is a major problem in clinical practice because it results in idiosyncratic drug reactions or lack of a "typical" response to normal doses. Individual tailoring of treatments is becoming increasingly important in medicine, both as a means to eliminate unnecessary and often highly toxic side effects and as a way to limit costly monitoring and treatment for drug toxicity. Traditional causes of drug response variability include disease type and severity, other concomitant illnesses, drug interactions, patient age, nutritional status, renal function and liver function. More recently, genetic polymorphism in drug metabolizing enzymes, transporters, receptors, and other targets is increasingly being recognized as a source of treatment failure and toxicity.[1]

As early as the 1920s, inter-ethnic variations in response to medication were observed. In 1920, Paskind investigated the effect of atropine sulphate on 20 Caucasians and 20 African American men in Cook County Hospital, Chicago, USA.[2] Initial slowing of the heart rate, reaching a maximum of 10-15 minutes, was observed frequently in Caucasian but not in African American subjects. Chen and Both (1929) measured the change in the transverse diameter of the pupil after the administration of various mydriatics.[3] The increase in diameter was greatest in Caucasians, intermediate in Chinese, and least in African Americans.

Studies after World War II related an ethnic difference in drug response to genetic differences between ethnic groups. A toxic hemolytic reaction to primaquine, an antimalarial, among

*Howard L. McLeod—UNC Institute for Pharacogenomics and Individualized Therapy, University of North Carolina, Chapel Hill, Campus Box 7360, 3203 Kerr Hall, Chapel Hill, NC 27599-7360, U.S.A. Email: hmcleod@unc.edu

Pharmacogenomics in Admixed Populations, edited by Guilherme Suarez-Kurtz.

black soldiers in the US Army was linked to a genetic deficiency of erythrocyte glucose 6-phosphate dehydrogenase (G6PD).[4] Later work linked the high frequency of G6PD in black populations to an ability to survive falciparum malaria. The deficiency conveyed a biological advantage in malaria-infested countries and, therefore, a higher frequency of G6PD-deficient individuals was present in populations originating from such countries.[4]

To date, all pharmacogenetic polymorphisms studied differ in frequency to some degree among ethnic and racial groups. Medication dosing recommendations, however, have historically been based on clinical drug trials conducted in Caucasian populations. As different ethnic groups have begun to be incorporated into clinical research studies, it has become clear that ethnic groups may differ in their response to drugs. If drug metabolism can differ between ethnic groups, then the data generated in one population cannot be directly extrapolated to another. When such differences exist, one ethnic group may be at increased risk of therapeutic failure or toxicity. Differences in allele frequency between different ethnic/racial groups are also commonplace.[5] Examining allele frequencies in different ethnic groups can help differentiate functional polymorphisms—those causing a change in phenotype—from marker polymorphisms—those that are in linkage disequilibrium with functional polymorphisms in a particular group of people. As the linkage disequilibrium of a marker polymorphism is unlikely to remain true across the breadth of human diversity, identifying the correct, causative polymorphism is important for designing accurate genetic tests for people of all ethnic backgrounds.

To date, several genetic polymorphisms of therapeutic relevance in African American subjects have been identified and characterized, and illustrative examples are shown below.

Thiopurine Methyltransferase

Thiopurine methyltransferase (TPMT) catalyzes the S-methylation of thiopurine drugs such as 6-mercaptopurine (6-MP), 6-thioguanine, and azathioprine, to inactive metabolites.[6-9] Thiopurines form part of the routine treatment for patients with acute lymphoblastic leukemia, rheumatoid arthritis, and autoimmune diseases such as systemic lupus erythematosus and Crohn's disease and are used as an immunosuppressant following organ transplantation.

Molecular pharmacogenetic studies of *TPMT* resulted in the discovery of a series of single nucleotide polymorphisms that have been associated with significantly decreased levels of TPMT activity.[10-13] Approximately 1 in 300 white subjects have low activity, 6-11% have intermediate activity, and 89-94% have high activity.[6,9,14] To date, eight variant alleles have been identified, including three alleles (*TPMT*2, TPMT*3A,* and *TPMT*3C*), which account for approximately 80-95% of low or intermediate TPMT activity in Caucasians.[13,14]

The association between low TPMT activity and excessive hematological toxicity is now well recognized.[8,12,14] Patients with low or undetectable levels of TPMT activity develop severe myelosuppression when treated with "standard" doses of thiopurines, while patients with very high TPMT levels are more likely to have a reduced clinical response to these agents.[15-20] A 1999 study in 180 children identified an important role for TPMT genotype on tolerance to 6-MP therapy.[21] Two of the patients were TPMT-deficient and tolerated a full dose of 6-MP for only 7% of weeks of the planned therapy. Heterozygous and homozygous wild type patients tolerated full doses for 65% and 84% of weeks of therapy over 2.5 years of treatment, respectively. The percentage of weeks in which 6-MP dosage had to be decreased to prevent toxicity was 2%, 16%, and 76% in wild type, heterozygous, and homozygous variant individuals respectively.[21]

Interethnic variability in RBC TPMT activity has been reported in several populations. Red blood cell TPMT was 29% higher in Saami subjects in Northern Norway compared to white subjects from the same geographic region.[22] African American subjects have 17-33% lower red blood cell TPMT activity than American white subjects.[11,23] The TPMT activity in African and white Americans was substantially lower than that reported in 119 Chinese subjects.[24]

The frequency and pattern of variant alleles is different among various ethnic populations. For example, Southwest Asians (Indian, Pakistani) have a lower frequency of variant TPMT alleles, and all variant alleles identified to date are *TPMT*3A*.[25] This is in contrast to Kenyans and Ghanaians where the frequency of variant alleles is similar to Caucasians but all variant alleles are *TPMT*3C*.[26,27] Conversely, in Caucasian populations *TPMT*3A* is the most common allele, but *TPMT*2* and *TPMT*3C* are also found.[25] Among African Americans *TPMT*3C* is most prevalent, but *TPMT*2* and *TPMT*3A* are also found.[28] This highlights the genetic admixture in the African American population that has been observed by others. It also points to the desire to perform genotype analysis in individual patients, in order to optimize the safety and efficacy of specific medications.

ABCB1

The human multidrug-resistance (*MDR1; ABCB1*) gene encodes an ATP-binding integral membrane transporter protein, P-glycoprotein (PGP).[29,30] PGP was originally identified by its ability to confer multidrug resistance on tumor cells against a variety of structurally unrelated anticancer agents. PGP limits the bioavailability of several commonly prescribed drugs such as cyclosporine A, paclitaxel, colchicines, doxorubicin, vinblastine, ivermectin, digoxin, several antipsychotics and antidepressants, and HIV-1 protease inhibitors.

PGP protein level is highly variable between subjects; however, the molecular basis for interpatient variation is not clear.[31] In 2000 fifteen single nucleotide polymorphisms (SNPs) were detected in the *ABCB1* gene. One of these SNPs, a C to T transition in exon 26 (C3435T), showed a correlation with PGP levels and function, with the homozygous T genotype associated with a greater than two-fold reduction in duodenal PGP levels relative the homozygous C genotype.[31,32]

The distribution of C and T allele frequencies is significantly different between the African/African American populations and the Caucasian/Asian populations (Fig. 1). The variant T allele, which results in decreased PGP levels, is relatively rare in populations of African

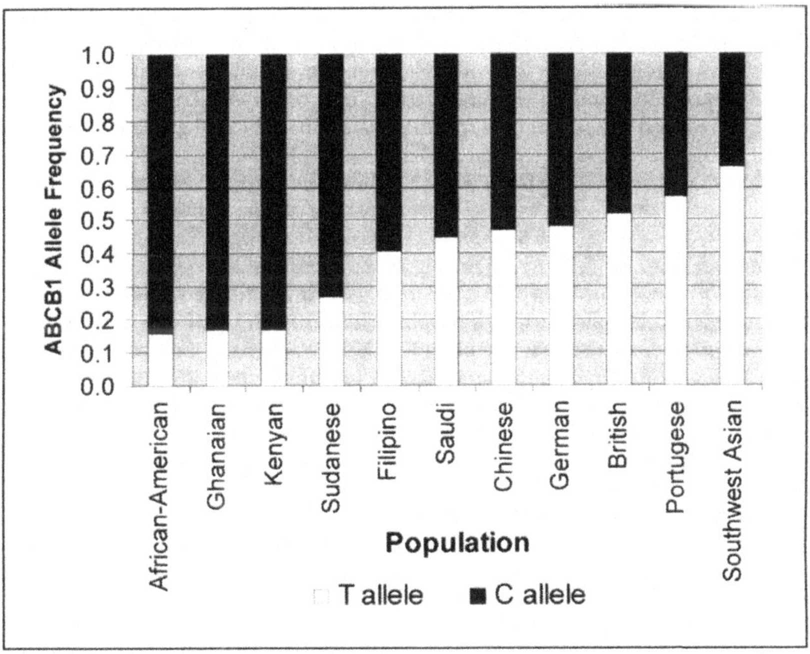

Figure 1. *ABCB1 3435C>T* variant allele frequencies in various ethnic populations. The variant allele was significantly less common among African populations than among Asians or Caucasians. A statistically significant difference was also seen between the Sudanese and Kenyan samples.[31,33]

ancestry, with allele frequencies between 0.16 and 0.27, but exists at higher frequencies (0.41 - 0.66) in Caucasian, Chinese, Filipino, Portuguese, and Saudi populations.[33] From the previous functional studies, this implies that populations of African ancestry will have higher PGP protein levels and drug efflux.

There is notable variation between African countries. While the Ghanaian and Kenyan subjects have an identical allele frequency, there is a significant difference between the Sudanese and geographically neighbouring Kenyan subjects ($p = 0.009$).[33] Also, the Southwest Asian subjects are significantly different from all other populations except the Portuguese; the Portuguese are significantly different from the Filipino population ($p = 0.02$), but similar to the other Caucasian and Asian subjects.[33]

PGP limits the bioavailability of many commonly prescribed medications, including anticancer agents and HIV-1 protease inhibitors. With a population of about 600 million (approximately 10% of the world's total), sub-Saharan Africa accounts for over two-thirds of the worlds HIV-infected persons and 80% of the world's HIV-infected women and children.[34] HIV infection is already the leading cause of death in many cities on the continent and has also increased child mortality in many countries. HIV-1 protease inhibitors are largely inaccessible in most of sub-Saharan Africa, but this may soon change. As most of the approved HIV-1 protease inhibitors are PGP substrates, bioavailability of these drugs may be limited in African patients as a result of high PGP levels, making the *ABCB1* genotype an important public health issue for health care providers in Africa. Prospective studies are now required to determine the utility of the *ABCB1* C3435T genotype for optimizing therapy for HIV, cancer, and other common diseases. The high frequency of the C allele in the African group may also contribute to the high incidence of drug resistance and the prevalence of more aggressive tumors, such as breast cancers, in individuals of African origin.[35]

Several new drugs in development are being targeted at reversal or prevention of the multidrug resistance mechanism caused by the expression of the *ABCB1* gene. Such drugs may be important in populations of African descent in order to improve the bioavailability of drugs that are PGP substrates. Information on the allele distribution of this functional *ABCB1* SNP will therefore be important for drug manufacturers, providing a tool to optimize the efficiency of commonly prescribed drugs. Current research examining *ABCB1* haplotypes may also prove important in examining variation in patient reactions to pharmacotherapy; however, the breakdown of haplotypes amongst different world populations remains to be assessed.[36-40]

TYMS

Thymidylate synthase (*TS, TYMS*) catalyses the intracellular transfer of a methyl group to deoxyuridine-5-monophosphate (dUMP) to form deoxythymidine-5-monophosphate (dTMP), which is anabolised in cells to the triphosphate (dTTP). This pathway is the only de novo source of thymidine, an essential precursor for DNA synthesis and repair. Thymidylate synthase has been of considerable interest as a target for cancer chemotherapeutic agents such as 5-fluorouracil and Raltitrexed.[41,42] Fluoropyrimidine resistance in several tumors, including colorectal cancer, has been shown to be mediated through increased mRNA and TYMS protein levels.[43] High levels of *TYMS* expression have been correlated with poor prognosis in breast cancer, gastric cancer, and colorectal cancer, possibly due to increased tumor cell proliferation as a result of increased TYMS levels.[44-48]

In 1995, Horie et al described a polymorphic tandem repeat found in the 5'-untranslated region of the thymidylate synthase gene.[49] In vitro studies have shown that increasing the number of repeats leads to stepwise increases of *TYMS* gene expression, with the presence of a triple repeat resulting in a 2.6-fold greater *TYMS* expression than a double repeat.[49,50] In vivo studies in human gastrointestinal tumors have shown a significant increase in *TYMS* protein levels and functional activity in patients with the triple repeat *TYMS TSER*3* polymorphism compared to individuals with the double repeat *TYMS TSER*2* polymorphism.[51,52]

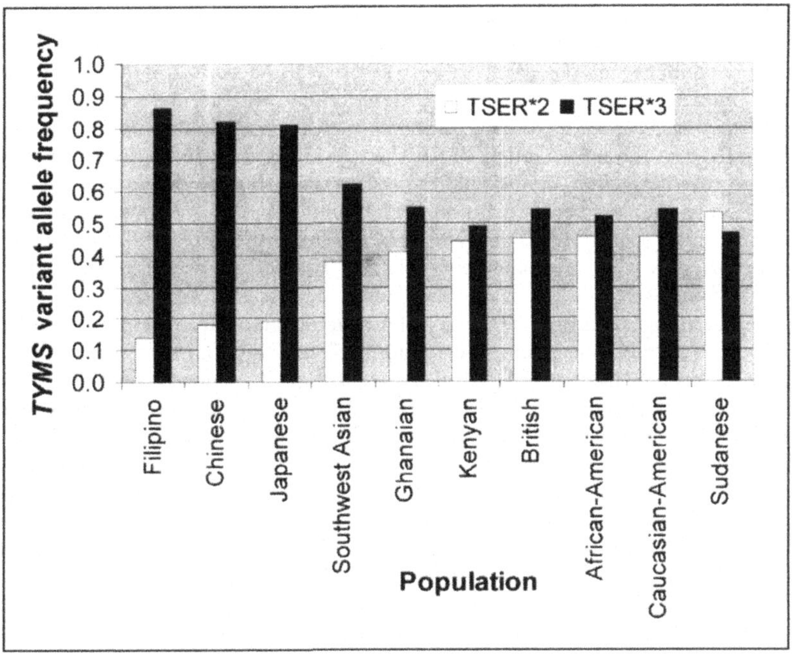

Figure 2. *TYMS TSER* variant allele frequencies in various ethnic populations. The frequency of the *TYMS TSER*3* variant is significantly higher in Asian populations than in the rest of the world, while the *TYMS TSER*2* variant is less frequent.[53,54,100]

In a pair of studies in 1999 and 2000, Marsh et al surveyed the *TYMS* allele frequencies in ten world populations. The allele frequencies of *TYMS TSER*2* ranged between 0.38-0.53 in Caucasian and African populations (Fig. 2).[53,54] However, Chinese and Filipino populations had a lower frequency of the *TYMS TSER*2* allele (0.18 and 0.14, respectively). The frequency for the *TYMS TSER*3* allele ranged between 0.47-0.62 in all populations except the Chinese and Filipino, where it was 0.82 and 0.86, respectively. The allele frequencies for *TYMS TSER*2* and *TSER*3* in the Chinese, Japanese, and Filipino were significantly different from all the other populations studied ($p < 0.001$).[53,54] The Sudanese were significantly different from the Southwest Asians ($p = 0.033$). No other significant difference in *TYMS TSER*2* and *TSER*3* allele frequency between populations were observed, including the allele frequencies of the Caucasian and African populations (>0.05 in all cases).[54]

Warfarin Dosing

Warfarin is the most commonly prescribed anticoagulant with over 21 million prescriptions in the United States alone in 2003.[55,56] Warfarin, a derivative of coumarin, is used to treat and prevent thromboembolic disorders including pulmonary embolism, stroke, atrial fibrillation, and heart attacks.[57] Although warfarin has been shown to prevent 20 strokes for every one bleeding episode that it causes, the Agency for Healthcare Policy and Research reported that warfarin is greatly underutilized for stroke prevention, in part because physicians are reluctant to prescribe a medication that is perceived as difficult to manage safely.[58] Warfarin has a narrow therapeutic index that varies widely between individuals; therefore, it requires constant monitoring and adjustment.[57,59] In a study following Taiwanese patients on warfarin, Chenhsu et al found 4.3 dose adjustments and 13.7 blood samples taken for monitoring per patient-year.[60]

Warfarin is an antagonist of Vitamin K, a necessary cofactor for the modification of glutamic acid to γ-carboxyglutamate in coagulation factors VIII, IX, and prothrombin by vitamin K-dependent γ-carboxylase. Between carboxylation cycles, vitamin K is recycled by vitamin K epoxide reductase (VKOR) in the endoplasmic reticulum membrane.[56,61] In 2004 *VKORC1* was identified as the gene encoding the catalytic subunit of VKOR.[56,61,62] Li et al showed that the expression of VKORC1 alone in insect cells could confer enzymatic activity, and that this activity could be inhibited by warfarin.[56] Warfarin is metabolized via hydroxylation in the liver by Cytochrome P450, subfamily IIC, polypeptide 9 (*CYP2C9*), the gene for which has two nonsynonymous polymorphisms, *CYP2C9*2* and *CYP2C9*3*, coding for enzymes with approximately 40% and 10% of the wild-type enzyme activity, respectively.[63-65]

There is a strong association between genetic factors and warfarin dose. D'Andrea et al reported a C to T polymorphism at nucleotide 1173 of *VKORC1* that significantly correlated with the average dose of warfarin. Patients with a CC genotype required 7.0 mg per day of warfarin, significantly higher than those with the CT genotype (5.1 mg per day) and those with the TT genotype (3.7 mg per day).[63,64] Yuan et al linked a G to A polymorphism in the *VKORC1* promoter to warfarin sensitivity among Chinese patients. Individuals with the AA phenotype had significantly lower dose requirements (2.61 mg/day) than those carrying a G allele (3.81 mg/day).[65] Similarly, patients who carry one *CYP2C9* variant allele have a 27% lower daily warfarin dose.[64] It has been estimated that polymorphisms in *VKORC1* predict approximately 25% of the inter-individual variance in warfarin dose, while *CYP2C9* polymorphisms account for and additional 10%.[59] Although the remaining 65% of warfarin dosing variation remains unexplained, gamma glutamyl carboxylase (*GGCX*) genotype, the enzyme which uses Vitamin K to activate several coagulation factors, and Apolipoprotein E4 genotype, which is involved in dietary Vitamin K absorption, have also been implicated; however, data remains limited.[66-69]

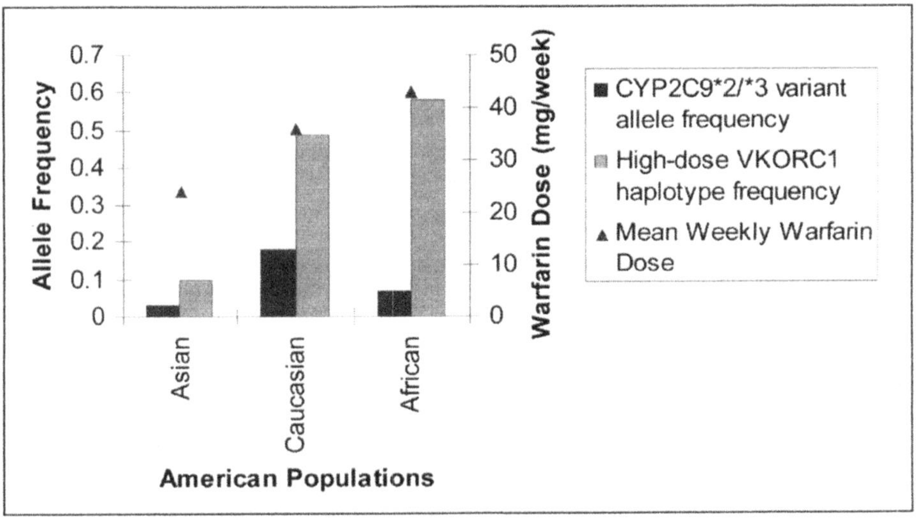

Figure 3. *CYP2C9* and *VKORC1* variant allele frequencies and mean weekly warfarin dose in Asian American, Caucasian American, and African American populations. Polymorphisms in *VKORC1* are thought to explain 25% of the interindividual difference in warfarin dose, while *CYP2C9* variants may explain an additional 10%.[59,70,101]

Known inter-ethnic differences in warfarin dosing correlate with the genetic data, as shown in Figure 3. Mean weekly warfarin doses vary from 24 mg per week among Asian-Americans to 31 mg per week among Hispanics, 36 mg per week for Caucasians, and 43 mg per week for African Americans ($p < 0.001$).[70] This fits with recent data showing that *VKORC1* haplotypes predicting a low dose of warfarin are more common among Asian-American populations (89%) and less common among African American populations (14%) than among European Americans (37%).[59,71] Likewise, low-dose linked *CYP2C9* variants are approximately five times more common among Caucasian Americans, with an allele frequency of 0.12 to 0.22, than among African Americans, with an allele frequency of 0.02 to 0.04.[72,73] However, as *CYP2C9* variants are very rare among Asians and Hispanics, they cannot explain the low warfarin dosing levels in these populations.[71]

The Need for "Resequencing" in African American Subjects

While there is only one human genome, the variants found therein have traveled under the influence of geographic origin. This was highlighted in our recent resequencing of 51 pharmacogenetic candidate genes, in which over 900 SNPs were discovered.[74] Of the novel polymorphisms identified by resequencing, 38% had estimated allele frequencies ≥10%, and 40% were found in more than one population group. In contrast, 75% of the variants that were already present in the databases had estimated allele frequencies ≥10%, and 72% were found in more than one ethnic group. Therefore, the novel variants identified by this resequencing study were more likely to be specific to one population group and have allele frequencies <10%, while the variants that were also found in a public database were more likely to be common to multiple populations and have allele frequencies ≥10%. In addition, the African-American samples contained the greatest number of variants and had the most population-specific variants, compared to the European-American and Asian-American samples (Fig. 4). Conversely, 25%, 31%, and 45% of the variants were not found in the African-American, European-American, and Asian-American samples, respectively, and 21-34% of the variants had allele frequencies <10% in those populations. 40% of the variants identified were "cosmopolitan" (both alleles were found in all 3 populations). Taken together, these results emphasize the importance of determining allele frequencies for candidate variants prior to selecting them for analysis in a specific ethnic group.

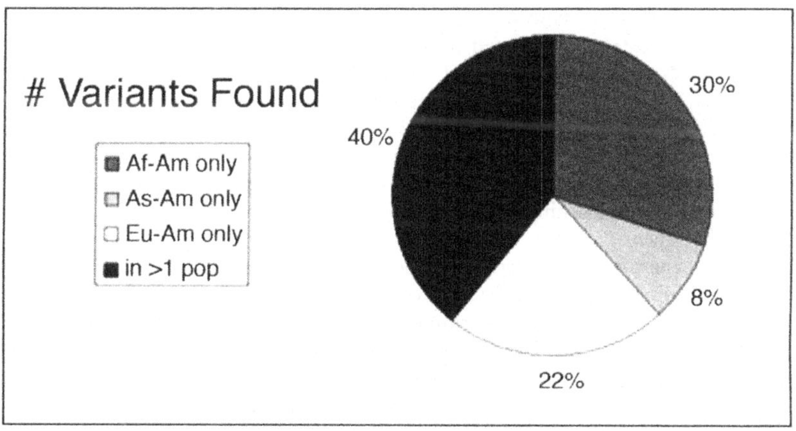

Figure 4. Distribution of SNP discovery in resequencing of 51 pharmacogenetic candidate genes.[74]

Who Is African American?

The approach most commonly favored for ethnic/racial pharmacogenetic studies has been the use of "representative" populations as surrogates for the entire population. This has been useful in extreme cases, such as comparing east Asian and African populations where allele frequency differences are great. However, this approach has been challenged in African American subjects, because of the high degree of genetic admixture and the vast geographic region from which the ancestral genome was derived. For example, there has been great interest in CYP2D6 pharmacogenetics as a predictor of efficacy for drugs ranging from the breast cancer agent tamoxifen to antipsychotic and antidepressant medications.[75] However, the frequency of poor metabolizer alleles in the African American population ranges from 0% to 8%, depending on the region of the USA in which the study was performed.[76] The same is true for the frequency of the CYP2D6 copy number variation. Similar data has been demonstrated for most of the cytochrome P450 genes and likely would be found for most any pharmacogenetic example. Therefore, it is imperative that studies do more than just rely on genotype frequency, but also begin to have more focus on clinically relevant biomarkers and actual clinical effects. These endpoints won't decrease the impact of geographic variation in allele frequency in African American subjects across the USA, but will at least allow focus on those variants with a high degree of clinical impact.

The Role of Ethnicity in Pharmacogenetics

The relationship between genetic makeup and ethnicity, defined by factors such as language, culture, and skin pigmentation, has long troubled biologists. While race can be an important factor in estimating the probability of a treatment's success or failure, there remains far more genetic difference among individuals than between ethnic groups.[77] It is estimated that about 90% of human genetic variation is within the continental groups typically used to assign ethnicity, such as Africa, European, or Asian.[78] Furthermore, an individual's self-described ethnicity is usually dependent upon cultural factors that may define reproductive patterns for only a few generations in the past.[79] When genetic markers such as microsattelites are used to assign individuals to inferred ancestral groups, the assignments often do not match self- or researcher-reported race.[80]

Complicating the relationship between genes and ethnicity is the oft-ignored issue of admixture. Science often assigns individuals to single ethnic groups, and then treats those ethnic groups as isolated, discontinuous populations; research done on one subpopulation ("Ghanaians" or "Inuit") is often extrapolated to a larger population ("Africans" or "Native Americans"). Individuals with so-called "mixed ancestry" are often excluded from research.[79,81] In reality, however, the human population appears to have been intermixing since time immemorial. When Serre and Pääblo collected 90 DNA samples to represent the geographic distribution of the human race, they found that most individuals were highly admixed between two or three of the continental populations, and that genetic makeup followed a gradient across distances.[79] Other research tracking the geographic distribution of allele frequencies has found a similar spatial gradient.[82-84]

Considering the uncertain relationship between ethnicity and genes, it is impossible to determine any individual's genotype or phenotype without performing a genetic test.[78,85] Language, culture, and skin pigmentation cannot adequately describe the hundreds of generations of human variety that have merged to form each individual's genetic makeup. Data from racial or ethnic groups may identify populations enriched for a toxicity or resistance genotype, such as that described above for *ABCB1* and Africans or *CYP2C9* and Asians. However, in any pharmacogenetically-informed medical practice, it will be necessary to test each individual's relevant genes; in research, ethnicity should be merely a crude placeholder, a tool to describe the immense human genetic variation.[86] Even if it was possible to know an individual's precise ancestry, it would not accurately predict whether he was one of the 60% of Caucasians who are slow acetylators of the tuberculosis drug isoniazid or one of the 40% who are not.[87]

The issue of race-based medicine is even more important with the recent, controversial Food and Drug Administration approval of BiDil, a heart failure therapy approved for one ethnicity.[88] BiDil's label will read that the drug is for self-identified black patients, although many cardiologists believe that the drug will be useful in treating patients of other ethnicities as well. It is unclear why, or even if, BiDil's two constituent generic drugs, hydralazine hydrochloride and isosorbide dinitrate, appear to work better in African-American patients, and a direct comparative study between black and white patients has not been performed.[86,88] It has been hypothesized that variants of the endothelial nitric oxide synthase (eNOS) gene and in the renin-angiotensin system may be responsible for higher rates of heart disease among African-Americans, but study results remain inconclusive.[89-94]

Proponents, including the Association of Black Cardiologists, hail BiDil as a way to remedy decades of ethnic inequalities in medical treatment and outcomes.[88] Critics point out that the data support a non-race based explanation for the effectiveness of BiDil and condemn NitroMed, the manufacturer of BiDil, for using race to speed FDA approval and extend its patent on the hydralazine-isosorbide dinitrate combination.[86,95,96] Meanwhile observers worry about the reification of race, the assumption that man-made ethnic categories represent actual divisions of the empirical world.[81]

Conclusion

While the promise of pharmacogenomics is enormous, it is likely to have the greatest initial benefit for patients in developed countries, due to expense, availability of technology, and the focus of initial research.[97] Pharmacogenomics should ultimately be useful to all world populations, but this is possible only if research includes participants from all world populations. There is clear evidence for ethnic variation in disease risk, disease incidence, and response to therapy. In addition, many polymorphic drug metabolizing enzymes will have qualitative and quantitative differences among racial groups.

One approach to using pharmacogenomics is in public health via SNP allele frequency analysis in defined populations.[98] For example, analysis in the five major tribes of Ghana found distinct differences in TPMT variant allele frequencies, ranging from 5.9% in the Ewe population to 13.2% in Ga individuals.[26,27] Even greater ethnic differences have been established for other polymorphic drug metabolizing enzymes (e.g., NAT2, CYP2D6, CYP2C19), and this will likely be the case for most pharmacogenomic traits, including drug transporters and targets.[99] This general approach needs to be more extensively evaluated, but does offer the potential for generating information that will have broad application to the development of clinical practice guidelines and national formularies in developing countries. Data from ethnic groups will not be as useful as analysis of individual patients. However, it will allow broader utility of pharmacogenetic principles while we wait for genotyping technology to be available for all. While knowledge of ethnic differences may be relevant to much of the world's populations, its usefulness significantly limited in situations of extensive genetic mixing. Therefore, great care must be made when applying pharmacogenomics to public health issues, and testing at the genetic level in each patient will remain the most definitive approach.

Acknowledgements

The author is supported in part by NIH grant UO1 GM63340, R21 CA102461, R21 CA113491, and P01 CA101937. The input and discussion of members of PGENI (PharmacoGenetics for Every Nation Initiative) are greatly appreciated.

References

1. Evans WE, Relling MV. Pharmacogenomics: Translating functional genomics into rational therapeutics. Science 1999; 286(5439):487-491.
2. Paskind HA. Some differences in response to atropine in white and coloured races. J Am Med Assoc 1921; 76:104-108.
3. Chenn KK, Poth EJ. Racial differences as illustrated by the mydriatic action of cocaine, euphthalmine, and ephedrine. J Pharmacol Exp Ther 1929; 36:429-445.
4. Alving AS, Carson PE, Flanagan CL et al. Enzymatic deficiency in primaquine-sensitive erythrocytes. Science 1956; 124(3220):484-485.
5. Park J, Hwang S, Lee YS et al. SNP@Ethnos: A database of ethnically variant single-nucleotide polymorphisms. Nucleic Acids Res 2006.
6. Krynetski EY, Evans WE. Pharmacogenetics as a molecular basis for individualized drug therapy: The thiopurine S-methyltransferase paradigm. Pharm Res 1999; 16(3):342-349.
7. McLeod HL, Krynetski EY, Relling MV et al. Genetic polymorphism of thiopurine methyltransferase and its clinical relevance for childhood acute lymphoblastic leukemia. Leukemia 2000; 14(4):567-572.
8. McLeod HL, Relling MV, Liu Q et al. Polymorphic thiopurine methyltransferase in erythrocytes is indicative of activity in leukemic blasts from children with acute lymphoblastic leukemia. Blood 1995; 85(7):1897-1902.
9. Weinshilboum RM, Sladek SL. Mercaptopurine pharmacogenetics: Monogenic inheritance of erythrocyte thiopurine methyltransferase activity. Am J Hum Genet 1980; 32(5):651-662.
10. Szumlanski C, Otterness D, Her C et al. Thiopurine methyltransferase pharmacogenetics: Human gene cloning and characterization of a common polymorphism. DNA Cell Biol 1996; 15(1):17-30.
11. Krynetski EY, Schuetz JD, Galpin AJ et al. A single point mutation leading to loss of catalytic activity in human thiopurine S-methyltransferase. Proc Natl Acad Sci USA 1995; 92(4):949-953.
12. Otterness D, Szumlanski C, Lennard L et al. Human thiopurine methyltransferase pharmacogenetics: Gene sequence polymorphisms. Clin Pharmacol Ther 1997; 62(1):60-73.
13. Otterness DM, Szumlanski CL, Wood TC et al. Human thiopurine methyltransferase pharmacogenetics: Kindred with a terminal exon splice junction mutation that results in loss of activity. J Clin Invest 1998; 101(5):1036-1044.
14. McLeod HL, Lin JS, Scott EP et al. Thiopurine methyltransferase activity in American white subjects and black subjects. Clin Pharmacol Ther 1994; 55(1):15-20.
15. McLeod HL, Miller DR, Evans WE. Azathioprine-induced myelosuppression in thiopurine methyltransferase deficient heart transplant recipient. Lancet 1993; 341(8853):1151.
16. Evans WE, Horner M, Chu YQ et al. Altered mercaptopurine metabolism, toxic effects, and dosage requirement in a thiopurine methyltransferase-deficient child with acute lymphocytic leukemia. J Pediatr 1991; 119(6):985-989.
17. Lennard L, Lilleyman JS, Van Loon J et al. Genetic variation in response to 6-mercaptopurine for childhood acute lymphoblastic leukaemia. Lancet 1990; 336(8709):225-229.
18. Lennard L, Gibson BE, Nicole T et al. Congenital thiopurine methyltransferase deficiency and 6-mercaptopurine toxicity during treatment for acute lymphoblastic leukaemia. Arch Dis Child 1993; 69(5):577-579.
19. Lilleyman JS, Lennard L. Mercaptopurine metabolism and risk of relapse in childhood lymphoblastic leukaemia. Lancet 1994; 343(8907):1188-1190.
20. Schutz E, Gummert J, Mohr F et al. Azathioprine-induced myelosuppression in thiopurine methyltransferase deficient heart transplant recipient. Lancet 1993; 341(8842):436.
21. Relling MV, Hancock ML, Rivera GK et al. Mercaptopurine therapy intolerance and heterozygosity at the thiopurine S-methyltransferase gene locus. J Natl Cancer Inst 1999; 91(23):2001-2008.
22. Klemetsdal B, Tollefsen E, Loennechen T et al. Interethnic difference in thiopurine methyltransferase activity. Clin Pharmacol Ther 1992; 51(1):24-31.
23. Jones CD, Smart C, Titus A et al. Thiopurine methyltransferase activity in a sample population of black subjects in Florida. Clin Pharmacol Ther 1993; 53(3):348-353.
24. Lee EJ, Kalow W. Thiopurine S-methyltransferase activity in a Chinese population. Clin Pharmacol Ther 1993; 54(1):28-33.
25. Collie-Duguid ES, Pritchard SC, Powrie RH et al. The frequency and distribution of thiopurine methyltransferase alleles in Caucasian and Asian populations. Pharmacogenetics 1999; 9(1):37-42.
26. Ameyaw MM, Collie-Duguid ES, Powrie RH et al. Thiopurine methyltransferase alleles in British and Ghanaian populations. Hum Mol Genet 1999; 8(2):367-370.
27. McLeod HL, Pritchard SC, Githang'a J et al. Ethnic differences in thiopurine methyltransferase pharmacogenetics: Evidence for allele specificity in Caucasian and Kenyan individuals. Pharmacogenetics 1999; 9(6):773-776.

28. Hon YY, Fessing MY, Pui CH et al. Polymorphism of the thiopurine S-methyltransferase gene in African-Americans. Hum Mol Genet 1999; 8(2):371-376.
29. Borst P, Evers R, Kool M et al. A family of drug transporters: The multidrug resistance-associated proteins. J Natl Cancer Inst 2000; 92(16):1295-1302.
30. Ueda K, Cornwell MM, Gottesman MM et al. The mdr1 gene, responsible for multidrug-resistance, codes for P-glycoprotein. Biochem Biophys Res Commun 1986; 141(3):956-962.
31. Hoffmeyer S, Burk O, von Richter O et al. Functional polymorphisms of the human multidrug-resistance gene: Multiple sequence variations and correlation of one allele with P-glycoprotein expression and activity in vivo. Proc Natl Acad Sci USA 2000; 97(7):3473-3478.
32. Hitzl M, Drescher S, van der Kuip H et al. The C3435T mutation in the human MDR1 gene is associated with altered efflux of the P-glycoprotein substrate rhodamine 123 from CD56+ natural killer cells. Pharmacogenetics 2001; 11(4):293-298.
33. Ameyaw MM, Regateiro F, Li T et al. MDR1 pharmacogenetics: Frequency of the C3435T mutation in exon 26 is significantly influenced by ethnicity. Pharmacogenetics 2001; 11(3):217-221.
34. Cock KM, Weiss HA. The global epidemiology of HIV/AIDS. Trop Med Int Health 2000; 5(7):A3-9.
35. Elmore JG, Moceri VM, Carter D et al. Breast carcinoma tumor characteristics in black and white women. Cancer 1998; 83(12):2509-2515.
36. Bercovich D, Friedlander Y, Korem S et al. The association of common SNPs and haplotypes in the CETP and MDR1 genes with lipids response to fluvastatin in familial hypercholesterolemia. Atherosclerosis 2005.
37. Brunner M, Langer O, Sunder-Plassmann R et al. Influence of functional haplotypes in the drug transporter gene ABCB1 on central nervous system drug distribution in humans. Clin Pharmacol Ther 2005; 78(2):182-190.
38. Colombo S, Soranzo N, Rotger M et al. Influence of ABCB1, ABCC1, ABCC2, and ABCG2 haplotypes on the cellular exposure of nelfinavir in vivo. Pharmacogenet Genomics 2005; 15(9):599-608.
39. Shon JH, Yoon YR, Hong WS et al. Effect of itraconazole on the pharmacokinetics and pharmacodynamics of fexofenadine in relation to the MDR1 genetic polymorphism. Clin Pharmacol Ther 2005; 78(2):191-201.
40. Hung CC, Tai JJ, Lin CJ et al. Complex haplotypic effects of the ABCB1 gene on epilepsy treatment response. Pharmacogenomics 2005; 6(4):411-417.
41. Rustum YM, Harstrick A, Cao S et al. Thymidylate synthase inhibitors in cancer therapy: Direct and indirect inhibitors. J Clin Oncol 1997; 15(1):389-400.
42. Leichman CG, Lenz HJ, Leichman L et al. Quantitation of intratumoral thymidylate synthase expression predicts for disseminated colorectal cancer response and resistance to protracted-infusion fluorouracil and weekly leucovorin. J Clin Oncol 1997; 15(10):3223-3229.
43. Berger SH, Jenh CH, Johnson LF et al. Thymidylate synthase overproduction and gene amplification in fluorodeoxyuridine-resistant human cells. Mol Pharmacol 1985; 28(5):461-467.
44. Pestalozzi BC, Peterson HF, Gelber RD et al. Prognostic importance of thymidylate synthase expression in early breast cancer. J Clin Oncol 1997; 15(5):1923-1931.
45. Lenz HJ, Leichman CG, Danenberg KD et al. Thymidylate synthase mRNA level in adenocarcinoma of the stomach: A predictor for primary tumor response and overall survival. J Clin Oncol 1996; 14(1):176-182.
46. Johnston PG, Fisher ER, Rockette HE et al. The role of thymidylate synthase expression in prognosis and outcome of adjuvant chemotherapy in patients with rectal cancer. J Clin Oncol 1994; 12(12):2640-2647.
47. Kornmann M, Link KH, Lenz HJ et al. Thymidylate synthase is a predictor for response and resistance in hepatic artery infusion chemotherapy. Cancer Lett 1997; 118(1):29-35.
48. Kaye SB. New antimetabolites in cancer chemotherapy and their clinical impact. Br J Cancer 1998; 78(Suppl 3):1-7.
49. Horie N, Aiba H, Oguro K et al. Functional analysis and DNA polymorphism of the tandemly repeated sequences in the 5'-terminal regulatory region of the human gene for thymidylate synthase. Cell Struct Funct 1995; 20(3):191-197.
50. Kawakami K, Salonga D, Omura K et al. Effects of polymorphic tandem repeat sequence on the in vitro translation of messenger RNA. Proc Am Assoc Cancer Res 1999; 40:436-437.
51. Kawakami K, Omura K, Kanehira E et al. Polymorphic tandem repeats in the thymidylate synthase gene is associated with its protein expression in human gastrointestinal cancers. Anticancer Res 1999; 19(4B):3249-3252.

52. Villafranca E, Okruzhnov Y, Dominguez MA et al. Polymorphisms of the repeated sequences in the enhancer region of the thymidylate synthase gene promoter may predict downstaging after preoperative chemoradiation in rectal cancer. J Clin Oncol 2001; 19(6):1779-1786.
53. Marsh S, Collie-Duguid ES, Li T et al. Ethnic variation in the thymidylate synthase enhancer region polymorphism among Caucasian and Asian populations. Genomics 1999; 58(3):310-312.
54. Marsh S, Ameyaw MM, Githang'a J et al. Novel thymidylate synthase enhancer region alleles in African populations. Hum Mutat 2000; 16(6):528.
55. Marketos M. The top 200 generic drugs in 2003 (by units). Accessed 2005.
56. Li T, Chang CY, Jin DY et al. Identification of the gene for vitamin K epoxide reductase. Nature 2004; 427(6974):541-544.
57. Ansell J, Hirsh J, Poller L et al. The pharmacology and management of the vitamin K antagonists: The seventh ACCP conference on antithrombotic and thrombolytic therapy. Chest 2004; 126(3 Suppl):204S-233S.
58. Horton JD, Bushwick BM. Warfarin therapy: Evolving strategies in anticoagulation. Am Fam Physician 1999; 59(3):635-646.
59. Rieder MJ, Reiner AP, Gage BF et al. Effect of VKORC1 haplotypes on transcriptional regulation and warfarin dose. N Engl J Med 2005; 352(22):2285-2293.
60. Chenhsu RY, Chiang SC, Chou MH et al. Long-term treatment with warfarin in Chinese population. Ann Pharmacother 2000; 34(12):1395-1401.
61. Goodstadt L, Ponting CP. Vitamin K epoxide reductase: Homology, active site and catalytic mechanism. Trends Biochem Sci 2004; 29(6):289-292.
62. Rost S, Fregin A, Ivaskevicius V et al. Mutations in VKORC1 cause warfarin resistance and multiple coagulation factor deficiency type 2. Nature 2004; 427(6974):537-541.
63. D'Andrea G, D'Ambrosio RL, Di Perna P et al. A polymorphism in the VKORC1 gene is associated with an interindividual variability in the dose-anticoagulant effect of warfarin. Blood 2005; 105(2):645-649.
64. Sanderson S, Emery J, Higgins J. CYP2C9 gene variants, drug dose, and bleeding risk in warfarin-treated patients: A HuGEnet systematic review and meta-analysis. Genet Med 2005; 7(2):97-104.
65. Yuan HY, Chen JJ, Lee MT et al. A novel functional VKORC1 promoter polymorphism is associated with inter-individual and inter-ethnic differences in warfarin sensitivity. Hum Mol Genet 2005; 14(13):1745-1751.
66. Wadelius M, Chen LY, Downes K et al. Common VKORC1 and GGCX polymorphisms associated with warfarin dose. Pharmacogenomics J 2005; 5(4):262-270.
67. Loebstein R, Vecsler M, Kurnik D et al. Common genetic variants of microsomal epoxide hydrolase affect warfarin dose requirements beyond the effect of cytochrome P450 2C9. Clin Pharmacol Ther 2005; 77(5):365-372.
68. Kohnke H, Scordo MG, Pengo V et al. Apolipoprotein E (APOE) and warfarin dosing in an Italian population. Eur J Clin Pharmacol 2005.
69. Kohnke H, Sorlin K, Granath G et al. Warfarin dose related to apolipoprotein E (APOE) genotype. Eur J Clin Pharmacol 2005; 61(5-6):381-388.
70. Dang MT, Hambleton J, Kayser SR. The influence of ethnicity on warfarin dosage requirement. Ann Pharmacother 2005; 39(6):1008-1012.
71. Marsh S, King CR, Porche-Sorbet RM et al. Population variation in VKORC1 haplotype structure. J Thromb Haemost 2006; 4(2):473-474.
72. Llerena A, Dorado P, O'Kirwan F et al. Lower frequency of CYP2C9*2 in Mexican-Americans compared to Spaniards. Pharmacogenomics J 2004; 4(6):403-406.
73. Lee SS, Kim KM, Thi-Le H et al. Genetic polymorphism of CYP2C9 in a Vietnamese Kinh population. Ther Drug Monit 2005; 27(2):208-210.
74. Freimuth RR, Xiao M, Marsh S et al. Polymorphism discovery in 51 chemotherapy pathway genes. Hum Mol Genet 2005; 14(23):3595-3603.
75. Marsh S, McLeod HL. Pharmacogenomics: From bedside to clinical practice. Hum Mol Genet 2006; 15(Spec No 1):R89-93.
76. Gaedigk A, Bradford LD, Marcucci KA et al. Unique CYP2D6 activity distribution and genotype-phenotype discordance in black Americans. Clin Pharmacol Ther 2002; 72(1):76-89.
77. Holden C. Race and medicine. Science 2003; 302(5645):594-596.
78. Bevan D. Genes, race and drugs. Clin Invest Med 2004; 27(1):5-6.
79. Serre D, Paabo S. Evidence for gradients of human genetic diversity within and among continents. Genome Res 2004; 14(9):1679-1685.
80. Wilson JF, Weale ME, Smith AC et al. Population genetic structure of variable drug response. Nat Genet 2001; 29(3):265-269.

81. Duster T. Medicine: Race and reification in science. Science 2005; 307(5712):1050-1051.
82. Krings M, Salem AE, Bauer K et al. mtDNA analysis of Nile River Valley populations: A genetic corridor or a barrier to migration? Am J Hum Genet 1999; 64(4):1166-1176.
83. Rosser ZH, Zerjal T, Hurles ME et al. Y-chromosomal diversity in Europe is clinal and influenced primarily by geography, rather than by language. Am J Hum Genet 2000; 67(6):1526-1543.
84. Ding YC, Wooding S, Harpending HC et al. Population structure and history in East Asia. Proc Natl Acad Sci USA 2000; 97(25):14003-14006.
85. McLeod HL. Pharmacogenetics: More than skin deep. Nat Genet 2001; 29(3):247-248.
86. Bloche MG. Race-based therapeutics. N Engl J Med 2004; 351(20):2035-2037.
87. Evans DA. N-acetyltransferase. Pharmacol Ther 1989; 42(2):157-234.
88. Saul SFDA. Approves a heart drug for African-Americans. NY Times (Print) 2005; C2.
89. Rao S, Austin H, Davidoff MN et al. Endothelial nitric oxide synthase intron 4 polymorphism is a marker for coronary artery disease in African-American and Caucasian men. Ethn Dis Spring 2005; 15(2):191-197.
90. Lapu-Bula R, Quarshie A, Lyn D et al. The 894T allele of endothelial nitric oxide synthase gene is related to left ventricular mass in African Americans with high-normal blood pressure. J Natl Med Assoc 2005; 97(2):197-205.
91. Malhotra S, Poole J, Davis H et al. Effects of NOS3 Glu298Asp polymorphism on hemodynamic reactivity to stress: Influences of ethnicity and obesity. Hypertension 2004; 44(6):866-871.
92. Kalinowski L, Dobrucki IT, Malinski T. Race-specific differences in endothelial function: Predisposition of African Americans to vascular diseases. Circulation 2004; 109(21):2511-2517.
93. Hooper WC, Lally C, Austin H et al. The relationship between polymorphisms in the endothelial cell nitric oxide synthase gene and the platelet GPIIIa gene with myocardial infarction and venous thromboembolism in African Americans. Chest 1999; 116(4):880-886.
94. Marroni AS, Metzger IF, Souza-Costa DC et al. Consistent interethnic differences in the distribution of clinically relevant endothelial nitric oxide synthase genetic polymorphisms. Nitric Oxide 2005; 12(3):177-182.
95. Fitzgibbons TP. Isosorbide dinitrate and hydralazine in blacks with heart failure. N Engl J Med 2005; 352(10):1041-1043, (author reply 1041-1043).
96. Moran AE, Cooper RS. Isosorbide dinitrate and hydralazine in blacks with heart failure. N Engl J Med 2005; 352(10):1041-1043, (author reply 1041-1043).
97. McLeod HL, Evans WE. Pharmacogenomics: Unlocking the human genome for better drug therapy. Annu Rev Pharmacol Toxicol 2001; 41:101-121.
98. Marsh S, Van Booven DJ, McLeod HL. Global pharmacogenetics: Giving the genome to the masses. Pharmacogenomics 2006; 7(4):625-631.
99. Kalow W, Bertilsson L. Interethnic factors affecting drug response. Adv Drug Res 1994; 25:1-53.
100. Marsh S, McLeod HL. Thymidylate synthase pharmacogenetics in colorectal cancer. Clin Colorectal Cancer 2001; 1(3):175-178, (discussion 179-181).
101. Kirchheiner J, Brockmoller J. Clinical consequences of cytochrome P450 2C9 polymorphisms. Clin Pharmacol Ther 2005; 77(1):1-16.

CHAPTER 5

Pharmacogenetics of Cytochrome P450 in Hispanic Populations

Pedro Dorado, Guilherme Suarez-Kurtz and Adrián LLerena*

Abstract

The present review focuses on the pharmacogenetics of the cytochrome P450 (CYP) enzymes in Hispanic populations, comprising the people living in Spanish speaking countries of the Americas as well as those categorized as Hispanics in the United States. We acknowledge the diversity of these peoples by their country of origin or residence, culture, as well as genetic composition, the latter resulting from centuries of inter-ethnic crosses between Amerindians, Europeans and Africans. This diversity is reflected in the frequency distribution of polymorphisms at the *CYP* genes that encode the main CYP enzymes involved in the biotransformation of xenobiotics, namely CYP1A1, CYP1A2, CYP2C9, CYP2C19, CYP2D6, CYP2E1, CYP3A4 and CYP3A5. Our review of the literature disclosed data for all these CYPs only in Mexicans or Mexican-Americans. For other populations, including extant Amerindian groups, scattered information was recovered on individual CYPs. For several Latin American countries, no information could be retrieved on any of these enzymes or, indeed other pharmacogenetic targets. With the purpose of fulfilling this information gap and to promote collaborative pharmacogenetic/genomic research in Spanish- and Portuguese-speaking peoples in the Americas and the Iberian peninsula, a network—the Iberian American Network of Pharmacogenetics and Pharmacogenomics—was recently created. This initiative represents a promising step towards the inclusion of Latin American populations among those who will benefit from the implementation of pharmacogenetic principles and tools in drug therapy.

Introduction

This chapter deals with Hispanic populations, comprising approximately 373 million individuals living in Spanish speaking countries of the American continent (Fig. 1) and 40,2 million, categorized as Hispanics in the United States (Fig. 2). The creation of the latter category evolved from a decision of the US Office of Management and Budget in 1978, which stated: "a person of Mexican, Puerto Rican, Cuban, Central or South American or other Spanish culture or origin, regardless of race" is described as a Hispanic. The operational and methodological problems with this definition have been repeatedly emphasized.[1,2] It is widely acknowledged that Hispanic groups within the continental US are heterogeneous by their country of origin, culture, as well as genetic composition. The genetic structure of Hispanics and other Spanish speaking peoples in the Americas—discussed by Parra in this book[3]—reflects five centuries of gender biased interethnic crosses between Amerindians, Europeans and Africans. Contribution of these ancestral roots to the genetic makeup of present day Hispanic individuals varies

*Corresponding Author: Dr. Adrián LLerena—University of Extremadura, Faculty of Medicine, Avda. de Elvas s/n. E-06071, Badajoz, Spain. Email: allerena@unex.es

Pharmacogenomics in Admixed Populations, edited by Guilherme Suarez-Kurtz.

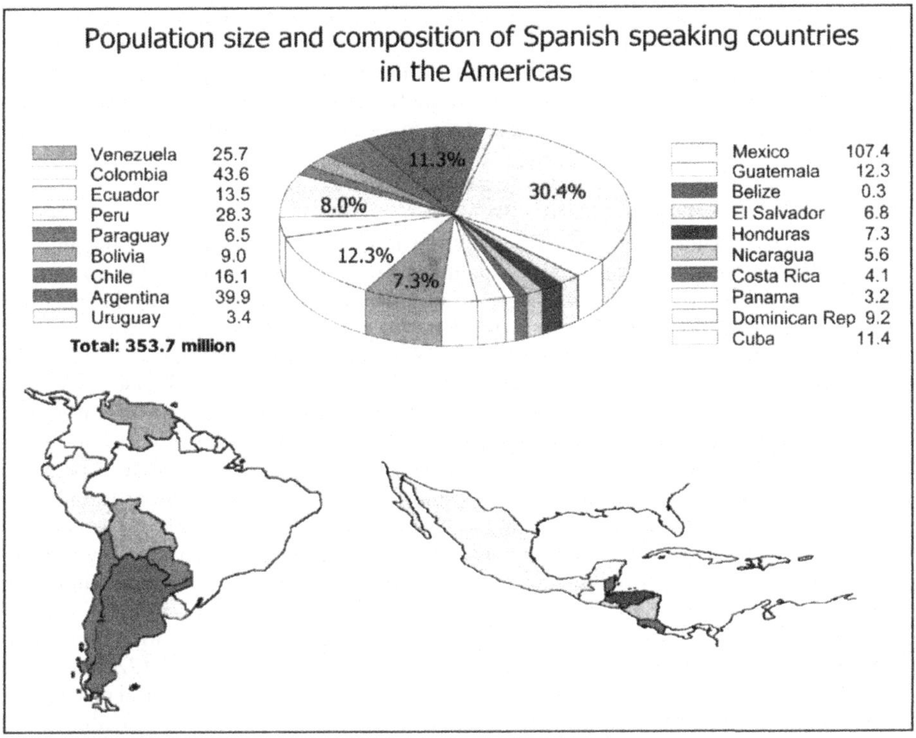

Figure 1. Population size and composition of Spanish speaking countries in the Americas. The figures next to the color codes represent the population in millions. The corresponding percentages for the five most populated countries are indicated in the pie graph. Data source: www.infoplease.com.

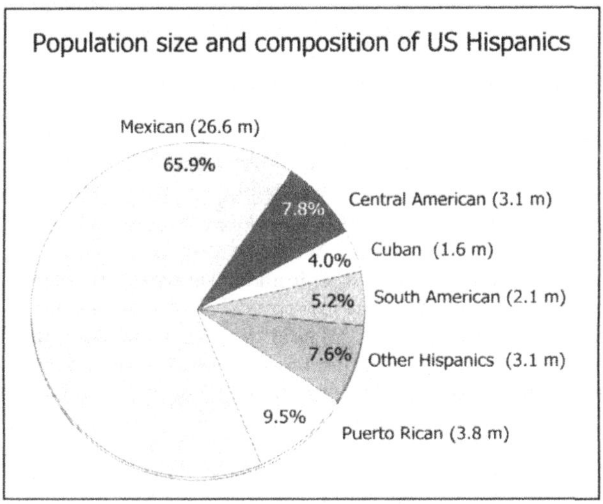

Figure 2. Population size and composition of Hispanic Americans. Data source: US Census Bureau (2004).

considerably both within and across populations and countries. The impact of such heterogeneity on the implementation of pharmacogenetics in clinical practice is evident, and may be highlighted by the recent report of significant differences in genetic polymorphisms at the β_2-adrenergic receptor, asthma severity and bronchodilator response to β_2 agonist albuterol, between individuals of Mexican or Puerto Rican ancestry, the two largest Hispanic groups in the US.[4]

Variability in Drug Metabolism

Inter-individual and interethnic differences in drug response are extensively documented. Ideally, the recommended drug dosage should be adapted to every individual in any population.[5,6] This is the goal of pharmacogenetics, a term coined by Vogel et al[7] in the late fifties. Since that time our knowledge of the importance of genetic factors in drug disposition and response has grown remarkably, in parallel with the understanding that nongenetic factors—physiological (age, gender, pregnancy, exercise, etc.), pathological (fever, diseases, infections, etc.), and environmental (diet, tobacco smoking, alcohol intake, xenobiotics)—also affect drug pharmacokinetics and pharmacodynamics, and thereby modulate therapeutic efficacy and/or side effects.[8,9] Nevertheless, it might be argued that genetic polymorphisms in drug metabolizing enzymes have proven to account for several of the most extreme variations in individual drug responses, whether beneficial or adverse. Many examples could be evoked to support this assertion, from the early demonstration of polymorphisms in butyrylcholinesterase as the cause of the succinylcholine-induced prolonged apnea[10] to the recognition of the major clinical impact of genetic polymorphisms in uridine diphosphate glucuronisyl transferase (UGT), particularly the UGT1A isoform, in cancer patients under treatment with irinotecan, which led the FDA to recently recommend genotyping for polymorphisms in *UGT1A* prior to administration of this drug.[11] These two examples refer, respectively, to phase 1 and phase 2 drug metabolism reactions. Polymorphisms have been described for several other enzymes that catalyse these reactions. Because phase 1 oxidative processes mediated by CYP enzymes seem to be the most important metabolic step for many, if not most clinically-relevant drugs, genetic polymorphisms of CYPs have been extensively studied since the original observations of metabolic deficiency of sparteine[12] and debrisoquine[13] were described. The enzyme involved in both cases, later identified as CYP2D6, is encoded by the extremely polymorphic *CYP2D* gene, which displays over 70 known variant alleles that account for a wide range of metabolic phenotypes affecting 20% of all therapeutic drugs in current clinical practice.

CYP Polymorphisms in Hispanic Populations

The present review is focused on the CYP enzymes in Hispanic populations. We report our own data, published and unpublished, as well as information retrieved from a literature search on the PubMed database, concluded in August 2006. We used as search terms the individual CYPs of pharmacogenetic relevance (CYP1A*, CYP2A*, CYP2C, CYP2D, CYP2E and CYP3A) and the individual Spanish-speaking populations/countries of the Americas (i.e., Amerindian*, Mexican*, Hispanic*, Chile*, Cuba* etc). We kept the original "ethnic" categorization of the retrieved studies, in which the terms Hispanic and Latino are often used interchangeably, the expression Mexican Americans designate "Hispanics" of Mexican ancestry, and the word Mestizo acknowledges admixture in the population samples. In the case of Amerindian (Native American) groups we indicate the regions and/or countries where they are presently living. Finally, data for Spaniards, who brought the most important European contribution to the genetic makeup of Hispanic Americans, are presented for comparative purposes.

CYP1A1

CYP1A1 is an inducible enzyme, important for the conversion of carcinogenic polycyclic aromatic hydrocarbons to epoxides. A phenotype polymorphism in inducibility was first described in 1973, when 10% of Caucasians showed much higher CYP1A1 activity in lymphocytes

Table 1. Frequency of* CYP1A1 *and* CYP1A2 *variant alleles in Hispanic populations

Group	N	CYP1A1 *2A	CYP1A1 *2C	References
Latinos	278	0.38	0.29	17
Mexican Mestizo	106	0.40	0.34	18
Mexican	136		0.28	19
Ecuadorian	108	0.50	0.70	20
Chilean	140	0.25		22
Chilean	128	0.26		23
Teenek (Mexico)	52	0.71	0.66	18
Mayos (México)	54	0.47	0.55	18
Aché (Paraguay)	67	1.0	1.0	24
Mapuche (Chile)	84	0.82	0.91	21
Spanish	142		0.04	25
		CYP1A2		
		*1C	*1F	
Mexican	46	0.53	0.77	31
Mexican	159	0.43		32
Spanish	192		0.67	34

after exposure to an inducer, than the rest of the studied group.[14] The molecular mechanism underlying this observation has not been determined, although it has been related to polymorphisms of the Ah receptor gene.[15] The most widely studied *CYP1A1* polymorphisms are in the 3′-flanking region *Msp*I site, namely 3801T>C (*CYP1A1*2A*) and in exon 7, a 2455A>G transition leading to Ile142Val substitution (*CYP1A1*2C*). When both polymorphisms are present, the variant allele is noted *CYP1A1*2B.*[16]

The frequency distribution of *CYP1A1*2A* and **2C* in Hispanic populations is shown in Table 1.[17-25] Both variant alleles are fixed in the Aché living in Paraguay and display consistently high frequencies in Amerindian groups from Mexico (Teenek and Mayos), Chile (Mapuche) as well as several groups from Brazil.[25] The unusually high frequency of *CYP1A1*2A* and **2C* in extant Amerindians throughout the Americas has been interpreted as resulting from genetic drift, caused by a founder effect in the settlement of Amerindian populations by immigrants from northern Asia through the Bering land bridge, between 30,000-120,000 years ago.[18,24] Interestingly, *CYP1A1*2A* and **2C* occur at relatively high frequencies in present-day Asians (e.g., 0.35 in Siberians), but are uncommon in Europeans (~0.04 in Spaniards) and rare or absent in sub-Saharan Africans.[26] The frequency of *CYP1A1*2A* and **2C* in rural and urban Spanish speaking populations of the Americas varies from 0.25 - 0.50 and 0.28 - 0.70, respectively, the upper limit for either variant allele being from a study in Ecuador (Table 1). This range of variation most likely reflects the different proportions of Amerindian admixture with Europeans and Africans in these trihybrid populations.[3]

Studies of *CYP1A1* polymorphisms as disease susceptibility factors, disclosed a significant association between the homozygous *CYP1A1*2C* (Val/Val) and the risk of acute lymphoblatic leukaemia in adult Mexicans.[19] In addition, *CYP1A1* polymorphisms appear to interact in a complex manner with smoking habits as risk factors for smooking-related lung cancer in Mexican-Americans and other "Latinos" living in Texas or in California.[17,27]

CYP1A2

The other CYP1A enzyme important for xenobiotic metabolism in humans is CYP1A2, which catabolizes therapeutic drugs such as amitriptyline, clozapine and fluvoxamine and catalyses the activation of several carcinogens, among them aromatic and heterocyclic amines, nitroaromatic compounds, mycotoxins, and oestrogens.[28]

The *CYP1A2* gene is located on chromosome 15 and shares 72% sequence identity with *CYP1A1*. The *CYP1A2* gene presents more than 20 polymorphic sites some of which have been related to changes in CYP1A2 enzymatic activity or inducibility in vivo.[16] One of these functional polymorphisms is in the 5'-flanking region, at the -3860 position (*CYP1A2*1C*) where a G>A transition leads to decreased CYP1A2 activity.[29] A C>A transversion at position -163 in intron 1 (*CYP1A2*1F*) has been associated with higher inducibility by smoking among Caucasians.[30] This -163C>A SNP is in strong linkage disequilibrium with other SNPs giving rise to different *CYP1A2* haplotypes, including *CYP1A2*1K* which is identified by a -729C>T transition and is related to decreased CYP1A2 activity.[16]

The *CYP1A2* genotypes and phenotypes have been little studied in the Spanish-speaking populations of the Americas, and our literature search disclosed only two studies, both in Mexicans, reporting frequencies of 0.77 for *CYP1A2*1F* and 0.43 - 0.53 for *CYP1A2*1C* (Table 1).[31,32] This variation in *CYP1A2*1C* frequency may be explained by the different sampling of the heterogeneous Mexican population in the two studies.[33] For comparison, the frequency of *CYP1A2*1F* in Spaniards is 0.67 (Table 1)[34] and to the best of our knowledge, that of *CYP1A2*1C* in healthy Spaniards has not been studied. No data on *CYP1A2* polymorphism in Amerindians is available for comparison with the Mexican results.

CYP2C9

CYP2C9 is involved in the metabolism of ca. 20% of currently used pharmaceutical drugs, including some with narrow therapeutic indices, e.g., S-warfarin, tolbutamide and phenytoin.[35,36] The CYP2C9 phenotype has been studied in vivo using different probe drugs such as tolbutamide, phenytoin, losartan, diclofenac and flurbiprofen either in isolation or as part of various "cocktails" for evaluation of drug metabolizing capacity of CYP family members.[37-40]

The human gene coding for the CYP2C9 protein has been mapped to chromosome 10q24.2, and is greater than 55 kb in length.[41] At present, there are at least seven *CYP2C9* alleles whose activity has been studied in vivo. The commonest allele and considered the wild-type, is denoted *CYP2C9*1*. Five *CYP2C9* variants codify proteins with decreased activity (*CYP2C9*2, *3, *4, *5* and **13*) whereas allele *CYP2C9*6*—a base pair deletion at position 818, which yields a premature stop codon—encodes a truncated, inactive protein.[16] Variant alleles denoted *CYP2C9*2* to **6* have been examined in Spanish speaking populations of the Americas (Table 2).[42-44] Mexican-Americans, West Mexicans and Bolivians display frequencies in the range of 0.05 - 0.08 for *CYP2C9*2* and 0.03 - 0.06 for *CYP2C9*3*. These ranges are below the corresponding frequency in Spaniards, i.e., 0.16 and 0.10, respectively.[38] There are no published data on the influence of *CYP2C9* genetic polymorphisms on the effects of therapeutic drugs in the populations under study in this chapter, differently from Brazilians where the impact of variant alleles *CYP2C9*2* and **3* on the pharmacokinetics and pharmacodynamics of nonsteroidal anti-inflammatory drugs has been described.[45-47]

CYP2C19

The *CYP2C19* genetic polymorphism is of clinical importance because the encoded enzyme catalyses the metabolism of several pharmacologically important drugs such us proton pump inhibitors, anticonvulsants, hypnosedatives, anti-infectives and antidepressants.[48] The anticonvulsant mephenytoin was the first effective probe that allowed discrimination of two different phenotypes (extensive metabolizers, EMs and poor metabolizers, PMs) of CYP2C19 activity. Mephenytoin exists as a racemic mixture of two active enantiomers, the (R)- and (S)-forms, and its elimination is stereo-selective: the catabolism of the (S)-enantiomer in humans is significantly faster due to rapid 4-OH hydroxylation, than that of (R)-mephenytoin, which undergoes slow N-demethylation. This difference in elimination kinetics, measured as the S-/R- mephenytoin ratio in urine samples collected for 8 hours following a single dose of racemic

Table 2. Frequency of* CYP2C9 *and* CYP2C19 *variant alleles in Hispanic populations

Group	N	*CYP2C9* *2	*CYP2C9* *3	References
Mexican American	98	0.08	0.06	42
Mexican Mestizo	109	0.06		43
Bolivian	778	0.05	0.03	44
Spanish	102	0.16	0.10	38
		***CYP2C19* *2**	***CYP2C19* *3**	
Mexican American	346	0.097	0.001	51
Bolivian	778	0.078	0.001	44

mephenytoin is the basis of the use of this compound as a phenotypic probe for CYP2C19. The reproducibility of this phenotyping test in healthy volunteers has been probed.[49] Alternatively, the omeprazole/5′-hydroxyomeprazol ratio in serum following a single oral dose of omeprazole can be used to discriminate CYP2C19 phenotypes.[41,43]

The frequency of CYP2C19 PMs varies across continental populations, being significantly higher (0.13 - 0.23) in East- and Southeast Asians than in Europeans and White Americans (0.01 - 0.06) or in Africans (0.03 - 0.05).[48] Despite the relatively high frequency of CYP2C19 PMs in Asians, no PMs were detected among 90 Panamanian Cuna Amerindians.[50] A recent study in Mexican-Americans disclosed a phenotype PM frequency of 0.032,[51] whereas a frequency of 0.06 was previously reported for West Mexicans.[52] Among the latter, five subjects (4%) had a log metabolic index for omeprazole/5′-hydroxyomeprazol below -0.9, which the authors interpreted as suggestive of the ultra-extensive (UM) phenotype. No mephenytoin UM was found among the Mexican Americans studied by Luo et al.[51] We found a frequency of 1.3% PMs of mephenytoin among Spaniards.[53] Later we showed that CYP2C19 activity evaluated with mephenytoin was not affected during treatment with antidepressant or antipsychotic drugs.[49]

The impaired activity of the CYP2C19 enzyme is inherited as an autosomal-recessive trait,[5,47,48] and the two most common defective alleles are *CYP2C19*2* - a G>A transition in exon 5 leading to an aberrant splice site—and *CYP2C19*3*—also a G>A transition, but in exon 4, creating a premature stop codon and a truncated protein. These two SNPs account for >99% of the defective alleles in Asians but only ~87% of the defective alleles in Caucasians.[54,55] Among the populations studied in this chapter, *CYP2C19*2* was detected at frequencies of 0.08 in Bolivians and 0.10 in Mexican Americans, whereas *CYP2C19*3* was rare (~0.01) in both these groups (Table 2).[44,51] We found no data on *CYP2C19* polymorphisms in Amerindian populations living in Spanish speaking countries in the Americas nor in Spaniards.

Alleles *CYP2C19*2* and **3* accounted for less than 1/3 of the PMs identified by Luo et al in Mexican Americans.[51] Thus, only 2 of the 7 PMs who were genotyped for *CYP2C19* polymorphisms carried two defective alleles (*2/*2 in both cases); the genotypes of the other 5 PMs were **1/*2* (n = 2) and **1/*1* (N = 3). The possibility that the rare defective alleles **4*, **5*, **6*, **7* and **8* might account for this discordance was excluded by genotyping.

CYP2D6

CYP2D6 is the major determinant of the plasma concentration of several clinically important drugs, such as beta-blockers, antidepressants, and neuroleptics.[56-58] Descriptions of the phenotypic polymorphism of CYP2D6 towards debrisoquine and sparteine are landmarks of pharmacogenetics and paved the way for the identification of the multiplicity of polymorphisms in the encoding gene, *CYP2D6*.[12,13] Phenotyping with debrisoquine, sparteine and other probes such as dextromethorphan allows the identification of two classes of CYP2D6 metabolizers: extensive (EM) and poor (PM). The distribution of these phenotypes varies markedly across continental populations: for example, PMs are more common in Europeans (e.g., 4.9% in Spaniards)[59] than in East Asians (~1%), whereas a wide range of frequencies (0 - 19%) has been reported for Africans.[60,61]

To our knowledge, assessment of CYP2D6 phenotypes has been carried out in Mexican Americans, Mexicans, Nicaraguans and Colombians. The frequency of PMs in these populations ranged from 3.2% to 10% (Table 3). This variation is not unexpected in view of the genetic heterogeneity of the trihybrid Latin-American peoples and the distinct frequencies of PMs in European, African and Amerindian populations.[60,61] Diversity in the frequency of PMs was also disclosed among Amerindian groups (Table 3): no PMs were detected in Mexican Tepehuano tested with dextromethorphan or in Panama Cuna tested with sparteine and debrisoquine. However, among Ngawbe and Embera living in Colombia and Panama, sparteine PMs presented frequencies of 4.4% and 2.2%, respectively. All the PMs in the latter two groups possessed nonfunctional *CYP2D6* alleles (either *CYP2D6*4* or *CYP2D6*6*, see below) and there were no disagreements between genotypic and phenotypic data. By contrast, genotype/phenotype discrepancies were reported in other studies carried out in Mexican Americans and Mexican Mestizos. Thus, Mendoza et al[62] identified three discordant subjects among eight Mexican Americans phenotyped as PMs whereas Sosa-Macias et al[63] reported phenotype/genotype discordance in two out of six Mestizo Mexicans PMs. It is noteworthy that one of the discordant PMs in each of these two studies was genotyped as wild-type *CYP2D6*1* homozygous, and that the discordant Mexican Mestizo had the lowest rate of dextromethorphan metabolism among all the subjects in that study.[63] Collectively, these discordant results might be due to one of the rare mutations previously known but not examined in the respective studies, or to yet unidentified mutations.

The CYP2D6 phenotypic variability results mainly from the extensive polymorphism in the encoding *CYP2D6* gene. Over 70 variant alleles are known, which encode enzymes with normal, decreased, null or increased activity. The frequency distribution of these alleles varies markedly amongst racial/ethnic groups, but a few variants, namely *CYP2D6*4, *5, *10, *17* and allele duplication respond for most of the variation in enzyme activity.[60] Alleles *CYP2D6*4* and **5* are nonfunctional, alleles **10* and **17* are associated with decreased activity, whereas duplication (or multiplication) of *CYP2D6*1* or **2* leads to the increased enzyme activity observed in some individuals (UMs).

The frequency distribution of variant *CYP2D6* alleles in Spanish speaking populations of the Americas is shown in Table 3.[21,62-74] Data are available for Mexican Americans, Cubans, Mexican Mestizos, Colombians, Nicaraguans as well as for Amerindians living in Mexico, Panama, Colombia and Chile. Among the variant polymorphisms investigated in these populations, the defective allele *CYP2D6*4* was found to predominate in most, displaying frequencies from 10 - 20%. However, Mapuche Amerindians and especially Tepehuano had much lower frequencies of *CYP2D6*4*: 3.6% and 0.6%, respectively. *CYP2D6*4* is the predominant variant allele in Europeans, occurring in 16.9% of Spaniards, but is rare or absent in sub-Saharan African and East Asians.[75] Different degrees of admixture among these three ancestral populations are a plausible explanation for the variable frequency of *CYP2D6*4*—and other variant *CYP2D6* alleles, as well—in the Americas, although genetic drift and/or founder effects must also be considered as affecting allele frequencies, especially in isolated Amerindian groups.

Allele *CYP2D6*10* was common (>5%) in some of the populations listed in Table 3, although wide variation was observed across, as well as within the groups, with ranges of 1.0 - 7.4% in Mexican Americans, 2.3 - 12.4% in Mexicans, and 0 - 17.5% among different Ameridian groups. A relatively high frequency of allele *CYP2D6*41* (9.5%) was reported in one study in Mexican Americans. This allele is the most common variant in Ethiopians (21%) but is relatively uncommon (1-3%) among Europeans, Asians and sub-Saharan Africans.[60]

Table 3. Distribution of CYP2D6 variant alleles and frequency of CYP2D6 poor metabolizers (PMs) in Hispanic populations

			Defective Alleles (%)				Reduced Activity Alleles (%)							
Populations	**N**	***2**	***3**	***4**	***5**	***6**	***9**	***10**	***17**	***29**	***41**	**Dupl.**	**% PMs ****	**Refs.**
Mexican American	349	22.8	<1	10.3	2.3			7.4	<1			1		62
	285												3.2 (DXT)	62
Mexican American	264	18	0.2	10	1.7	0.4	1.1	2.8	0.2	0.4	9.5	0.8		64
	236												10 (DXT)	64
Mexican American	50			17	2			1	2			3	6 (DXT)	65
Mexican Mestizo	243	19.3	1.4	11.2	2.7			12.4	1.7			12.8		66
	100												10 (DXT)	66
Mexican Mestizo	110		0.9	13.1		0		2.3						63
	88												6.8 (DXT)	63
Nicaraguan	137		1.8	15.7	3.6			3.3				1.1		67
Nicaraguan	125												5.6 (DBQ)	74
Colombian	121	37	1.2	19.4	0.8	0			1.6			1.2		68
Mapuche (Chile)	84	18.5	0	3.6	4.2		0	1.8						21
Cubans	254		0	14.6	1.9	1.0		0.6	6.4			4.6		69
	260												4.8 (DBQ)	69
Cuna (Panama)	170												0 (SPT)	71
Cuna (Panama)	89												0 (DBQ)	72
Ngawbe (Panama, Colombia)	344		0	17.1	0	0.5	0	17.5					4.4 (SPT)	70
Embera (Panama, Colombia)	153		0	14	0	1.1	0	6.9					2.2 (SPT)	70
Tepehuano (Mexico)	85		0	0.6		0		0						63
	58												0 (DXT)	63
Spanish	925												4.9 (DBQ)	59
Spanish	142		0	16.9	1.4	6.7		2.1	0			3.9*		73

*Duplications of functional *CYP2D6* alelles. ** CYP2D6 phenotyping test-drug: DBQ (debrisoquine); DXT (dextromethorphan); SPT (sparteine).

The duplication/multiplication of *CYP2D6*1* or **2* alleles has been identified in some Hispanic populations (Table 3), while the *CYP2D6*4* multiplication has only been evaluated in Spaniards, Cubans, and Colombians.[68,69,73] The percentages of multiplication of the inactive *CYP2D6*4* allele were similar in all studied populations (0.4-0.7%).

CYP2D6 provides the major metabolic pathway for ca. half of 100 top selling pharmaceutical drugs worldwide.[76] Thus the wide variation in this enzyme's activity has important therapeutic consequences, both in terms of efficacy and toxicity. Phenotyping and/or genotyping individuals receiving drugs that are CYP2D6 substrates may help clinicians to better design therapeutic regimens comprising such drugs. In view of the large number of known functional *CYP2D6* alleles and their distinct frequency distribution across populations, a relevant question here is: Which alleles must be tested in order to inform properly a therapeutic decision? For Mexican Americans, it has been recently recommended testing for four null alleles (*CYP2D6*3, *4, *5* and **6)*, five reduced function alleles *(*9, *10, *17, *29* and **41*), as well as duplications of alleles **1, *2* and **4*.[76]

While there are several strategies and methods for *CYP2D6* genotyping, such us SSCP real-time PCR, microarrays for DNA analysis, TaqMan real-time PCR, they are expensive and hence unaffordable in many locations. The XL-PCR and PCR-RFLP procedure for *CYP2D6* genotyping in a new published method allows the rapid, straightforward, and inexpensive identification of 90-95% of clinically relevant *CYP2D6* genotypes based on an already existing biotechnological method, and suitable for worldwide use.[77]

CYP2E1

CYP2E1 is an ethanol-inducible enzyme, involved in the catabolism of ethanol itself and of therapeutic drugs such as paracetamol, albendazole, artesunate, primaquine, and chloroquine. CYP2E1 is also responsible for the metabolic activation of aliphatic, aromatic, and halogenated hydrocarbons, many of which are industrial solvents and monomers, and some of which are pro-carcinogenic agents.[78]

Phenotyping studies with chlorzoxazone have shown a 4 - 5 fold variation in this drug's oral clearance among European Americans, and, on average, a 25 - 40% higher clearance in this group compared to Japanese.[79] Both environmental and genetic factors are thought to account for this diversity. Indeed, a number of genetic polymorphisms in the *CYP2E1* gene are known,[16] the most widely studied being allele **5B*, defined by the restriction enzymes *Pst*I and *Rsa*I, and allele **6*, defined by resistance to *Dra*I digestion. Both these alleles are found at relatively high frequency (>0.25, Table 4) among Amerindians living in Chile[21] and Paraguay.[24] For comparison, the frequency of *CYP2E1*5B* varied widely (0.02 - 0.33) among six Amerindian groups living in Brazil.[24] Among Spanish-speaking populations of the Americas frequencies of 0.13 - 0.30 for *CYP2E1*5B* and 0.17 - 0.21 for *CYP2E1*6* were reported (Table 4).[80-83] By contrast, the corresponding values in Spaniards are 0.04 and 0.11.[84] Taken together, these data are consistent with the high impact of genetic admixture between Native Americans and Europeans on the *CYP2E1*5* and **6* polymorphisms. These polymorphisms have been investigated as risk factors for cancer[80,85] and alcoholism in Spanish speaking populations of the Americas.[81,86] Wu et al[81] reported significant association of the *Dra*I and *Pst*I polymorphisms and lung cancer risk in Mexican Americans whereas Sierra-Torres et al[85] argued that there was tantalized evidence that genetic differences in the metabolism of wood smoke carcinogens that are CYP2E1 substrates, confer susceptibility to cervical neoplasia among women in Colombia. Konishi et al[86] reported much higher frequencies of both *CYP2E1*5* and **6* in alcoholic than in nonalcoholic Mexican Americans, and proposed that these polymorphisms might independently contribute to the development of alcoholism in this population. This observation must be weighed against the results of a recent meta-analysis, that disclosed no evidence whatsoever for *CYP2E1* polymorphisms being a risk factor for alcoholism and alcoholic liver disease.[87]

Table 4. Frequency of* CYP2E1 *variant alleles in Hispanic populations

Group	N	*CYP2E1* *5B	*CYP2E1* *6	References
Mexican American	69		0.17	80
Mexican American	251	0.14	0.17	81
Mexican	159	0.30		31
Mexican	97	0.20		82
Nicaraguan	137	0.17		83
Chilean	299	0.13 - 0.16		21
Chilean	148	0.16	0.21	22
Aché (Paraguay)	67	0.43		24
Mapuche (Chile)	84	0.25	0.26	21
Spanish	255	0.04	0.11	84

CYP3A4 and 3A5

The human CYP3A subfamily plays a dominant role in the metabolism of more drugs than any other biotransformation enzyme. In adult humans, CYP3A4 and 3A5 are expressed in liver and small intestine and thus modulate both first-pass and systemic drug metabolism. There is large variation in the expression and the activity of these CYP3A isoforms, which can lead to either toxicity, due to drug accumulation, or therapeutic failure, due to insufficient dosage. Phenotyping of CYP3A by orally administered probes is compounded by enzyme activity in the small intestine and the liver, and a parenteral route is required to ascertain phenotypes of hepatic drug metabolism. Nifedipine, erythromycin, midazolam and omeprazole are some of the probes currently used for CYP3A phenotyping, which can also be accomplished by measurements of endogenous CYP3A substrates, such as the cortisol metabolic ratio.[37,39,40] A study using both oral and intravenous midazolam disclosed no differences in the disposition of this benzodiazepine between Mexicans and European Americans.[88] This was taken as an indication that, from a pharmacokinetic standpoint, dosages of drugs metabolized by CYP3A need not be different between those two population groups. In another study, 80% of West Mexicans phenotyped with omeprazole proved to be extensive metabolizers of CYP3A4, and their mean log metabolic index (omeprazole/omeprazole sulfone) was found "to be higher than the data described for Caucasians and lower than that for Asians".[52]

There is at present no definitive in vivo evidence for a *CYP3A4* genetic polymorphism related to catalytic activity in humans.[89] Few studies have addressed the pharmacogenetic status of *CYP3A4* in the Spanish-speaking populations of the Americas. Discrepant results were observed in the frequency of allele *CYP3A4*1B*, a A>G transition at position -392, in the 5′-upstream region of the gene (Table 5): frequencies in the range of 0.06 - 0.09 were reported for Mexicans[90] and a broadly defined sample of Hispanic Americans,[91] whereas a much higher frequency (0.52) was detected in a relatively small sample Hispanic girls.[92] It was suggested that differences in "ancestry"—the girls in the latter study were primarily of Puerto Rican background—might account for the discordant *CYP3A4*1B* frequencies among Hispanics.[92] For comparison, the allele frequency of *CYP3A4*1B* in Spaniards was reported to be 0.04.[93] Allele *CYP3A4*2* (673T>C) was not detected in Mexicans, whereas several other variant alleles were investigated in a sample of only 20 Mexicans.[91]

Of the various known *CYP3A5* polymorphisms, only the variant allele *CYP3A5*3*, a SNP (6896A>G) within intron 3, which exhibits lower expression phenotype, was investigated in Hispanic populations (Table 5).[92,93] The reported frequency among girls of primarily Puerto Rican background (0.30) is intermediate between those of Spaniards (0.91) and sub-Saharan

Table 5. Frequency of* CYP3A4*1B *and* CYP3A5*3 *alleles in Hispanic populations

Group	N	*CYP3A4*1B*	*CYP3A5*3*	References
Hispanic	188	0.09		91
Hispanic	57	0.52	0.30	92
Mexican	69	0.06		90
Spanish	177	0.04	0.91	93

Africans (0.06 - 0.19),[60] and comparable to other populations in the Americas that have European and African ancestral roots, such as black Brazilians (0.33)[94] and African-Americans (0.27).[95]

Concluding Remarks

This overview of the pharmacogenetics of CYP enzymes and genes in Hispanic populations revealed that Mexicans and Mexican-Americans are the most extensively studied groups, whereas for other populations, including extant Amerindian groups, information is either limited to some CYPs or not available at all. This disparity is likely to involve multiple factors. Among these, one might consider that Mexican Americans account for the major proportion (65.9%) of the 40.4 million US Hispanics, and that Mexico has the largest population (107 million; Fig. 1) and pharmaceutical market (world's 9th at $7.9 billion/year as of July 2006)[96] among the Spanish-speaking countries of the Americas. However, the latter two parameters are not consistently associated with the availability of CYP data throughout Latin America, and it is likely that the research interests and efforts of individual groups as well as the financial support available for their research are more relevant for the local development of pharmacogenetics/genomics.

With the purpose of promoting pharmacogenetic and pharmacogenomic training and research in Spanish- and Portuguese-speaking countries in the Americas and the Iberian Peninsula, a network—the Iberian American Network of Pharmacogenetics and Pharmacogenomics—was recently created. The network includes research groups from Argentina, Brazil, Chile, Colombia, Cuba, Mexico, Nicaragua, Peru, Portugal and Spain. It is hoped that collaboration among these groups will contribute to the implementation of the principles and goals of pharmacogenetics/-genomics in Iberian American populations.

Acknowledgements and Financial Support

The present review is coordinated in the Iberian American Network of Pharmacogenetics and Pharmacogenomics (CYTED206RT0290). We thank Macarena C. Cáceres (Spain), Idilio González (Cuba), and Ruth E. Alanis (México) for assistance in collecting data. Work in the authors labs is supported by grants from Plan Nacional de Investigación Científica, Desarrollo e Innovación Tecnológica (I+D+I) and Fondo Social Europeo from European Union (FEDER), Instituto Carlos III-FIS (PI06/1681), and Ministerio de Educación y Ciencia (SAF2006-13589). PD is supported by a fellow from Consejería de Infraestructura y Desarrollo tecnologico, European Union, Fondo Social Europeo (REI05A003); and Conselho Nacional de Desenvolvimento Científico e Tecnológico, Fundação de Amparo à Pesquisa do Estado do Rio de Janeiro (Brazil) and the Swiss Bridge Foundation to GSK.

References

1. Chakraborty BM, Fernandez-Esquer ME, Chakraborty R. Is being Hispanic a risk factor for noninsulin dependent diabetes mellitus (NIDDM)? Ethn Dis 1999; 9:278-283.
2. Suarez-Kurtz G, Pena SDJ. Pharmacogenomics in the Americas: Impact of genetic admixture. Curr Drug Targets 2006, (in press).

3. Parra E. Admixture in North America. In: Suarez-Kurtz G, ed. Pharmacogenetics in Admixed Populations. Austin: Landes Bioscience, 2007.
4. Choudhry S, Ung N, Avila PC et al. Pharmacogenetic differences in response to albuterol between Puerto Ricans and Mexicans with asthma. Am J Respir Crit Care Med 2005; 171:563-70.
5. Kalow W. Pharmacogenetics: Past and future. Life Sci 1990; 47:1385-1397.
6. Meyer UA. Molecular genetics and the future of pharmacogenetics. Pharmacol Ther 1990; 46:349-355.
7. Vogel F. Moderne Probleme der Humangenetik. Ergeb Inn Med Kinderheilk 1959; 12:52-125.
8. LLerena A, Cobaleda J, Martinez C et al. Interethnic differences in drug metabolism: Influence of genetic and environmental factors on debrisoquine hydroxylation phenotype. Eur J Drug Metab Pharmacokinet 1996; 21:129-138.
9. Pelkonen O, Sotaniemi EA. Environmental factors of enzyme induction and inhibition. Pharmacol Ther 1987; 33:115-120.
10. Kalow W, Gunn DR. The relation between dose of succinylcholine and duration of apnea in man. J Pharmacol Exp Ther 1957; 120:203-214.
11. Maitland ML, Vasisht K, Ratain MJ. TPMT, UGT1A1 and DPYD: Genotyping to ensure safer cancer therapy? Trends Pharmacol Sci 2006; 27:432-7.
12. Eichelbaum M. Ein nueentdeckter Defekt im Arzeinmittel-stoffwechsel dês Menschen: Dir fehlende N-Oxidation dês Spartein. Bonn: Medizinische Kakultat Rheinishen Friedrich-Wilhelmsd Universitat, 1975.
13. Smith RL. Introduction: Human genetic variations in oxidative drug metabolism. Xenobiotica 1986; 16:361-365.
14. Kellermann G, Shaw CR, Luyten-Kellerman M. Aryl hydrocarbon hydroxylase inducibility and bronchogenic carcinoma. N Engl J Med 1973; 289:934-937.
15. Fujii-Kuriyama Y, Ema M, Mimura J et al. Polymorphic forms of the Ah receptor and induction of the CYP1A1 gene. Pharmacogenetics 1995; 5(Spec No):S149-53.
16. (http://www.cypalleles.ki.se).
17. Wrensch MR, Miike R, Sison JD et al. CYP1A1 variants and smoking-related lung cancer in San Francisco Bay area Latinos and African Americans. Int J Cancer 2005; 113:141-147.
18. Fragoso JM, Juarez-Cedillo T, Hernandez-Pacheco G et al. Cytochrome P4501A1 polymorphisms in the Amerindian and Mestizo populations of Mexico. Cell Biochem Funct 2005; 23(3):189-193.
19. Gallegos-Arreola MP, Batista-Gonzalez CM, Delgado-Lamas JL et al. Cytochrome P4501A1 polymorphism is associated with susceptibility to acute lymphoblastic leukemia in adult Mexican patients. Blood Cells Mol Dis 2004; 33:326-329.
20. Paz-y-Mino C, Arevalo M, Munoz GMJ et al. CYP1A1 genetic polymorphisms in Ecuador, South America. Dis Markers 2005; 21:57-59.
21. Muñoz S, Vollrath V, Vallejos MP et al. Genetic polymorphisms of CYP2D6, CYP1A1 and CYP2E1 in the South-Amerindian population of Chile. Pharmacogenetics 1998; 8(4):343-351.
22. Quiñones L, Lucas D, Godoy J et al. CYP1A1, CYP2E1 and GSTM1 genetic polymorphisms. The effect of single and combined genotypes on lung cancer susceptibility in Chilean people. Cancer Lett 2001; 174:35-44.
23. Acevedo C, Opazo JL, Huidobro C et al. Positive correlation between single or combined genotypes of CYP1A1 and GSTM1 in relation to prostate cancer in Chilean people. Prostate 2003; 57:111-7.
24. Gaspar PA, Hutz MH, Salzano FM et al. Polymorphisms of CYP1A1, CYP2E1, GSTM1, GSTT1, and TP53 Genes in Amerindians. Amer J Phys Anthropol 2002; 119:249-246.
25. Ladona MG, Izquierdo-Martinez M, Posada de la Paz MP et al. Pharmacogenetic profile of xenobiotic enzyme metabolism in survivors of the Spanish toxic oil syndrome. Environ Health Perspect 2001; 109:369-75.
26. Garte S, Gaspari L, Alexandrie AK et al. Metabolic gene polymorphism frequencies in control populations. Cancer Epidemiol Biomarkers Prev 2001; 10:1239-1248.
27. Ishibe N, Wiencke JK, Zuo ZF et al. Susceptibility to lung cancer in light smokers associated with CYP1A1 polymorphisms in Mexican- and African-Americans. Cancer Epidemiol Biomarkers Prev 1997; 6:1075-1080.
28. Landi MT, Sinha R, Lang NP et al. Human cytochrome P4501A2. IARC Sci Publ 1999; 148:173-195.
29. Nakajima M, Yokoi T, Mizutani M et al. Genetic polymorphism in the 5'-flanking region of human CYP1A2 gene: Effect on the CYP1A2 inducibility in humans. J Biochem (Tokyo) 1999; 125(4):803-808.
30. Sachse C, Brockmoller J, Bauer S et al. Functional significance of a C>A polymorphism in intron 1 of the cytochrome P450 CYP1A2 gene tested with caffeine. Br J Clin Pharmacol 1999; 47:445-449.

31. Mendoza-Cantu A, Castorena-Torres F, Bermudez M et al. Genotype and allele frequencies of polymorphic cytochromes P450 CYP1A2 and CYP2E1 in Mexicans. Cell Biochem Funct 2004; 22:29-34.
32. Castorena-Torres F, Mendoza-Cantu A, de Leon MB et al. CYP1A2 phenotype and genotype in a population from the Carboniferous Region of Coahuila, Mexico. Toxicol Lett 2005; 156:331-339.
33. Arnulfo Albores, personal communication.
34. Dorado P, Llerena A. Unpublished observations.
35. Lee CR, Goldstein JA, Pieper JA. Cytochrome P450 2C9 polymorphisms: A comprehensive review of the in-vitro and human data. Pharmacogenetics 2002; 12:251-63.
36. Schwarz UI. Clinical relevance of genetic polymorphisms in the human CYP2C9 gene. Eur J Clin Invest 2003; 33(Suppl 2):23-30.
37. Streetman DS, Bertino Jr JS, Nafziger AN. Phenotyping of drug-metabolizing enzymes in adults: A review of in-vivo cytochrome P450 phenotyping probes. Pharmacogenetics 2000; 10:187-216.
38. Dorado P, Berecz R, Norberto MJ et al. CYP2C9 genotypes and diclofenac metabolism in Spanish healthy volunteers. Eur J Clin Pharmacol 2003; 59:221-225.
39. Christensen M, Andersson K, Dalen P et al. The Karolinska cocktail for phenotyping of five human cytochrome P450 enzymes. Clin Pharmacol Ther 2003; 73:517-28.
40. Blakey GE, Lockton JA, Perrett J et al. Pharmacokinetic and pharmacodynamic assessment of a five-probe metabolic cocktail for CYPs 1A2, 3A4, 2C9, 2D6 and 2E1. Br J Clin Pharmacol 2004; 57:162-169.
41. Goldstein JA, de Morais SM. Biochemistry and molecular biology of the human CYP2C subfamily. Pharmacogenetics 1994; 4:285-299.
42. LLerena A, Dorado P, O'Kirwan F et al. Lower frequency of CYP2C9*2 in Mexican-Americans compared to Spaniards. Pharmacogenomics J 2004; 4:403-406.
43. Machorro-Lazo MV, Flores-Martínez SE, García-Zapién AG et al. Genetic frequency of the variant CYP2C9*2 in Western Mexican Mestizos. HGM2003 Posters Abstracts Cancún 2003, (http://hgm2003.hgu.mrc.ac.uk).
44. Bravo-Villalta HV, Yamamoto K, Nakamura K et al. Genetic polymorphism of CYP2C9 and CYP2C19 in a Bolivian population: An investigative and comparative study. Eur J Clin Pharmacol 2005; 61:179-184.
45. Vianna-Jorge R, Perini JA, Rondinelli E et al. CYP2C9 genotypes and the pharmacokinetics of tenoxicam in Brazilians. Clin Pharmacol Ther 2004; 76:18-26.
46. Perini JA, Vianna-Jorge R, Brogliato AR et al. Influence of CYP2C9 genotypes on the pharmacokinetics and pharmacodynamics of piroxicam. Clin Pharmacol Ther 2005; 78:362-369.
47. Perini JA, Suarez-Kurtz G. Impact of CYP2C9*3/*3 genotype on the pharmacokinetics and pharmacodynamics of piroxicam. Clin Pharmacol Ther 2006, (in press).
48. Desta Z, Zhao X, Shin JG et al. Clinical significance of the cytochrome P450 2C19 genetic polymorphism. Clin Pharmacokinet 2002; 41:913-958.
49. Llerena A, Valdivielso MJ, Benitez J et al. Reproducibility over time of mephenytoin and debrisoquine hydroxylation phenotypes. Pharmacol Toxicol 1993; 73(1):46-48.
50. Inaba T, Jorge LF, Arias TD. Mephenytoin hydroxylation in the Cuna Amerindians of Panama. Br J Clin Pharmacol 1988; 25(1):75-79.
51. Luo HR, Poland RE, Lin KM et al. Genetic polymorphism of cytochrome P450 2C19 in Mexican Americans: A cross-ethnic comparative study. Clin Pharmacol Ther 2006; 80:33-40.
52. Gonzalez HM, Romero EM, Peregrina AA et al. CYP2C19- and CYP3A4-dependent omeprazole metabolism in West Mexicans. J Clin Pharmacol 2003; 43:1211-1215.
53. Reviriego J, Bertilsson L, Carrillo JA et al. Frequency of S-mephenytoin hydroxylation deficiency in 373 Spanish subjects compared to other Caucasian populations. Eur J Clin Pharmacol 1993; 44(6):593-595.
54. De Morais SM, Wilkinson GR, Blaisdell J et al. The major genetic defect responsible for the polymorphism of S-mephenytoin metabolism in humans. J Biol Chem 1994; 269:15419-15422.
55. De Morais SM, Wilkinson GR, Blaisdell J et al. Identification of a new genetic defect responsible for the polymorphism of (S)-mephenytoin metabolism in Japanese. Mol Pharmacol 1994; 46:594-598.
56. Llerena A, Cobaleda J, Martinez C et al. Interethnic differences in drug metabolism: Influence of genetic and environmental factors on debrisoquine hydroxylation phenotype. Eur J Drug Metab Pharmacokinet 1996; 21:129-38.
57. Eichelbaum M, Ingelman-Sundberg M, Evans WE. Pharmacogenomics and individualized drug therapy. Annu Rev Med 2006; 57:119-137.
58. Dorado P, Berecz R, Peñas-LLedó EM et al. Clinical implications of CYP2D6 genetic polymorphism during treatment with antipsychotic drugs. Current Drug Target 2006, (In press).

59. LLerena A, Edman G, Cobaleda J et al. Relationship between personality and debrisoquine hydroxylation capacity. Suggestion of an endogenous neuroactive substrate or product of the cytochrome P4502D6. Acta Psychiatr Scand 1993; 87:23-28.
60. Aklillu E, Dandara C, Bertilsson L et al. Pharmacogenetics of cytochrome P450s in African populations - Clinical and molecular evolutionary implications. In: Suarez-Kurtz G, ed. Pharmacogenetics in Admixed Populations. Austin: Landes Bioscience, 2007.
61. Bernard S, Neville KA, Nguyen AT et al. Interethnic differences in genetic polymorphisms of CYP2D6 in the U.S. population: Clinical implications. Oncologist 2006; 11:126-135.
62. Mendoza R, Wan YJ, Poland RE et al. CYP2D6 polymorphism in a Mexican American population. Clin Pharmacol Ther 2001; 70:552-560.
63. Sosa-Macias M, Elizondo G, Flores-Perez C et al. CYP2D6 genotype and phenotype in Amerindians of Tepehuano origin and Mestizos of Durango, Mexico. J Clin Pharmacol 2006; 46:527-536.
64. Luo HR, Gaedigk A, Aloumanis V et al. Identification of CYP2D6 impaired functional alleles in Mexican Americans. Eur J Clin Pharmacol 2005; 61:797-802.
65. Casner PR. The effect of CYP2D6 polymorphisms on dextromethorphan metabolism in Mexican Americans. J Clin Pharmacol 2005; 45(11):1230-1235.
66. Lopez M, Guerrero J, Jung-Cook H et al. CYP2D6 genotype and phenotype determination in a Mexican Mestizo population. Eur J Clin Pharmacol 2005; 61:749-754.
67. Agundez JA, Ramirez R, Hernandez M et al. Molecular heterogeneity at the CYP2D gene locus in Nicaraguans: Impact of gene-flow from Europe. Pharmacogenetics 1997; 7:337-340.
68. Isaza CA, Henao J, Lop… …ıg metabolizer
CYP2D6 gene in the Col… …); 22:695-705.
69. Gonzalez I et al. CYP2D…
…z AM et al. Isolation, sequence and genotyping of the dr…
70. Jorge LF, Eichelbaum M, … …s of CYP2D6
in Ngawbe and Embera … ombian population. Methods Find Exp Clin Pharmacol 200… …versus drift in
world populations. Pharm… …6 genotypes in a Cuban population. 2006, (unpublished).
Griese EU et al. Comparative evolutionary pharmacogeneti…
71. Arias TD, Jorge LF, Lee … …Amerindians
Amerindians of Panama and Colombia: Role of selection
of Panama: Absence of evidence for deficient metabolizers. Clin Pharmacol Ther 1988; 43:456-465.
72. Jorge LF, Arias TD, Inaba T et al. Unimodal distribution of the metabolic ratio for debrisoquine in Cuna Amerindians of Panama. Br J Clin Pharmacol 1990; 30:281-285.
73. LLerena A, Dorado P, Cáceres MC. CYP2D6 genotypes in a Spanish population. 2006, (unpublished).
74. LLerena A, Ramirez R. Debrisoquine hydroxylation phenotypes (CYP2D6) in a Nicaraguan population. 2006, (unpublished).
75. Bradford LD. CYP2D6 allele frequency in European Caucasians, Asians, Africans and their descendants. Pharmacogenomics 2002; 3:229-243.
76. Flores DL, Alvarado I, Wong ML et al. Clinical implications of genetic polymorphism of CYP2D6 in Mexican Americans. Ann Intern Med 2004; 140:939.
77. Dorado P, Caceres MC, Pozo-Guisado E et al. Development of a PCR-based strategy for CYP2D6 genotyping including gene multiplication of worldwide potential use. Biotechniques 2005; 39:571-574.
78. Guengerich FP, Kim DH, Iwasaki M. Role of human cytochrome P-450 IIE1 in the oxidation of many low molecular weight cancer suspects. Chem Res Toxicol 1991; 4:168-179.
79. Daly AK. Molecular basis of polymorphic drug metabolism. J Mol Med 1995; 73:539-553.
80. Wu X, Amos CI, Kemp BL et al. Cytochrome P450 2E1 DraI polymorphisms in lung cancer in minority populations. Cancer Epidemiol Biomarkers Prev 1998; 7:13-18.
81. Konishi T, Smith JL, Lin KM et al. Influence of genetic admixture on polymorphisms of alcohol-metabolizing enzymes: Analyses of mutations on the CYP2E1, ADH2, ADH3 and ALDH2 genes in a Mexican-American population living in the Los Angeles area. Alcohol Alcohol 2003; 38:93-94.
82. Mendoza-Cantu A, Castorena-Torres F, Bermudez de Leon M et al. Occupational toluene exposure induces cytochrome P450 2E1 mRNA expression in peripheral lymphocytes. Environ Health Perspect 2006; 114:494-499.
83. Martinez C, Agundez JA, Olivera M et al. Influence of genetic admixture on polymorphisms of drug-metabolizing enzymes: Analyses of mutations on NAT2 and C gamma P2E1 genes in a mixed Hispanic population. Clin Pharmacol Ther 1998; 63:623-628.
84. Vidal F, Lorenzo A, Auguet T et al. Genetic polymorphisms of ADH2, ADH3, CYP4502E1 Dra-I and Pst-I, and ALDH2 in Spanish men: Lack of association with alcoholism and alcoholic liver disease. J Hepatol 2004; 41:744-750.

85. Sierra-Torres CH, Arboleda-Moreno YY, Orejuela-Aristizabal L. Exposure to wood smoke, HPV infection, and genetic susceptibility for cervical neoplasia among women in Colombia. Environ Mol Mutagen 2006; 47:553-561.
86. Konishi T, Luo HR, Calvillo M et al. ADH1B*1, ADH1C*2, DRD2 (-141C Ins), and 5-HTTLPR are associated with alcoholism in Mexican American men living in Los Angeles. Alcohol Clin Exp Res 2004; 28:1145-1152.
87. Zintzaras E, Stefanidis I, Santos M et al. Do alcohol-metabolizing enzyme gene polymorphisms increase the risk of alcoholism and alcoholic liver disease? Hepatology 2006; 43:352-361.
88. Poland RA, Lin KM, Nuccio C et al. Cytochrome P450 2E1 and 3A activities do not differ between Mexicans and European Americans. Clin Pharmacol Ther 2002; 72:288-293.
89. Lamba JK, Lin YS, Schuetz EG et al. Genetic contribution to variable human CYP3A-mediated metabolism. Adv Drug Deliv Rev 2002; 54:1271-1294.
90. Reyes-Hernandez OD, Arteaga-Illan G, Elizondo G. Detection of CYP3A4*1B and CYP3A4*2 polymorphisms by RFLP. Distribution frequencies in a Mexican population. Clin Genet 2004; 66:166-168.
91. Ball SE, Scatina J, Kao J et al. Population distribution and effects on drug metabolism of a genetic variant in the 5' promoter region of CYP3A4. Clin Pharmacol Ther 1999; 66:288-294.
92. Kadlubar FF, Berkowitz GS, Delongchamp RR et al. The CYP3A4*1B variant is related to the onset of puberty, a known risk factor for the development of breast cancer. Cancer Epidemiol Biomarkers Prev 2003; 12:327-331.
93. Gervasini G, Vizcaino S, Gasiba C et al. Differences in CYP3A5*3 genotype distribution and combinations with other polymorphisms between Spaniards and Other Caucasian populations. Ther Drug Monit 2005; 27:8198-21.
94. Suarez-Kurtz G, Pena SDJ. Pharmacogenetic studies in the Brazilian population. In: Suarez-Kurtz G, ed. Pharmacogenetics in Admixed Populations. Austin: Landes Bioscience, 2007.
95. Hustert E, Zibat A, Presecan-Siedel E et al. Natural protein variants of pregnane X receptor with altered transactivation activity toward CYP3A4. Drug Metab Dispos 2001; 29:1454-1459.
96. (www.imshealth.com).

Chapter 6

Pharmacogenetic Studies in the Brazilian Population

Guilherme Suarez-Kurtz* and Sergio D.J. Pena

Abstract

By virtue of being the product of the genetic admixture of three ancestral roots: Europeans, Africans and Amerindians, the Brazilian population displays very high levels of genomic diversity and several peculiarities in its genetic structure that are reviewed in this chapter. After painting this background we then move to the pharmacogenetic/genomic arena and review the data available for the Brazilian population, including extant Amerindian groups, for phase I (cytochrome P450 superfamily, butyrylcholinesterase and aldehyde dehydrogenases) and phase II (gluthatione-S-transferases, thiopurine S-methyltransferase and N-acetyltransferases) drug metabolizing enzymes, the ABCB1 transmembrane drug transporter and various drug receptors/targets (adrenergic beta-receptors, the G-protein sub-unit 3, the renin-angiotensin system, modulators of drug effects on lipid metabolism and methylenetetrahydrofolate reductase). Finally, we take examples from available Brazilian data to document the challenges and advantages created by population admixture for the study and the implementation of pharmacogenetic/genomic methodology and personalized drug therapy.

Introduction

Brazil is the 5th largest country in the world and occupies an area of 8.5 million km^2. Its present population exceeds 184 million people, who speak Portuguese—differently from all other Latin Americans countries, which speak Spanish. The language reflects the colonization of Brazil by the Portuguese, initiated in the year 1500. At that time, the indigenous population living in the area of what is now Brazil, was estimated at ~2.5 million.[1,2] The Portuguese-Amerindian admixture started soon after the arrival of the first colonizers. Mating between European men and indigenous women became commonplace and later (after 1755) was even encouraged as a strategy for population growth and colonial occupation of the country. From the middle of the 16th century, Africans were brought to Brazil to work on sugarcane farms and, later, in the gold and diamond mines and on coffee plantations. Historical records suggest that between 1551 and 1850 (when the slave trade was abolished), ~3.5 million Africans arrived in Brazil.[1,3] As to the European immigration, it is estimated that ~500,000 Portuguese arrived in the country between 1500 and 1808.[1] From then on, after the Brazilian ports were legally opened to all friendly nations, Brazil received increasing numbers of immigrants from several parts of the world. Portugal remained by far the most important source of migrants, followed by Italy, Spain, and Germany. In the 20th century, Asian immigration took place, mainly from Japan, as well as from Lebanon and Syria. According to Callegari-Jacques

*Corresponding Author: Guilherme Suarez-Kurtz—Division of Pharmacology, Instituto Nacional de Câncer, Rio de Janeiro, RJ 22290-290, Brazil. Email: kurtz@inca.gov.br.

Pharmacogenomics in Admixed Populations, edited by Guilherme Suarez-Kurtz.

and Salzano,[4] 58% of the immigrants who arrived in Brazil between 1500 and 1972 were Europeans, 40% were Africans, and 2% were Asians. In this sense, Brazil might be seen as representing a "meeting point" for the three major historical geographical components of humanity [Africans, Asians (represented by their Native American descendants) and Europeans]. It is therefore not surprising that, after five centuries of mating between these distinct populations, Brazilians form one of the most heterogeneous peoples of the world, a fact that has far reaching consequences for the implementation of pharmacogenetics in practice.

Genetic Variation in Brazilians

In the past few years one of us (SDJP) has been using different molecular tools to characterize the ancestry of Brazilians and the formation and structure of Brazilian populations. We will describe briefly these studies, starting with uniparental markers because they can be useful to bring forth the concept of haplotype genealogies.

Uniparental Genetic Markers in Brazilians

In our initial studies we examined DNA polymorphisms in the nonrecombining portion of the Y-chromosome to investigate the contribution of distinct patrilineages to the present-day White Brazilian population. Twelve unique-event polymorphisms were typed in 200 unrelated males from four geographical regions of Brazil and in 93 Portuguese males.[5] These polymorphisms identified the most important region-specific Y haplogroups of mankind. Of particular interest were haplogroup E3a, which is typical of sub-Saharan Africa and haplogroup Q3, which is a marker of Amerindian ancestry. In our Brazilian sample, the vast majority of Y-chromosomes proved to be of European origin. Only 2% of the Y-chromosome lineages belonged to the African haplogroup E3a and none typed as the Amerindian haplogroup Q3. Indeed, there were no significant differences when the haplogroup frequencies in Brazil and Portugal were compared by means of an exact test of population differentiation. Likewise, there was no population differentiation among the four geographical regions of Brazil. Nevertheless, by typing with fast evolving NRY markers we later could uncover a higher within-population haplotype diversity in Brazil than in Portugal, explainable by the input of diverse European Y chromosomes.[6]

We also studied the same White Brazilians for mtDNA, revealing a very different picture from the NRY. Considering Brazil as a whole, 33%, 28% and 39% of matrilineages were of Amerindian, African and European origin, respectively.[7] As expected, the frequency of different regions reflected their genealogical histories: most matrilineal lineages in the Amazonian region had Amerindian origin, while African ancestrality was preponderant in the Northeast (44%) and the European haplogroups in the South (66%).

In summary, these phylogeographical studies with White Brazilians revealed that the vast majority of patrilineages have European origin, while most matrilineages (>60%) were Amerindian or African. Together, these results configure a picture of strong directional mating between European males and Amerindian and African females, which agrees with the known history of the peopling of Brazil since 1500. An example of the relevance of these phylogeographical studies to pharmacogenetics, is our demonstration[8] that the occurrence in a White Brazilian subject of the *CYP2C9*5* allele—previously detected exclusively in Sub-Saharan Africans and African-Americans—derived from his matrilineal, African ancestry (see below).

Biparental Genetic Markers in Brazilians

As mentioned above, the population of Brazil, formed by extensive admixture between Amerindians, Europeans and Africans, is one of the most variable in the world. In 2003 we published a study whose objective was to ascertain to what extent the physical appearance of a Brazilian individual was predictive of the degree of genomic African or European ancestry.[9] We used a panel of 10 ancestry-informative markers (AIMs), i.e., genetic polymorphisms that displayed large differences in allelic frequencies (>0.40) between Europeans and Africans[10] to estimate, on an individual level, the ancestry of Brazilians. With the purpose of verifying the individual discrimination power of this set of 10 AIMs we genotyped two samples of individuals from Northern Portugal and from the island of São Tomé, located on the west coast of Africa. Since this island was a Portuguese entrepôt for assembly of slaves before shipment to Brazil, its population had geographical origin in several regions of Africa.[11] From the genotype information we calculated for each individual the African Ancestry Index (AAI) that is the logarithm of the product over all loci of the ratio of the likelihood of a given genotype being from African origin to the likelihood of it being of European origin.[9] There was no overlap between the AAI values obtained for the two groups, permitting a complete individual discrimination between the European and African genomes. We then studied a Brazilian sample composed of 173 individuals from a Southeastern rural community, phenotypically classified according to their Color (White, Black, or Intermediate) with a multivariate evaluation based on skin pigmentation in the medial part of the arm, hair color and texture, and the shape of the nose and lips. When we compared the AAI values for these individuals, we observed that the groups had much wider ranges than those of Europeans and Africans and that there was very significant overlap between them. In other words, the differences in AAI values of the group of Brazilian Blacks compared with Brazilian Whites were very discrete and several orders of magnitude smaller than those observed between Africans and Europeans. We next embarked on a study of 200 unrelated Brazilian White males who originated from cosmopolitan centers of the four major geographic regions of the country. The results showed AAI values intermediate between Europeans and Africans, even in southern Brazil, a region predominantly peopled by European immigrants.

Altogether, our data strongly suggested that in Brazil, at an individual level, color, as determined by physical evaluation, was a poor predictor of genomic European or African ancestry, estimated by molecular markers. To corroborate these findings with other AIMs, we undertook an investigation based on data from 12 commercially available forensic microsatellites that were utilized to estimate the personal genomic origin for each of 752 individuals (275 Whites, 192 Intermediates and 285 Blacks) from the city of São Paulo.[12] The genotypes permitted the calculation of a personal likelihood-ratio estimator of African or European ancestry. Although the 12 marker set proved capable of discriminating between European and African individuals, we observed very significant overlaps among the three color categories of Brazilians. This was confirmed quantitatively using a Bayesian analysis of population structure that did not demonstrate significant genetic differentiation between the three color groups. These results corroborate and validate our previous conclusions using AIMs that in Brazil at the individual level there is significant dissociation of color and genomic ancestry.

The two studies mentioned above did not specifically analyze the Amerindian contribution to the Brazilian population. To achieve that we needed new polymorphic markers that would be sensitive to the three ancestralities. We then turned to a set of 40 insertion-deletion (indel) polymorphisms that proved to be exquisitely discriminating in that regard.[13] We genotyped 59 Europeans, 36 Africans and 54 Amerindians with the 40-indel set and used the Structure software[14] to obtain individual proportion of the three ancestral populations. The results are shown as a triangle plot (Fig. 1A). Clearly this indel set is capable of separating the three populations very well. In Figure 1B we show the results for 200 self-declared White Brazilians superimposed on the three ancestral populations. It is quite evident that many of White Brazilians have significant degrees African and/or Amerindian ancestry. Even more striking are the results with 100 Black Brazilians sampled in the city of São Paulo (Fig. 1C).

It is obvious then that, regardless of their skin color, the overwhelming majority of Brazilians have a significant degree of African ancestry. Likewise it could be easily demonstrated that, regardless of their skin color, the overwhelming majority of Brazilians have a significant degree of European ancestry. Finally, although we have not calculated the exact numbers, we can safely predict that regardless of their skin color, a sizeable proportion of Brazilians have a significant degree of Amerindian ancestry!

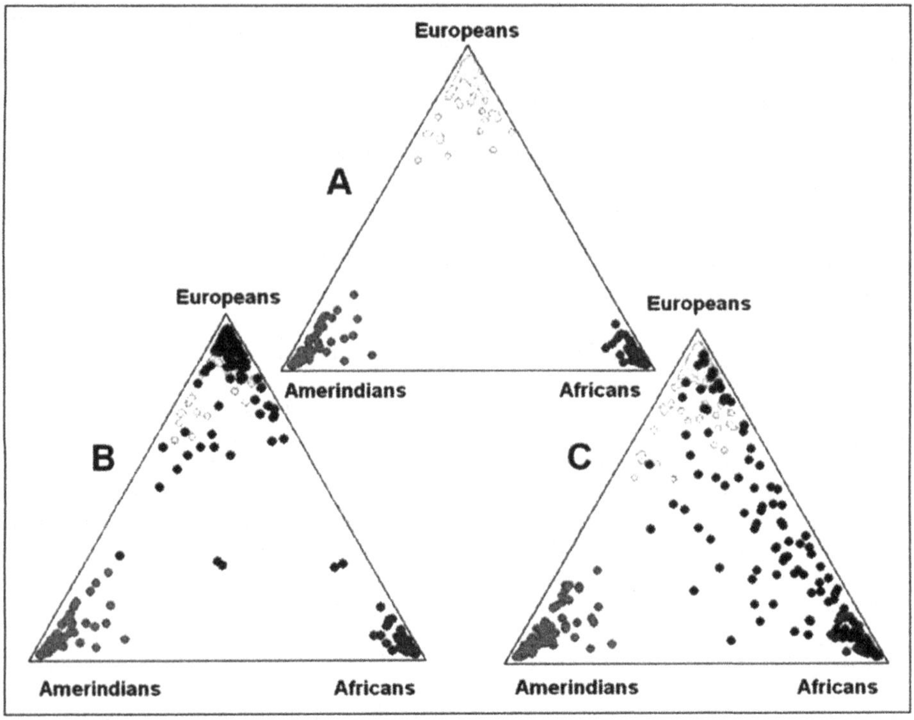

Figure 1. Triangular plots of ancestral proportions based on a set of 40 insertion-deletion polymorphisms used as ancestry-informative markers (L. Bastos-Rodrigues, J. Pimenta, S.P. Bydlowski and S.D.J. Pena, manuscript in preparation). The data were analyzed with the *Structure 2.1* software.[14] A) 220 Europeans (green), 159 Amerindians (red) and 161 Africans (blue). B) 200 White self-declared Brazilians. C) 100 Black self-declared Brazilians. A color version of this figure is available online at www.eurekah.com.

It thus makes no sense talking about "populations" of "White Brazilians" or "Black Brazilians" because of the poor correlation between color and ancestry. Also, it does not make sense talking about African-Brazilians or European-Brazilians because most Brazilians will have significant proportions of African and of European (and of Amerindian) ancestry. Thus, the only possible basis to deal with genetic variation in Brazilians is on a person-by-person basis, according with the Variable Mosaic Genome" paradigm,[15] which allows any individual to have different ancestries in different genomic segments.

This introduction serves to show the ethnosemantic confusion and the perplexities that emerge from the naïve application of paradigms established for populations with, at face value, little genetic admixture to the interpretation of genetic variation in highly admixed populations such as the Brazilians. We also would like to highlight hazards of equating color or "race" with geographical ancestry and of using interchangeably terms such as White, Caucasian and European in one hand, and Black, Negro or African in the other, as is often done in the scientific and medical literature.

Pharmacogenetics in Brazilians

The heterogeneity and extensive admixture of the Brazilian population has important implications for pharmacogenetics because extrapolation of data derived from well-defined ethnic groups is clearly not applicable to the majority of Brazilians. Nevertheless, only recently recognition of this fact translated into pharmacogenetic research on prescribed drugs, such as thiopurines,[16,17] nonsteroidal anti-inflammatory agents (NSAIDs)[18,19] and lipid-lowering statins.[20-22] By comparison, much more information has accumulated over the last 20 years on genetic variation in metabolic pathways for environmental pro-carcinogens, and its impact on cancer risk. Other pharmacogenetic targets, such as adrenergic beta-receptors, components of the renin-angiotensin system (RAS) and methylene-tetrahydrofolate reductase (MTHFR) have been investigated in relation to disease susceptibility and phenotypes, rather than drug response. The following sections present an overview of these different aspects of pharmacogenetics in Brazilians, based on a PubMed-based literature search, completed on June 15, 2006. Allele, genotype and/or haplotype frequencies of polymorphic genes of pharmacogenomic interest were compiled from studies including a minimum of 50 healthy individuals/group. The data are presented according to the pharmacological function of the encoded proteins, i.e., drug metabolism, transport or receptor. Because diverse racial/ethnic categorization criteria and terminology were used in different studies—authored, in some cases, by the same researchers—comparison of the data for distinct strata of the population is not straightforward. We attempted to deal with this challenge by adopting the classification scheme used in the 2000 Brazilian Census[23] which recognizes five categories: "branco" (white), "pardo" (brown), "preto" (black), "amarelo" (yellow, indicating Asians) and "indígena" (Amerindian or Native American). Accordingly, data from individuals classified in the original reports as white, Caucasian, Caucasoid, European-derived or of European descent are grouped under the category "White", data from those described as "Mulatto", "interethnic-admixed" or of mixed-ancestry are grouped under "Intermediate" and data from those reported as black, African-derived or African-Brazilian, under Black. Because several studies lumped Intermediate and Black subjects as "nonwhite", we included this category in some tables. One fundamental caveat should be made at this point. The poor correlation between color and ancestry in Brazilians (see above) casts a shadow of uncertainty over studies that have used only self-reported color (or "race") or investigator assessment of color without genomic ancestry analysis. Moreover, the different color terminologies used have different meaning according to the region of Brazil. For instance, a person who is considered Black in Southern Brazil might be classified as Intermediate in the Northeast.

Pharmacogenetics of Drug Metabolic Pathways

Phase I Enzymes

The cytochrome P450 (CYP) enzymes are the most extensively studied phase I enzymes in Brazilian populations,[24] which is not surprising in view of the major role of these enzymes in the metabolism of xenobiotics in humans. Other phase I enzymes investigated in Brazilians include the alcohol-dehydrogenases (ALDs) and butyrylcholinesterase (BChE).

CYPs

In humans, over 55 *CYP* genes have been identified to date, but only a relatively small number of the encoded proteins, mainly in the CYP1, CYP2 and CYP3 families, are involved in drug metabolism. Genetic polymorphisms in these CYP enzymes contribute to the pronounced inter-individual and inter-ethnic variation in the capacity of an individual to metabolize prescribed drugs and other xenobiotics. The *CYP* genes studied in Brazilians and the reported frequency of the variant alleles in the different population strata are shown in Table 1.[25-37]

CYP1A1

Two common polymorphisms in *CYP1A1* have been studied in Brazilians: one is 2455A>G, which alters the protein structure by replacing an isoleucine with a valine (I462V) and the other is 3801T>C, in the 3′ noncoding region (*Msp*I). The single 3801C variant is termed *CYP1A1**2A and the single 2455G variant is termed *CYP1A1**2C, whereas 3801C linked to

Table 1. Frequency of variant CYP alleles in the population of Brazil

Gene	Allele	Population Groups*						Reference
		General Pop.	White	Intermediate	Black	Nonwhite	Amerindian**	
CYP1A1	*2A	0.22 (112)						25
		0.27 (256)	0.20 (121)			0.32 (135)		26
							0.72 - 0.96 (190)	27
		0.17 (221)						28
			0.22 (472)					29
	*2C	0.09 (108)	0.22 (90)					25
			0.12 (86)		0.14 (231)			30
							0.54 - 0.97 (190)	27
			0.14 (472)					29
		0.10 (212)						31
CYP2A6	*1B	0.30 (412)	0.38 (147)	0.30 (142)	0.20 (123)			32
		0.27 (289)	0.29 (151)	0.22 (63)	0.25 (75)	0.24 (138)		33
	*2	0.02 (412)	0.03 (147)	0.01 (142)	0.01 (123)			32
		0.01 (289)	0.01 (151)	0.02 (63)	<0.01 (75)	0.01 (138)		33
	*4	0.01 (412)	0.01 (147)	0.0 (142)	<0.01 (123)			32
	*9	0.06 (412)	0.05 (147)	0.08 (142)	0.04 (123)			32
CYP2C9	*2	0.09 (331)	0.12 (136)	0.07 (118)	0.04 (77)			18
	*3	0.07 (331)	0.08 (136)	0.07 (118)	0.03 (77)			18
CYP2D6	*3		0.01 (99)					34
	*4		0.20 (99)					34
	*5		0.03 (99)					34
	*2N		0.04 (99)					34

continued on next page

Table 1. Continued

Gene	Allele	Population Groups*						Reference
		General Pop.	White	Intermediate	Black	Nonwhite	Amerindian**	
CYP2E1	*5B	0.05 (108)						25
			0.05 (206)	0.06 (86)				36
							0.02 - 0.33 (190)	27
		0.05 (221)						28
		0.05 (150)						37
		0.06 (212)						31
		0.05 (289)	0.07 (151)	0.03 (63)	0.03 (75)	0.03 (138)		33
	*6		0.08 (151)			0.08 (138)		34
CYP3A4	*1B	0.03 (110)						20
		0.08 (472)						29
CYP3A5	*3		0.92 (107)					20
		0.63 (339)	0.78 (128)	0.74 (111)	0.33 (100)			37

* Number of individuals in brackets; ** Amerindian groups living in Brazil.

2455G is labeled *CYP1A1**2B.[38] Both variant alleles are associated with increased enzymatic activity and their frequency distribution shows marked inter- and intra-ethnic variation. For example, *CYP1A1**2C is virtually absent in Sub-Saharan Africans,[39] relatively uncommon (frequency <0.05) in Europeans,[40] but is the major allele (0.82) in Mapuche from Chile.[41] Amerindian groups living in Brazil also exhibit very high frequencies of both the *CYP1A1**2A and *CYP1A1*2C* polymorphisms (Table 1), and both variant alleles appear to be more common in White and nonWhite Brazilians than in Europeans and Africans, respectively.[40] Gaspar et al[27] argued that genetic drift or interethnic admixture could not explain the surprisingly high frequency (0.14) of *CYP1A1**2C in Black Brazilians, and suggested that the *CYP1A1**2C allele although originated from Africa, was restricted to some ethnic groups not yet investigated, but that may have been present in parts of Africa before the slave traffic to South America began. *CYP1A1 *2A* and **2C* polymorphisms have been investigated in Brazilians as risk factors for cancer [25,26,31,42] cirrhosis and pancreatites,[18] as modulators of estrogen metabolism,[29] but not in relation to individual response to prescribed drugs. Two other *CYP1A1* polymorphisms have been reported in Brazilians, each one in a single individual: a novel silent mutation in exon 7 (2445C>T) and the rare *CYP1A1*9* allele (2461C>T).[26]

CYP1A2

This cytochrome metabolizes a relatively small number of prescribed drugs (e.g., acetaminophen and theophylline) plus many carcinogenic environmental aromatic amines. The *CYP1A2* gene is highly inducible and polymorphic.[38] Only one polymorphism (295C>T) has been studied in Brazilians: its frequency in White women was similar to those reported for European populations and no correlation with hormonal status or lipid metabolism was observed (Table 1).[29]

CYP2A6

This enzyme inactivates prescribed drugs (e.g., coumarin and Tegafur), activates a number of pro-carcinogens, especially tobacco-specific nitrosamines, and provides the main pathway for nicotine catabolism. Over 20 polymorphisms in the *CYP2A6* gene have been described[38] some of which exhibit marked ethnic/racial-dependence. For example, the *CYP2A6*4* allele (null CYP2A6) is relatively common in Asians (frequency ~0.20 in Japanese) and rare (<0.04) in Europeans, Africans and their descendants.[43] Some of the *CYP2A6* genetic polymorphisms (e.g., homozygous *CYP2A6*2* or *CYP2A6*4*) are associated with virtually absent nicotine metabolism, whereas others (e.g., *CYP2A6*9*) result in reduced enzymatic activity.[34] *CYP2A6*2, *4* and *9 are rare in Brazilians (Table 1), but allele **1B* (identified as a gene conversion in the 3′flanking region) is relatively common (0.29 in the overall population). A statistically significant trend for decreasing frequency from White, to Intermediate and to Black persons was detected in one study[32] but not in another.[30] Although the different criteria used for categorization might account for this discrepancy, it is noteworthy that data from other populations consistently show higher frequency of allele **1B* in Europeans compared to Africans and their respective descendants.[43] An association between the *CYP2A6*1B* allele and smoking status was reported for Brazilians, such that individuals having one or two copies of *CYP2A6*1B* were over-represented among nonsmokers, as compared to ever-smokers.[32] This association was markedly influenced by population structure, being significant in individuals self-identified as White and Intermediate, but not Black. When this study was performed, *CYP2A6*1B* was thought to codify for a normal phenotype, and therefore its association with tobacco dependence could not be explained by impaired nicotine metabolism—as previously proposed for *CYP2A6* alleles encoding enzymes with null- or reduced activity.[43] It was suggested that *CYP2A6*1B* might be in linkage with other mutation(s), which are the true cause of the observed association with smoking behavior in Brazilians, and that this linkage disequilibrium (LD) is sensitive to population structure.[32]

CYP2C9

The human CYP2C family metabolizes a wide range of therapeutic agents through three polymorphic enzymes, CYP2C9, CYP2C8 and CYP2C19. CYP2C9 provides the major metabolic pathway for anticoagulants, anti-convulsants, anti-diabetics and NSAIDs. Over 50 single nucleotide polymorphisms (SNPs) have been described in the regulatory and coding regions of *CYP2C9*,[38] of which *CYP2C9*2*, *CYP2C9*3* and *CYP2C9*5* have been investigated In Brazilians. The frequency distribution of *CYP2C9*2* and *3 alleles and genotypes display a population structure: both variant alleles are 2.5-3 times more frequent in White than in Black Brazilians, with Intermediate individuals displaying frequencies midway those observed in White and Black persons (Table 1).[18] The observation of *CYP2C9*5* in a self-identified White Brazilian[45] provided a distinguished example of the peculiarities associated with pharmacogenetics in admixed populations.[46] Prior to this observation, attempts to detect *CYP2C9*5* in Europeans and Asians had repeatedly failed, leading to suggestions that this allele was private to African-derived populations. In the case of the White Brazilian carrier of *CYP2C9*5*, the relative contributions of European, African and Amerindian roots to his genetic pool were 92.0%, 7.5% and 0.5%, respectively, whereas his mtDNA haplotype was L3d, characteristic of Western African populations. Genotyping the proband's mother for the *CYP2C9* gene, revealed that she was carrier of allele *5, configuring the inheritance of the *CYP2C9*5* allele via the matrilineal, African ancestry.[45]

*CYP2C9*3* and, to a lesser extent *CYP2C9*2* encode enzymes with impaired activity. We have reported the impact of these variants on the pharmacokinetics and/or pharmacodynamics of the NSAIDs, tenoxicam and piroxicam.[8,18,19] Compared to *CYP2C9*1/*1* homozygous, individuals heterozygous for either *CYP2C9*2* or *CYP2C9*3 had significantly lower oral clearance and increased area under the plasma concentration-time curve (AUC) of either drug, administered as single or repeated doses. The increased exposure to piroxicam in heterozygous for either *CYP2C9*2* or *3 was accompanied by a significantly greater inhibition of the drug's target, cyclooxygenase-1, measured by ex-vivo inhibition of TxB_2 generation. Homozygosity for *CYP2C9*3*, a rare genotype in the Brazilian population (~0.6%), had a remarkable impact on both the pharmacokinetics and pharmacodynamics of piroxicam: The oral clearance was reduced 8-fold, whereas the terminal half-life of elimination, the AUC and the area above the effect-time curves for the inhibition of COX-1 and COX-2 were several fold higher in a carrier of the *CYP2C9*3/*3* genotype, compared to individuals homozygous for the wild-type allele.[8] We suggested that the protracted inhibition of COX-1 and COX-2 in homozygous carriers of the *CYP2C9*3* might be advantageous in relation to cancer chemoprevention with NSAIDs, by allowing the use of lower daily doses of these drugs and/or administration at longer time intervals. Indeed, our data indicate that COX-2 activity was maintained below 50% of its control value for over a month and the COX-1 activity below 20% for over two months following a single 20 mg oral dose of piroxicam to a homozygous *CYP2C9*3* individual! These results highlight the potential advantage of pharmacogenetic polymorphisms associated with impaired drug metabolism for long-term strategies in disease prevention or treatment.

CYP2C8 plays important roles in the metabolism of therapeutic drugs such as paclitaxel, amiodarone and pioglitazone. Several variants in the *CYP2C8* gene have been identified,[38] of which the most extensively studied is *CYP2C8*3* (416G>A; 1196A>G). This allele is associated with impaired enzyme activity and is strongly linked to *CYP2C9*2* in Europeans and Africans.[48] This linkage has prompted suggestions that the presence of *CYP2C8*3* is the true cause for impaired metabolism among carriers of *CYP2C9*2*.[48] We confirmed the strong LD between *CYP2C9*2* and *CYP2C8*3* in Brazilians, but could not verify whether *CYP2C8*3* or *CYP2C9*2* was responsible for the impaired metabolism of piroxicam, since *CYP2C8*3* occurred in all *CYP2C9*2* carriers and in none of the *CYP2C9*1/*1* individuals.[19]

CYP2D6

This cytochrome plays a major role in the metabolism of several therapeutic drugs, including the conversion of codeine and tramadol into their active metabolite, morphine, and the inactivation of betablockers, neuroleptics and antidepressants. The *CYP2D6* gene contains in excess of 60 allelic variations, and such variability generates a wide range of phenotypes, from "poor metabolizers" with no CYP2D6 enzyme activity to "ultrarapid metabolizers" who have multiple copies of the entire *CYP2D6* gene. Ethnic differences in the distribution of genetic polymorphisms and the ensuing CYP2D6 activity are well known. For example, the most common variant *CYP2D6* alleles in Europeans, Asians and Africans are, respectively, **4*, **10* and **17*.[49] Some functional *CYP2D6* polymorphisms were investigated in relation to the lipid-lowering response to simvastatin in White Brazilians (Table 1). The frequencies of the defective alleles *3, *4 and *5 and of the duplicated variant were within the range reported for Europeans and their descendants, and no influence of these polymorphisms on the response to simvastatin was discerned.[34]

CYP2E1

The human CYP2E1 metabolizes drugs such as the muscle relaxant chlorzoxazone, plays a minor role in the oxidation of ingested alcohol but is of major importance in the bioactivation of several promutagens and procarcinogens. Two polymorphisms in the *CYP2E1* gene, namely *CYP2E1*5B* (-1053C>T linked to -1293G>C) and *CYP2E1*6* (7632T>A) have been investigated as genetic markers for cancer risk,[25,26,31,35,36] chronic gastritis,[36] cirrhosis and pancreatites[28] in Brazilians. The frequency of these polymorphisms in healthy controls varied over a narrow range (0.03 - 0.08) with no difference between Whites and nonWhites (Table 1) By contrast, among Amerindians living in Brazil, the *CYP2E1*5B* allele frequency varies over 15-fold, some groups (Gavião, Surui and Zoro) displaying high frequencies (~0.30; Table 1) in the range reported for other Native South-Americans (0.25 - 0.42).[27,41]

CYP3A4 and CYP3A5

The CYP3A subfamily provides the most important pathway for human xenobiotic metabolism. CYP3A4, the dominant CYP3A in adult liver, exhibits marked inter-individual variability in expression and activity. The *CYP3A4* gene is highly polymorphic and over 30 genetic variants have been described, although most show no association with clinical phenotypes. One variant that has been extensively studied is *CYP3A4*1B*, a SNP in the 5′-regulatory region of the gene (-392A>G), which is in strong LD with a SNP within intron 3 of *CYP3A5*3* (6896A>G). The frequency of these linked SNPs varies extensively among different ethnicities.[50] Self-identified White Brazilians display allele frequencies of *CYP3A4*1B* (0.03 - 0.08) and *CYP3A5*3* (0.78 - 0.92) in the range of European-derived populations. The frequency of *CYP3A5*3* in Black Brazilians (0.33) is significantly lower than in White or Intermediate subjects,[37] but not quite as low as in African Americans.[50] A clinical trial in White Brazilians found no association of *CYP3A4*1B* or *CYP3A5*3* polymorphisms and the efficacy or tolerability of simvastatin.[20]

ADH

In humans, ethanol elimination occurs mainly in the liver, through oxidation to acetaldehyde and acetate by the successive action of ADHs (types 1-4) and aldehyde dehydrogenase (ALDH). *ALDH2*, the most important gene that affects predisposition to alcoholism in Asians, is virtually absent in nonAsians populations. Regarding the ADH gene cluster, a recent meta-analysis disclosed significant associations of the *ADH2*1* and *ADH3*2* less active coding enzymatic alleles and increased risk of alcoholism in East Asians and in Caucasians.[51] *ADH4* is another polymorphic gene in the ADH cluster, which contributes to ethanol oxidation at intoxicating levels of ethanol. Guindalini et al[52] studied the association of three polymorphisms in the promoter of *ADH4* with alcohol dependence in Brazilians and disclosed positive associations of the *-75C* and the *-159A* alleles, but not the *-192A>T* polymorphism, with the risk of developing dependence. Furthermore, haplotypes AAC and TAC (at positions -192, -159 and

-75, respectively) were more frequent among alcohol-dependent subjects, and were considered risk haplotypes. Conversely, haplotypes TAA and AGC were over-represented among controls and were considered protective.

BChE

This plasma enzyme is involved in the metabolism of drugs of abuse (e.g., heroin and cocaine), local anesthetics and the short-acting muscle relaxants succinylcholine and mivacurium. The demonstration, in the early 1950's, that inherited deficiency of BChE activity was responsible for the prolonged apnea following the administration of succinylcholine is one of the landmarks of pharmacogenetics.[53] Genetic variability of BChE in Brazilian populations was first reported in 1984, and over the years several thousands individuals, including various Amerindian populations, have been genotyped and/or phenotyped, allowing the identification of novel *BChE* gene mutations.[54-56] BChE polymorphisms have not been studied in relation to drug response in Brazilians, but were found to be positively associated with weight, stature and body mass index.[57]

Phase 2 Enzymes

Among these enzymes, glutathione-S-transferases (GSTs), thiopurine-S-methyltransferase (TPMT) and N-acetyl-transferase type 2 (NAT2) were studied in Brazilians.

GSTs

This superfamily comprises a number of ubiquitous, multifunctional enzymes which play a key role in cellular detoxification, protecting macromolecules from attack by reactive electrophyles. Polymorphisms in the GSTs have the potential to alter an individual's susceptibility to carcinogens and toxins and influence therapeutic drug response. In humans, the *GST* family comprises 16 genes in six subfamilies, of which the mu (*GSTM*), pi (*GSTP*) and theta (*GSTT*) have been the most extensively investigated. Several polymorphisms have been described in members of these subfamilies, with variable consequences on enzyme activity. Four functional GST polymorphisms have been investigated in Brazilian populations: the *GSTM1-null* and the *GSTT1-null* genotypes (homologous deletions), the *GSTM3*B* polymorphism—a 3-pb deletion in intron 6, postulated to increase the expression levels of GSTM3—and the 313A>G transition (Ile105Val) in exon 5 of the *GSTP1* gene, associated with reduced enzymatic activity (Table 2).[58-68]

GSTT1

The *GSTT1-null* genotype occurs at similar frequencies (0.20 - 0.25) in Europeans, Caucasian-Americans, African-Americans, African Vendas and Zimbabweans, but is considerably more common among South African Xhosa and Coloured groups (0.41 - 0.57).[40,69] The frequencies reported in healthy White and nonWhite Brazilians vary within a relatively narrow range (0.18 - 0.26; Table 2), except in two studies in which higher frequencies (0.36 and 0.42) were detected. These apparent discrepancies were tentatively ascribed to regional variations in ethnic-admixture[62] and to different patterns of GST expression in various tissues.[64] The distribution of the *GSTT1-null* genotype among Amerindian populations living in Brazil is highly heterogeneous, regardless of whether they are of the same geographic region or linguistic group.[30,58]

GSTM1

The range of *GSTM1-null* frequencies in White and in Black Brazilians (Table 2), with one exception,[65] overlaps with the corresponding ranges in populations of predominantly European (0.42 - 0.60) or African (0.16 - 0.36) ancestry.[40,70] Accordingly, there is a trend for declining frequency of the *GSTM1-null* genotype from White, to Intermediate to Black Brazilians. Similar to the *GSTT1-null* genotype, the frequency of the *GSTM1-null* genotype varies markedly among extant Amerindian populations. Possible reasons for this heterogeneity, including environmental and evolutionary factors were discussed by Gaspar et al.[30]

Table 2. ***Frequency of variant* GST *alleles and genotypes in the population of Brazil***

		Population Groups*						
Gene	**Allele/Genotype**	**General Pop.**	**White**	**Intermediate**	**Black**	**Nonwhite**	**Amerindian****	**Reference**
GSTT1	GSTT1-null		0.19 (130)		0.19 (117)		0.11 (79)	58
		0.18 (276)						59
							0 - 0.30 (190)	30
		0.22 (300)						60
		0.25 (591)	0.25 (319)	0.24 (140)	0.28 (132)	0.26 (272)		61
		0.19 (221)						28
		0.19 (150)	0.16 (135)					36
		0.21 (666)	0.20 (432)	0.19 (138)	0.36 (96)			62
			0.22 (233)	0.17 (87)	0.26 (137)			63
		0.42 (81)						64
		0.26 (285)	0.26 (106)	0.26 (91)	0.26 (90)			65
GSTM1	GSTM1-null		0.55 (130)		0.33 (117)		0.20 (79)	58
		0.37 (276)						59
							0.04 - 0.43 (190)	30
		0.42 (300)						60
		0.42 (591)	0.49 (319)	0.40 (140)	0.28 (132)	0.34 (272)		61
		0.46 (221)						28
		0.41 (150)	0.44 (135)					36
		0.46 (666)	0.49 (492)	0.47 (138)	0.35 (96)			62
			0.55 (233)	0.41 (87)	0.33 (137)			63
		0.37 (278)	0.38 (196)			0.36 (82)		66
		0.40 (82)						64
		0.38 (285)	0.47 (106)	0.41 (91)	0.24 (90)			65

continued on next page

Table 2. Continued

		Population Groups*						
Gene	**Allele/Genotype**	**General Pop.**	**White**	**Intermediate**	**Black**	**Nonwhite**	**Amerindian****	**Reference**
GSTM3	*B	0.31 (212)	0.21 (131)					31
		0.37 (286)	0.28 (106)	0.34 (90)	0.52 (90)			65
GSTP1	313G (105Val)	0.31 (591)	0.32 (123)	0.31 (140)	0.31 (132)			61
		0.31 (221)						28
		0.04 (157)						68
TPMT	*2	0.8% (306)	0.6% (83)	0.8% (204)				16
		2.2% (204)						17
	*3A	1.6% (306)	1.8% (83)	2.0% (204)				16
		1.5% (204)						17
	*3B	0.2% (204)						17
	*3C	2.1% (306)	1.8% (83)	2.5% (204)				16
		1.0% (204)						17

* Number of individuals in brackets; ** Amerindian groups living in Brazil.

GSTM3

Our studies[65] revealed significant trends for increasing frequency of the *GSTM3*B* allele and the *GSTM3*B/*B* genotype from White to Intermediate to Black Brazilians, in agreement with the data reported by Marques et al[31] despite the relatively small numbers (<50) of Intermediate and Black individuals in the latter study. The frequency of the **B* allele and of the homozygous *B/*B genotype in White Brazilians (0.21 - 0.28 and 0.07, respectively) are similar to that reported for Caucasian Americans (0.24 and 0.05).[71] By contrast, the **B* allele and *B/*B genotype frequencies in Black Brazilians (0.52 and 0.30) are considerably lower than in African-Americans (0.68 and 0.48, respectively)[71] and even more so than in South Africa Bantus (0.80 and 0.64).[72] These data are consistent with the larger European contribution to the gene pool of Black Brazilians, compared to African-Americans.[15]

GSTP1

The 313A variant allele (Table 2) was detected at an overall frequency of 0.31 in two studies in healthy Brazilians, with no apparent influence of population structure,[28,61] whereas a surprisingly 8-fold lower frequency was reported for healthy controls in a *GSTP1* profiling study for thyroid malignancies.[68] This discrepancy might derive from different genotyping procedures,[24] but nevertheless casts a shadow of uncertainty over the conclusion that carriers of the *GSTP1* 313A allele have a 7 - 10-fold greater risk for thyroid carcinomas.[68]

TPMT

TPMT degrades purine antimetabolites (e.g., 6-mercaptopurine and azathioprine) used to treat childhood ALL and as immunosupressant agents. The TPMT activity displays important inter-individual variability in all populations hitherto investigated, a common pattern in European-derived populations being a trimodal distribution, with *ca.* 89%, 11% and <0.5% of individuals displaying high, intermediate and low or undetectable activity, respectively.[73] Adjusting the dose of 6-mercaptopurine according to the TPMT phenotype had a major impact on the 5-year ALL survival rate[74] and is now routinely performed in many pediatric centers, providing an outstanding example of the clinical value of pharmacogenomics. The variation in TPMT activity is largely due to SNPs in the *TPMT* gene, the three most common defective alleles being *TPMT*2* (238G>C), *TPMT*3A* (460G>A and 719A>G) and *TPMT*3C* (719A>G).[73] The frequency distribution of these alleles varies markedly across, as well as within, continental populations.[46,50] *TPMT*3A* is the most common allele in Europeans but has not been detected in East Asians or West Africans; the latter also do not carry the *TPMT*2* allele. Nevertheless, *TPMT*2* and *TPMT*3A* are present in African-Americans and in Black Brazilians, which is consistent with European admixture in these groups. Notably, the relative frequencies of *TPMT*3A* and *TPMT*3C* in Brazilians, White or nonWhite, are not significantly different (Table 2), in striking contrast with the wide differences observed in all other ethnic groups hitherto studied, including the neighbouring Argentinians.[75] The latter observation is a good example of the pharmacogenetic variability within South America.[15]

NAT2

The acetylation polymorphism of isoniazid mediated by arylamine-N-acetyltransferases (NATs) is one of the landmarks of pharmacogenetics.[76] In humans, NATs are encoded by two adjacent genes, *NAT1* and *NAT2* in chromosome 8p22. Alleles *NAT1*4* and *NAT2*4* are designated as "wildtype", although they are not the most common in all ethnic groups. Because NATs are involved in catabolism of several environmental carcinogens, polymorphisms of their encoding genes have been extensively investigated as cancer risk factors. One of these polymorphisms, a *481C>T* transition present in alleles *NAT2*5A, *5B, *5F, *6E, *11, *12C* and **14* and associated with the slow acetylator phenotype[77] was studied in Brazilians.[31] Two main observations were reported: a trend for declining frequency of the *481T* variant from White to Intermediate and to Black subjects, consistent with data from other populations,[40] and an over representation of this variant in control subjects as compared to oral cancer patients, in agreement with reports of positive association between rapid acetylator phenotype and increased oral cancer risk in Caucasians.[78]

Drug Transporters

Our literature search on pharmacogenetics of drug transporters in Brazilians disclosed results only for ABCB1, a member of the ABC (ATP-binding cassette) family.

ABCB1

Also known as P-glycoprotein or Pgp, this multidrug transporter is encoded by the *ABCB1* gene, which displays over 50 SNPs and 3 insertion/deletion polymorphisms. Among the commonly observed SNPs (frequency >10%), only 2677G>T/A leads to amino acid changes (Ala893Ser/Thr), whereas 1236C>T in exon 12 and 3435C>T in exon 26 are silent. The frequency of these SNPs and derived haplotypes varies considerably across populations.[79] Accordingly, we found a statistically significant effect of stratification on the distribution of these polymorphisms among healthy Brazilians, with a trend for decreasing frequency of the T allele at each locus from White, to Intermediate to Black individuals (Table 3).[80] Data from hypercholesterolemic White subjects recruited at two other centers[20,22] were in good agreement with our results, except for the higher frequency of the 3435T allele in White individuals recruited in the city of Porto Alegre. This might result from the relatively higher proportion of European ancestry in the South region of Brazil, where Porto Alegre is located.[9] For haplotype analysis, we grouped the A and T alleles at locus 2677 as non-G, and observed a significant effect of population structure on haplotype distribution, despite the fact that of the eigth possible haplotypes, four (G/C/G, T/nonG/T, C/G/T and T/G/C) accounted for 88 - 95% of the

Table 3. *ABCB1* polymorphisms in Brazilians*

	General Population	White	Intermediate	Black	Reference
Allele		**Allele Frequency**			
1236T		0.46			20
		0.41	0.34	0.33	80
2677A		0.03			22
		<0.01	0.02	<0.01	80
2677T		0.38			22
		0.39	0.28	0.18	80
2677 nonG		0.49			20
3435T	0.39				81
		0.46			22
		0.56			20
		0.41	0.34	0.31	80
Haplotype**		**Haplotype Frequency**			
C/G/C	0.52	0.41, 0,48	0.52	0.57	20, 80
T/nonG/T	0.25	0.41, 0.35	0.21	0.16	20, 80
C/G/T	0.08	0.08, 0.08	0.07	0.09	20, 80
T/G/C	0.07	0.01, 0.04	0.08	0.11	20, 80
T/G/T	0.03	0.01, <0.01	0.03	0.06	20, 80
C/nonG/T	0.02	0.03, 0.02	0.04	0.01	20, 80
T/nonG/C	0.02	0.01, 0.01	0.03	0.01	20, 80
C/nonG/C	0.01	0.02, 0.01	0.02	0.01	20, 80

* N = 315 (ref. 80) and 105 (ref. 81) healthy individuals, and 99 (ref. 20) and 69 (ref. 22) hypercholesterolemic individuals. ** derived from SNPs at loci 1236, 2677 and 3435

genetic variability in all three population categories. However, carriers of haplotype T/nonG/T were twice as frequent in Whites than in Blacks, while a reverse pattern was seen for haplotype T/G/C; the frequency of these two haplotypes in the Intermediate group was midway between White and Black subjects. Polymorphisms in *ABCB1* were investigated in relation to the efficacy and/or toxicity of statins in Brazilians, and the reported results are reviewed below, under "Pharmacogenetics of statins".

Drug Receptors and Targets

Beta-Adrenergic Receptors

Three nonsynonymous, and functionally important polymorphisms in the coding region of the beta-2 receptor (*ADBR2*) have been studied in relation to blood pressure levels and obesity in Brazilians: 46A>G (Arg16Gly), 79C>G (Gln27Glu) and 491C>T (Thr164Ile) (Table 4). An "ethnic-related genetic structure" was disclosed in the general population, such that the frequency of Glu27 and Ile164 was lower in nonWhite, compared to White individuals. Of the eight haplotypes that could be theoretically resolved with these SNPs, three accounted for >97% of the variability in Brazilians.[82] The most common haplotype was Arg16/Gln27/Thr164 (frequency 0.43), which also predominates among Sub-Saharan Africans (0.50 - 0.56) and Asians (0.58), but not in white Americans (0.35).[83] Association studies of *ADBR2* polymorphisms revealed that individuals harboring the Gly16/Gly16 genotype had a 1.48-fold higher risk of hypertension.[82] Regarding *ADRB2* polymorphisms and obesity, contradictory results were reported: in one study both Arg16 and Gln27 were found to increase the risk of obesity in the general population,[82] whereas in another study obesity was associated with Gly16 (rather than Arg16), but the association was restricted to males.[84] The latter study investigated also association of obesity and a common missense polymorphism at codon 64 (Trp64Arg) of the beta-3 adrenergic receptor gene (*ADRB3*). The Arg64 allele was detected in 11% of White Brazilians (Table 4) and had no effect on the risk of obesity in this group.[84]

GNB3

A common C825T polymorphism in the gene *GNB3*, which encodes the beta-3 subunit of heterotrimeric G proteins, has been associated with hypertension, obesity, and atherosclerosis in humans.[85] The 825T allele is ca. 3-fold more frequent in Africans than in Europeans,[86] which probably explains the trend for increasing frequency of this allele from White, to Intermediate to Black Brazilians (Table 4). The C825T polymorphism was not independently associated with obesity, cholesterol metabolism or blood pressure in Brazilians.[84] However a significant interaction between this polymorphism and obesity was observed for the systolic blood pressure (SBP), such that within the obese population, the SBP was significantly higher in carriers of the TT genotype, compared to CC homozygotes.[87]

RAS

Genes that encode components of the RAS are thought to modulate susceptibility to cardiovascular disease and drug effects on this system, although no clear picture has yet emerged despite extensive investigation.[88] Studies in Brazilians have focused on the cardiovascular risk—rather than on pharmacogenetics per se—associated with polymorphisms in three primary genes of the RAS (Table 4): the angiotensinogen (*AGT*) M235T SNP, the insertion/deletion (I/D) of 287 bp within intron 16 in the angiotensin I-converting enzyme gene (*ACE*), and the A1166C polymorphism in the gene encoding the angiotensin II subtype 1 receptor (*AGTR1*).[89-93] The frequency of polymorphisms I/D in *ACE* and M285T in *AGT* was significantly higher in Black, than in White Brazilians,[90] A similar situation prevails between Africans or African Americans *versus* European-derived populations.[95,96] Pereira et al[97] studied the effect of the I/D *ACE* polymorphism on the serum ACE activity and observed different influence of this polymorphism according to racial categorization: A higher level of association exists in White than in Black Brazilians, with the Intermediate subjects in between these two groups, consistent with a "distinct ethnic difference in how serum ACE activity is regulated regarding the *ACE* I/D polymorphism".[97] The observation of this phenomenon in different strata from the same population highlights the unique opportunities provided by admixed populations for gene-association studies in disease susceptibility and pharmacogenomics.[15,46]

eNOS

Nitric oxide (NO) is a highly diffusible and reactive molecule that plays crucial roles in cardiovascular homeostasis. Endothelial nitric oxide synthase (eNOS) is the primary physiological source of NO and polymorphisms in the *eNOS* gene have been associated with a variety of pathophysiological conditions. Three polymorphisms have been studied in Brazilians: SNPs in the promoter region (-786T>C) and in exon 7 (Glu298Asp) and a variable number of tandem repeats (VNTR) in intron 4 (Table 4). The frequency of all three polymorphisms varied considerably between two samples of Black individuals studied at the same center,[98,99] possibly the result of stratification within the population samples. An effect of population structure on the frequency of the Asp298 allele was recently reported in a large multi-ethnic Brazilian sample and evidence was presented for a role of the Glu298Asp polymorphism in modulating blood pressure through a relationship with serum blood levels.[101] However, other authors detected no difference in the allele and genotype frequency of the Glu298Asp, the -786T>C and the intron 4 VNTR polymorphisms between normotensive and hypertensive Brazilians.[99-100]

Pharmacogenetics of Statins

The association of genetic polymorphisms with the efficacy and toxicity of statins has been focused in several studies in Brazilians. Hutz and collaborators[20,21,102] investigated the effects of polymorphisms in *ABCB1* and in genes encoding apolipoprotein E (*APOE*), cholesterol ester transfer (*CEPT*), hepatic lipase (*LIPC*), sterol regulatory element-binding factors 1-a and 2 (*SREBF-1a* and *SREBF-2*) and SREBF cleavage-activating (*SCAP*) proteins on the response to simvastatin in hypercholesterolemic White Brazilians. The data were analysed separately for each gene tested, and significant associations between changes in plasma concentration of low-density lipoprotein cholesterol (LDL-C)—the primary phenotypic response to statin therapy—and genetic polymorphisms were disclosed only for *ABCB1*. Simvastatin caused a greater reduction in LDL-C (and total cholesterol) in carriers of the 1236T, 2677A or 2677T (2677 non-G) and the T/non-G/T haplotype.[20] By contrast, no significant association was detected between the lipid lowering effect of atorvastatin and *ABCB1 p*olymorphisms in another study of hypercholesterolemic White individuals.[22] Whether differences in the recruited patients or the prescribed statin account for these discrepant results is not known. Data for the other polymorphic genes studied by Hutz's group revealed greater decrease in total cholesterol (but not LDL-C) in carriers of the *SCAP* 2386G allele and a greater HDL-C increase in *CETP B2B2* homozygotes compared to B1B1 or B1B2 subjects.[102]

MTHFR

MTHFR is a key enzyme regulating intracellular folate concentrations, which in turn modulate the sensitivity to antifolate drugs such as methotrexate and 5-fluorouracil. Sixty-five common polymorphisms are known in the *MTHFR* gene,[103] two of which have been widely studied as disease risk factors and as pharmacogenetic targets: 677T>C - a missense (alanine to valine) associated with a thermolabile enzyme of reduced activity - and 1298C>A - a nonsynonimous variant that causes substitution of glutamate for alanine, predicted to be benign -. In Brazilians, the 677T>C and 1298C>A polymorphisms have been investigated as risk factors for congenital malformations, childhood leukemia and reduced homocystein plasma levels[104-111] but not as modulators of drug response. Contradictory results have been published regarding the effect of population structure on the distribution of the 677T and the 1298C alleles and genotypes among Brazilians, possibly because of the application of different criteria for population stratification (Table 4).

Table 4. Frequency of polymorphisms in drug targets in Brazilians

		Population Groups*						
Gene	**Allele**	**General Pop.**	**White**	**Intermediate**	**Black**	**Nonwhite**	**Amerindian****	**Reference**
ADRB2	Gly16	0.56 (1576)						82
			0.56 (335)					84
	Glu27	0.28 (1576)			Glu27			82
			0.32 (335)					84
	Ile164	<0.01 (1576)			Ile164			82
ADRB3	Arg64		0.11 (335)		Arg64			84
GNB3	825T	0.45 (1552)	0.36 (548)	0.49 (795)	0.63 (122)			87
			0.33 (335)					84
	814A		0.04 (335)					84
ACE	I/D (Del)	0.61 (137)						89
			0.54 (150)	0.57 (142)	0.65 (90)			90
		0.64 (525)						91
		0.53 (71)						92
					0.58 (103)			93
		0.60 (104)						94
AGT	235T		0.57 (150)	0.56 (142)	0.69 (90)			90
		0.34 (104)						94
AT1R	1166C (3´UTR)	0.25 (104)						94
eNOS	-786C		0.42 (154)		0.19 (136)			98
			0.41 (113)		0.29 (87)			99
		0.38 (102)						100

continued on next page

Table 4. Continued

Gene	Allele	Population Groups*						Reference
		General Pop.	White	Intermediate	Black	Nonwhite	Amerindian**	
	Asp298		0.33 (154)		0.15 (136)			98
			0.29 (113)		0.28 (87)			99
		0.26 (102)						100
		0.26 (1577)						101
	4a (VNTR intron 4)		0.18 (154)		0.32 (136)			98
			0.14 (113)		0.19 (87)			99
		0.14 (102)						100
MTHFR	677T		0.37(107)		0.20 (137)		0.11 (83)	104
			0.37 (75)					105
		0.26 (252)	0.26 (103)			0.25 (149)		106
		0.31 (220)						107
		0.23 (843)						108
		0.29 (564)	0.29 (474)			0.30 (90)		109
			0.35 (87)	0.24 (82)	0.12 (40)			110
		0.31 (198)	0.29 (119)			0.33 (79)		111
	1298C		0.25 (75)					108
		0.22 (252)	0.29 (103)			0.17 (149)		106
		0.24 (220)						107
			0.24 (87)	0.22 (82)	0.13 (40)			110
		0.25 (199)	0.28 (120)			0.22 (79)		111

* Number of individuals in brackets. ** Amerindian groups living in Brazil.

Concluding Remarks

Sir John Gaddum once described the pharmacologist as a "Jack of all trades" referring to the multiple practical and scientific talents required for devising and performing experiments in classical pharmacology around the time when the term pharmacogenetics was first used.[112] The global expansion of pharmacogenetics over the last fifty years required that pharmacologists learned new "trades" or scientific disciplines, but also witnessed the opposite, or rather complementary trend of specialists from various disciplines becoming interested and making significant contributions to pharmacogenetics/genomics. Both scenarios can be recognized in the development of pharmacogenetics/genomics in Brazil. This is reflected in the data reviewed in this chapter as well as in the membership of the Brazilian pharmacogenetics/pharmacogenomics network (Rede Nacional de Farmacogenética/farmacogenômica or REFARGEN).[113,114] The caleidoscopic heterogeneity of individual genetic ancestry in the tri-rooted Brazilian population exacerbates the challenges to the pharmacogeneticist studying admixed populations, and cannot be reconciled with the notion of "race/ethnicity-targeted therapy".[115]

References

1. Salzano FM, Freire-Maia N. Problems in human biology: A study of Brazilian populations. Detroit: Wayne State University Press, 1970:1-200.
2. Salzano FM, Bortolini MC. The Evolution and genetics of Latin American populations. Cambridge, UK: Cambridge University Press, 2002:1-512.
3. Curtin PD. The Atlantic slave trade: A census. Madison: University of Wisconsin Press, 1956.
4. Callegari-Jacques SM, Salzano FM. Brazilian Indian/non Indian interactions and their effects. Ciência e Cultura 1999; 51:166-174.
5. Carvalho-Silva DR, Santos FR, Rocha J et al. The phylogeography of Brazilian Y-chromosome lineages. Am J Hum Genet 2001; 68:281-286.
6. Carvalho-Silva DR, Tarazona-Santos E, Rocha J et al. Y chromosome diversity in Brazilians: Switching perspectives from slow to fast evolving markers. Genetica 2006; 126:251-260.
7. Alves-Silva J, da Silva Santos M, Guimarães PE et al. The ancestry of Brazilian mtDNA lineages. Am J Hum Genet 2000; 67:444-461.
8. Perini JA, Suarez-Kurtz G. Impact of the CYP2C9*3/*3 genotype on the pharmacokinetics and pharmacodynamics of piroxicam. Clin Pharmacol Ther 2006; 80:549-551.
9. Parra FC, Amado RC, Lambertucci JR et al. Color and genomic ancestry in Brazilians. Proc Natl Acad Sci 2003; 100:177-182.
10. Shriver,MD, Smith MW, Jin L et al. Ethnic-affiliation estimation by use of population-specific DNA markers. Am J Hum Genet 1997; 60:957-964.
11. Tomas G, Seco L, Seixas S et al. The peopling of Sao Tome (Gulf of Guinea): Origins of slave settlers and admixture with the Portuguese. Hum Biol 2002; 74:397-411.
12. Pimenta JR, Zuccherato LW, Debes AA et al. Color and genomic ancestry in Brazilians: A new study with forensic microsatellites. Human Heredity 2006; 62:190-195.
13. Bastos-Rodrigues L, Pimenta JR, Pena SD. The genetic structure of human populations studied through short insertion-deletion polymorphisms. Ann Hum Genet 2006; 70:658-65.
14. Pritchard JK, Stephens M, Donnelly P. Inference of population structure using multilocus genotype data. Genetics 2000; 155:945-959.
15. Suarez-Kurtz G, Pena SDJ. Invited Review: Pharmacogenomics in the Americas: Impact of genetic admixture. Curr Drug Targets 2006; 7:1649-1658.
16. Reis MM, Santoro A, Suarez-Kurtz G. Thiopurine methyltransferase phenotypes and genotypes in Brazilians. Pharmacogenetics 2003; 13:371-173.
17. Boson WL, Romano-Silva MA, Correa H et al. Thiopurine methyltransferase polymorphisms in a Brazilian population. Pharmacogenomics J 2003; 3:178-82.
18. Vianna-Jorge R, Perini JA, Rondinelli E et al. CYP2C9 genotypes and the pharmacokinetics of tenoxicam in Brazilians. Clin Pharmacol Ther 2004; 76:18-26.
19. Perini JA, Vianna-Jorge R, Brogliato AR et al. Influence of CYP2C9 genotypes on the pharmacokinetics and pharmacodynamics of piroxicam. Clin Pharmacol Ther 2005; 78:362-369.
20. Fiegenbaum M, Silveira FR, Van der Sand CR et al. The role of common variants of ABCB1, CYP3A4, and CYP3A5 genes in lipid-lowering efficacy and safety of simvastatin treatment. Clin Pharmacol Ther 2005; 78:551-558.
21. Fiegenbaum M, Silveira FR, Van der Sand CR et al. Determinants of variable response to simvastatin treatment: The role of common variants of SCAP, SREBF-1a and SREBF-2 genes. Pharmacogenomics J 2005; 5:359-364.

22. Rodrigues AC, Rebecchi IM, Bertolami MC et al. High baseline serum total and LDL cholesterol levels are associated with MDR1 haplotypes in Brazilian hypercholesterolemic individuals of European descent. Braz J Med Biol Res 2005; 38:1389-1397.
23. http://www.ibge.gov.br/home/estatistica/populacao/censo2000/.
24. Filho MB, Albano RM, Rossini A et al. Pharmacogenomics of xenobiotic metabolizing enzymes in South American populations. Curr Pharmacogenomics 2006; 4:9-18.
25. Sugimura H, Hamada GS, Suzuki I et al. CYP1A1 and CYP2E1 polymorphism and lung cancer, case-control study in Rio de Janeiro, Brazil. Pharmacogenetics 1995; 5:S145-148.
26. Amorim LM, Lotsch PF, Simao TD et al. Analysis of CYP1A1 exon 7 polymorphisms by PCR-SSCP in a Brazilian population and description of two novel gene variations. Mutat Res 2004; 547:35-40.
27. Gaspar PA, Kvitko K, Papadopolis LG et al. High frequency of CYP1A1*2C allele in Brazilian populations. Hum Biol 2002; 74:235-242.
28. Burim RV, Canalle R, Martinelli ALC et al. Polymorphisms in glutathione S-transferases GSTM1, GSTT1 and GSTP1 and cytochromes P450 CYP2E1 and CYP1A1 and susceptibility to cirrhosis or pancreatitis in alcoholics. Mutagenesis 2004; 19:291-298.
29. Almeida S, Zandona MR, Franken N et al. Estrogen-metabolizing gene polymorphisms and lipid levels in women with different hormonal status. Pharmacogenomics J 2005; 5:346-351.
30. Gaspar PA, Hutz MH, Salzano FM et al. Polymorphisms of CYP1A1, CYP2E1, GSTM1, GSTT1, and TP53 Genes in Amerindians. Amer J Phys Anthropol 2002; 119:249-246.
31. Marques CFS, Koifman S, Koifman RJ et al. Influence of CYP1A1, CYP2E1, GSTM3 and NAT2 genetic polymorphisms in oral cancer susceptibility: Results from a case-control study in Rio de Janeiro. Oral Oncol 2006, (in press).
32. Vasconcelos GM, Struchiner CJ, Suarez-Kurtz G. CYP2A6 genetic polymorphisms and correlation with smoking status in Brazilians. Pharmacogenomics J 2005; 5:42-48.
33. Rossini A, Lima SS, Rapozo DC et al. CYP2A6 and CYP2E1 polymorphisms in a Brazilian population living in Rio de Janeiro. Braz J Med Biol Res 2006; 39:195-201.
34. Fiegenbaum M. A farmacogenética da sinvastatina: Investigação de variantes em genes candidatos e sua influência sobre a resposta ao tratamento. PhD Dissertation. Universidade Federal do Rio Grande do Sul, 2005.
35. Gattas GJ, Soares-Vieira JA. Cytochrome P450-2E1 and glutathione S-transferase mu polymorphisms among Caucasians and mulattoes from Brazil. Occup Med (Lond) 2000; 50:508-511.
36. Colombo J, Rossit ARB, Caetano A et al. GSTT1, GSTM1 and CYP2E1 genetic polymorphisms in gastric cancer and chronic gastritis in a Brazilian population. World J Gastroenterol 2004; 10:1240-1245.
37. Perini JA, Suarez-Kurtz G. Unpublished observations.
38. http://www.cypalleles.ki.se.
39. Dandara C, Sayi J, Masimirembwa CM et al. Genetic polymorphism of cytochrome P450 1A1 (Cyp1A1) and glutathione transferases (M1, T1 and P1) among Africans. Clin Chem Lab Med 2002; 40:952-957.
40. Garte S, Gaspari L, Alexandrie AK et al. Metabolic gene polymorphism frequencies in control populations. Cancer Epidemiol Biomarkers Prev 2001; 10:1239-1248.
41. Munoz S, Vollrath V, Vallejos MP et al. Genetic polymorphisms of CYP2D6, CYP1A1 and CYP2E1 in the South-Amerindian population of Chile. Pharmacogenetics 1998; 8:343-351.
42. Canalle R, Burim RV, Tone LG et al. Genetic polymorphisms and susceptibility to childhood acute lymphoblastic leukemia. Environ Mol Mutagen 2004; 43:100-109.
43. Schoedel KA, Hoffmann EB, Rao Y et al. Ethnic variation in CYP2A6 and association of genetically slow nicotine metabolism and smoking in adult Caucasians. Pharmacogenetics 2004; 14:615-626.
44. Malaiyandi V, Sellers EM, Tyndale RF. Implications of CYP2A6 genetic variation for smoking behaviors and nicotine dependence. Clin Pharmacol Ther 2005; 77:145-158.
45. Suarez-Kurtz G, Vianna-Jorge R, Perini JA et al. Detection of CYP2C9*5 in a White Brazilian. Clin Pharmacol Ther 2005; 77:587-588.
46. Suarez-Kurtz G. Pharmacogenetics in admixed populations. Trends in Pharmacological Sciences 2005; 26:196-201.
48. Garcia-Martin E, Martinez C, Tabares B et al. Interindividual variability in ibuprofen pharmacokinetics is related to interaction of cytochrome P450 2C8 and 2C9 amino acid polymorphisms. Clin Pharmacol Ther 2004; 76:119-127.
49. Xie HG, Kim RB, Wood AJ et al. Molecular basis of ethnic differences in drug disposition and response. Annu Rev Pharmacol Toxicol 2001; 41:815-50.
50. http://pgeni.im.wustl.edu.

51. Zintzaras E, Stefanidis I, Santos M et al. Do alcohol-metabolizing enzyme gene polymorphisms increase the risk of alcoholism and alcoholic liver disease? Hepatology 2006; 43:352-361.
52. Guindalini C, Scivoletto S, Ferreira RG et al. Association of genetic variants in alcohol dehydrogenase 4 with alcohol dependence in Brazilian patients. Am J Psychiatry 2005; 162:1005-1007.
53. Kalow W, Gunn DR. The relation between dose of succinylcholine and duration of apnea in man. J Pharmacol Exp Ther 1957; 120:203-214.
54. Chautard-Freire-Maia EA, Carvalho RD, da Silva MC et al. Frequencies of atypical serum cholinesterase in a mixed population of northeastern Brazil. Hum Hered 1984; 34:364-370.
55. Alcantara VM, De Lourenco MA, Salzano FM et al. Butyrylcholinesterase polymorphisms (BCHE and CHE2 loci) in Brazilian Indian and admixed populations. Hum Biol 1995; 67:717-726.
56. Souza RL, Fadel-Picheth C, Allebrandt KV et al. Possible influence of BCHE locus of butyrylcholinesterase on stature and body mass index. Am J Phys Anthropol 2005; 126:329-334.
57. Souza RL, Mikami LR, Maegawa RO et al. Four new mutations in the BCHE gene of human butyrylcholinesterase in a Brazilian blood donor sample. Mol Genet Metab 2005; 84:349-353.
58. Arruda VR, Grignolli CE, Gonçalves MS et al. Prevalence of homozygosity for the deleted alleles of glutathione S-transferase mu (GSTM1) and (GSTT1) among distinct ethnic groups from Brazil: Relevance to environmental carcinogenesis? Clin Genet 1998; 54:210-214.
59. Arruda VR, Lima CSP, Grignoli CRE et al. Increased risk for acute myeloid leukaemia in individuals with glutathione S-transferase mu 1 (GSTM1) and theta 1 (GSTT1) gene defects. Eur J Haematol 2001; 66:383-388.
60. Morari EC, Leite JLP, Granja F et al. The null genotype of glutathione S-transferase M1 and T1 locus increases the risk for thyroid cancer. Cancer Epidemiol Biomarkers Prev 2002; 11:1485-1488.
61. Rossini A, Rapozo DCM, Amorim LMF et al. Frequencies of GSTM1, GSTT1, and GSTP1 polymorphisms in a Brazilian population. Genet Mol Res 2002; 3:233-240.
62. Hatagima A, Marques CFS, Krieger H et al. Glutathione S-transferase M1 (GSTM1) and T1 (GSTT1) polymorphisms in a Brazilian mixed population. Human Biology 2004; 6:937-942.
63. Gattás GJF, Kato M, Soares-Vieira JA et al. Bydlowski SP.Ethnicity and glutathione S- transferase (GSTM1/GSTT1) polymorphisms in a Brazilian population. Braz j Med Biol Res 2004; 37:451-458.
64. Drummond SN, De Marco L, Noronha JCM et al. GSTM1 polymorphism and oral squamous cell carcinoma. Oral Oncol 2004; 40:52-55.
65. Suarez-Kurtz G, Vargens DD, Struchiner C et al. Skin color, genomic ancestry and the distribution of GST polymorphisms. Pharmacogenetics/genomics 2007; (in press).
66. Linhares JJ, Silva IDCG, Souza NCN et al. Genetic polymorphism of GSTM1 in women with breast cancer and interact with reproductive history and several clinical pathologies. Biol Res 2005; 38:273-281.
67. Drummond SN, Gomez RS, Noronha JCM et al. Association between GSTT-1 gene deletion and the susceptibility to oral squamous cell carcinoma in cigarette-smoking subjects. Oral Oncol 2005; 41:515-519.
68. Granja F, Morari J, Morari EC et al. GST profiling may be useful in the screening for thyroid nodule malignancy. Cancer Letters 2004; 209:129-137.
69. Adams CH, Werely CJ, Victor TC et al. Allele frequencies for glutathione S-transferase and N-acetyltransferase 2 differ in African population groups and may be associated with oesophageal cancer or tuberculosis incidence. Clin Chem Lab Med 2003; 41:600-605.
70. Ye Z, Song H, Higgins JP et al. Five glutathione s-transferase gene variants in 23,452 cases of lung cancer and 30,397 controls: Meta-analysis of 130 studies. PLoS Med 2006; 3:91.
71. Park LY, Muscat JE, Kaur T et al. Comparison of GSTM polymorphisms and risk for oral cancer between African-Americans and Caucasians. Pharmacogenetics 2000; 10:123-131.
72. Tetlow N, Robinson A, Mantle T et al. Polymorphism of human mu class glutathione transferases. Pharmacogenetics 2004; 14:359-368.
73. Weinshilboum R, Sladek SL. Mercaptopurine pharmacogenetics: Monogenic inheritance of erythrocyte thipurine methyltransferase activity. Am J Hum Genet 1980; 32:651-662.
74. Evans WE, Relling MV. Moving towards individualized medicine with pharmacogenomics. Nature 2004; 429:464-468.
75. Larovere LE, de Kremer RD, Lambooy LH et al. Genetic polymorphism of thiopurine S-methyltransferase in Argentina. Ann Clin Biochem 2003; 40:388-393.
76. Evans DA, Manley KA, McKusick VA. Genetic control of isoniazid metabolism in man. Br Med J 1960; 13:485-491.
77. http://www.louisville.edu/medschool/pharmacology.html.
78. Chen C, Ricks S, Doody DR et al. N-Acetyltransferase2 polymorphisms, cigarette smoking and alcohol consumption, and oral squamous cell cancer risk. Carcinogenesis 2001; 22:1993-1999.

79. Leschziner G, Zabaneh D, Pirmohamed M et al. Exon sequencing and high resolution haplotype analysis of ABC transporter genes implicated in drug resistance. Pharmacogenet Genomics 2006; 16:439-450.
80. Estrela RCE, Suarez-Kurtz G. Unpublished observations.
81. Calado RT, Franco RF, Zago MA et al. MDR1 gene C3435T polymorphism and the risk of acquired aplastic anaemia. Br J Haematol 2002; 117:769.
82. Pereira AC, Floriano MS, Mota GF et al. Beta2 adrenoceptor functional gene variants, obesity, and blood pressure level interactions in the general population. Hypertension 2003; 42:685-692.
83. Maxwell TJ, Ameyaw MM, Pritchard S et al. Beta-2 adrenergic receptor genotypes and haplotypes in different ethnic groups. Int J Mol Med 2005; 16:573-580.
84. Mattevi VS, Zembrzuski VM, Hutz MH. Impact of variation in ADRB2, ADRB3, and GNB3 genes on body mass index and waist circumference in a Brazilian population. Am J Hum Biol 2006; 18:182-186.
85. Siffert W. G protein polymorphisms in hypertension, atherosclerosis, and diabetes. Annu Rev Med 2005; 56:17-28.
86. Rosskopf D, Manthey I, Siffert W. Identification and ethnic distribution of major haplotypes in the gene GNB3 encoding the G-protein beta3 subunit. Pharmacogenetics 2002; 12:209-220.
87. Danoviz ME, Pereira AC, Mill JG et al. Hypertension, obesity and GNB 3 gene variants. Clin Exp Pharmacol Physiol 2006; 33:248-252.
88. Baudin B. Polymorphism in angiotensin II receptor genes and hypertension. Exp Physiol 2005; 90:277-282.
89. Mansur AP, Annicchino-Bizzacchi J, Favarato D et al. Angiotensin-converting enzyme and apolipoproteins genes polymorphism in coronary artery disease. Clin Cardiol 2000; 23:335-340.
90. Pereira AC, Mota GA, Bensenor I et al. Effect of race, genetic population structure, and genetic models in two-locus association studies: Clustering of functional renin-angiotensin system gene variants in hypertension association studies. Braz J Med Biol Res 2001; 34:1421-1428.
91. Da Cruz IB, Oliveira G, Taufer M et al. Angiotensin I-converting enzyme gene polymorphism in two ethnic groups living in Brazil's southern region: Association with age. J Gerontol A Biol Sci Med Sci 2003; 58:851-856.
92. Galão AO, de Souza LH, da Costa BE et al. Angiotensin-converting enzyme gene polymorphism in preeclampsia and normal pregnancy. Am J Obstet Gynecol 2004; 191:821-824.
93. Sakuma T, Hirata RD, Hirata MH. Five polymorphisms in gene candidates for cardiovascular disease in Afro-Brazilian individuals. J Clin Lab Anal 2004; 18:309-316.
94. Araujo MA, Goulart LR, Cordeiro ER et al. Genotypic interactions of renin-angiotensin system genes in myocardial infarction. Int J Cardiol 2005; 103:27-32.
95. Ludwig EH, Borecki IB, Ellison RC et al. Associations between candidate loci angiotensin-converting enzyme and angiotensinogen with coronary heart disease and myocardial infarction: The NHLBI Family Heart Study. Ann Epidemiol 1997; 7:3-12.
96. Tiago AD, Nkeh B, Candy GP et al. Association study of eight candidate genes with renin status in mild-to-moderate hypertension in patients of African ancestry. Cardiovasc J S Afr 2001; 12:75-80.
98. Marroni AS, Metzger IF, Souza-Costa DC et al. Consistent interethnic differences in the distribution of clinically relevant endothelial nitric oxide synthase genetic polymorphisms. Nitric Oxide 2005; 12:177-182.
98. Pereira AC, Sposito AC, Mota GF et al. Endothelial nitric oxide synthase gene variant modulates the relationship between serum cholesterol levels and blood pressure in the general population: New evidence for a direct effect of lipids in arterial blood pressure. Atherosclerosis 2006; 184:193-200.
99. Sandrim VC, Coelho EB, Nobre F et al. Susceptible and protective eNOS haplotypes in hypertensive black and white subjects. Atherosclerosis 2006; 186:428-432.
100. Sandrim VC, de Syllos RW, Lisboa HR et al. Endothelial nitric oxide synthase haplotypes affect the susceptibility to hypertension in patients with type 2 diabetes mellitus. Atherosclerosis 2006, (in press).
101. Pereira AC, Morandini Filho AA, Heimann AS et al. Serum angiotensin converting enzyme activity association with the I/D polymorphism in an ethnically admixtured population. Clin Chim Acta 2005; 360:201-204.
102. Fiegenbaum M, da Silveira FR, Van der Sand CR et al. Pharmacogenetic study of apolipoprotein E, cholesteryl ester transfer protein and hepatic lipase genes and simvastatin therapy in Brazilian subjects. Clin Chim Acta 2005; 362:182-188.
103. Martin YN, Salavaggione OE, Eckloff BW et al. Human methylenetetrahydrofolate reductase pharmacogenomics: Gene resequencing and functional genomics. Pharmacogenet Genomics 2006; 16:265-277.

104. Arruda VR, Siqueira LH, Goncalves MS et al. Prevalence of the mutation C677 —> T in the methylene tetrahydrofolate reductase gene among distinct ethnic groups in Brazil. Am J Med Genet 1998; 78:332-335.
105. Cunha AL, Hirata MH, Kim CA et al. Metabolic effects of C677T and A1298C mutations at the MTHFR gene in Brazilian children with neural tube defects. Clin Chim Acta 2002; 318:139-143.
106. Perez AB, D'Almeida V, Vergani N et al. Methylenetetrahydrofolate reductase (MTHFR): Incidence of mutations C677T and A1298C in Brazilian population and its correlation with plasma homocysteine levels in spina bifida. Am J Med Genet A 2003; 119:20-25.
107. Alessio AC, Annichino-Bizzacchi JM, Bydlowski SP et al. Polymorphisms in the methylenetetrahydrofolate reductase and methionine synthase reductase genes and homocysteine levels in Brazilian children. Am J Med Genet A 2004; 128:256-260.
108. Couto FD, Adorno EV, Menezes JF et al. C677T polymorphism of the MTHFR gene and variant hemoglobins: A study in newborns from Salvador, Bahia, Brazil. Cad Saude Publica 2004; 20:529-533.
109. Gaspar DA, Matioli SR, de Cassia Pavanello R et al. Maternal MTHFR interacts with the offspring's BCL3 genotypes, but not with TGFA, in increasing risk to nonsyndromic cleft lip with or without cleft palate. Eur J Hum Genet 2004; 12:521-526.
110. Pereira AC, Xavier-Neto J, Mesquita SM et al. Lack of evidence of association between MTHFR C677T polymorphism and congenital heart disease in a TDT study design. Int J Cardiol 2005; 105:15-18.
111. Zanrosso CW, Hatagima A, Emerenciano M et al. The role of methylenetetrahydrofolate reductase in acute lymphoblastic leukemia in a Brazilian mixed population. Leuk Res 2006; 30:477-481.
112. Dollery CT. Clinical Pharmacology - The first 75 years and a view of the future. Brit J Clin Pharmacol 2006; 61:650-665.
113. Suarez-Kurtz G. The Brazilian National Pharmacogenomics/pharmacogenetics network. Pharmacogenomics 2004; 4:347-348.
114. http://www.refargen.org.br.
115. Suarez-Kurtz G. Pharmacogenetics, pharmacogenomics and population admixture: Implications for drug development and prescription. Nat Rev Genet 2005, (doi:10.1038/nrg1559-c1).

Chapter 7

Pharmacogenetics of Cytochrome P450s in African Populations:

Clinical and Molecular Evolutionary Implications

Eleni Aklillu,* Collet Dandara, Leif Bertilsson and Collen Masimirembwa

Abstract

Though the pharmacogenetics of drug metabolism had its origins in the 1960s, it is only during the past 10 years that it began to have clinical impact and pharmaceutical industry recognition. The delay in clinical application was due to both lack of convincing clinical data as to the relevance of phenotyping/genotyping in making drug prescription decisions and the perceived expense and difficulty of the techniques to do so. In the 1990s, molecular biology techniques become cheaper and simpler, facilitating efforts to evaluate the potential clinical applications of drug metabolism pharmacogenetics. Starting with phenotyping and genotyping studies in Caucasian populations in Europe, interest in other major populations in Asia and Africa also increased. The polymorphisms of *CYP2D6*, *CYP2C19* and *NAT-2* clearly demonstrated presence of major interethnic differences in the genotype and enzyme activity. Interestingly, phenotype data from populations of African origin reflected reduced CYP2D6 activity compared to Caucasians due to an African specific variant, *CYP2D6*17*, which exists in these populations at high frequency (14-34%) and is associated with reduced affinity for CYP2D6 substrates. On the contrary more people are deficient of the CYP2D6 activity in Caucasians than in populations of African origin due to the high prevalence (>20% allele frequency) of the defective *CYP2D6*4* variant in Caucasians as compared to African populations (<2%). The study of polymorphisms of other cytochrome P450s, CYP1A2, 2C8, 2C9, and 2C19 in African populations demonstrated other interethnic differences in allelic variant frequencies compared to the Caucasian and/or Oriental populations. In this review, we give an update of the phenotype and genotype status of the polymorphisms of the major cytochrome P450s, 1A2, 2C8, 2C9, 2C19, 2D6 and CYP3A. Since clinical data is still partly lacking, we will give postulations of drug treatments these polymorphism might affect. The studies which had initially been motivated by clinical interests are beginning to yield data that could add knowledge to the evolutionary relatedness of the major continental populations in both time and space. We conclude the review with a preliminary effort to use the allele frequency data in molecular evolutionary studies of the major ethnic groups.

Introduction

The science and technology of pharmacogenetics has benefited greatly from activities around the Human Genome Project (http://www.sanger.ac.uk/HGP/). This project was conceived and

*Corresponding Author: Eleni Aklillu—Division of Clinical Pharmacology, Department of Laboratory of Medicine, Karolinska Institutet, Karolinska University Hospital Huddinge, C-168, SE-141 86 Stockholm, Sweden. Email: eleni.aklillu@ki.se

Pharmacogenomics in Admixed Populations, edited by Guilherme Suarez-Kurtz.

popularized on the promise of better understanding of human disease and discovery of new drug targets in the pharmacogenomic era. Spin off international efforts to understand and exploit the human genome include the human SNP Consortium (http://snp.cshl.org), the International HapMap Project (http://www.hapmap.org/) and the recently launched Human Variome Project (http://www.humanvariomeproject.org/), which all directly and/or indirectly reinforced the potential importance of pharmacogenetics. Pharmacogenetics addresses genetic variability in drug response at both pharmacodynamic and pharmacokinetic levels. It therefore seems that it is through pharmacogenetics that the Human Genome, the SNP project, the HapMAP and the Human Variome Project are going to score their first success in positively influencing human health. This is because the basic work has already been done in how genetic variability affects the pharmacokinetics of drugs to rapidly find application in the clinic. Pharmacogenetic databases (http://www.cypalleles.ki.se), journals (http://www.jpharmacogenetics.com), and associations (http://www.pharmgkb.org) are dedicated to the work of better defining the current status of the subject.

The pharmacogenetics of drug absorption, distribution, metabolism and elimination, due to polymorphisms of drug metabolizing enzymes and drug transporters, has been most studied and is showing the greatest promise for possible clinical application. This is evidenced by the massive publication portfolio in this area, the existence of guidelines on pharmacogenetics in drug discovery, development and labeling of pharmaceutical products by major drug regulatory agents (FDA, EMEA) and recognition of the subject by WHO. FDA has approved a genotyping chip from IVD Technology for two important drug metabolising enzymes, CYP2C19 and CYP2D6 (http://www.devicelink.com/ivdt/archive/05/03/007.html).

There are increasing efforts to demonstrate the clinical utility of preprescription genotyping to optimize drug dosages.[1] Studies on the role of CYP2B6 polymorphism on the metabolism and elimination of efavirenz promises to be a convincing clinical case that genotyping could increase the safety and efficacy of this important antiretroviral drug.[2] Pharmacogenetics is getting into clinical practice and is being integrated in the drug discovery process in developed countries. Where is Africa in all this brewing excitement? To address this question, this review gives an update of the status of pharmacogenetics of drug metabolism with particular focus on CYP450 in African populations.

History of the African People

Africa is the home to the first human species from about 2.5 million years ago which gave rise to the modern human species (*Homo sapiens)*, thought to have spread to other parts of the world.[3-5] Presently, Africa is inhabited by aboriginal groups; Caucasoids in the north down to the southern border of Sahara, and Negroids in sub-Saharan Africa. The populations of sub-Saharan Africa have been greatly affected by the Bantu expansions that originated near the confluence of the Niger and Benue rivers. African populations show the largest amount of genetic diversity. Linguistic and genetic studies show that most sub-Saharan populations are closely related to each other whereas Pygmy, Khoisan, and eastern African populations are the most differentiated.[3]

Africa, the second largest continent hosting about 10% of the world population is home to more than 1500 languages that are thought to represent as much genetic variability. These languages belong to four major families; the Afro-Asiatic family which is spoken in countries like Mauritania, Morocco, Libya, Egypt, Mali, Niger, parts of Ethiopia, Somalia and Eritrea; the Nilo-Saharan family which is spoken in Sudan, Chad, parts of Kenya and parts of Tanzania; the Niger-Congo family which is spoken from Senegal to South Africa, also in East, West, Central and Southern Africa; the last and smallest being the Khoisan family spoken in parts of Namibia and parts of Botswana.[3-5] The African population may not be considered homogenous because they might not be more similar to each other genetically than they are to nonAfrican populations.

Linguistic barriers strengthen genetic isolation between groups speaking different languages; hence, people speaking the four major language families are thought to be genetically different. In addition to local genetic mixing of the different African groups, the slave trade that saw more than 12 million African people shipped to America, Arabia and Europe resulted in genetic admixture of genes from Africans with other population groups. There is a need to understand the extent of this admixture and its implication to evolutionary studies and to pharmacogenetics. Despite all this mixing, there are still concentrations of particular alleles in different African populations groups, for example, the high prevalence (29%) of *CYP2D6*2xN* among Ethiopians—who are classified as Cushitics under the Afro-Asiatics language family—compared to the low prevalence (1-2%) among Bantu Africans of the Niger-Congo language family (e.g., Ghanaians, Zimbabweans, Tanzanians).[6] Generally, Africans are genetically more diverse compared to Asians and Caucasians.[7,8]

Overview of the Polymorphic Status of Major Cytochrome Enzymes in African Populations

Multiple isozymes of cytochrome P450 have evolved from a common ancestral gene via divergent evolution through gene duplication, conversion, amplification and SNPs as defense mechanisms to protect organisms from environmental toxicants.[9] Based on amino acid sequence homology, the cytochrome P450 enzymes have been classified by family, subfamily, and isoform.[10] The P450s in families 1-3 are manly involved in the metabolism of xenobiotics including drugs and display genetic polymorphism to cause interindividual and interethnic variations in enzyme activity.[11] The major forms of importance for drug metabolism are CYP2C9, CYP2C19, CYP2D6 and CYP3A4/3A5, whereas CYP1A1, CYP1A2, CYP1B1, CYP2E1 and CYP3A4 are the most important isoforms responsible for metabolic activation of precarcinogens.

CYP1A2

CYP1A2, a hepatic enzyme inducible by smoking, metabolises various chemical procarcinogens and N-hetrocyclics found in tobacco s

moke to reactive carcinogens.[12] It metabolizes several clinically important drugs, such as olanzapine, theophylline and clozapine.[13] There is wide interindividual and interethnic variations in CYP1A2 enzyme activity, constitutive expression, enzyme inducibility and polymorphic metabolism of pro-carcinogens by human liver microsomes.[14] More than 60-fold differences in the constitutive levels of hepatic CYP1A2 between human individuals is reported.[15]

The human *CYP1A2* gene is located on chromosome 15, spans about 7.8 kb, contains seven exons and the coding region starts at nucleotide 10 of exon 2.[14] Several polymorphisms in the promoter, regulatory and noncoding regions of the CYP1A2 gene have been reported (http://www.imm.ki.se/CYPalleles/cyp1a2.htm). At least 14 SNPs in and near the human CYP1A2 gene have been described to date, of which three have been associated with decreased enzyme activity and one with increased inducibility. However, none of the variant alleles have yet been shown unequivocally to account for the striking variations in enzyme activity and basal expression level.[16]

Genetic Polymorphism of CYP1A2 in Africans

Genetic polymorphism of CYP1A2 is very well studied among Caucasian and Asian population in search of a genetic basis of the wide variation in phenotype status. Analysis of CYP1A2 genetic polymorphism in 113 individuals from 3 major continental regions, Africa, Asia, and Europe, indicated that the African population had the highest level of nucleotide diversity, the lowest level of linkage disequilibrium, and two distinct haplotype clusters with broadly overlapping geographic distributions.[8] Haplotypes found outside of Africa were mostly a subset of those found within Africa. These patterns were all consistent with the African origin of modern humans. Few studies investigated genetic polymorphism of CYP1A2 in Africans. The -163C>A SNP is suggested to be associated with higher enzyme inducibility by smoking among Caucasians,[17] however this remains controversial.[18,19] The frequency of -163C>A appears to be similar in different populations. It occurs in Egyptians (68%),[20] Ethiopians (60%),[21] Tanzanians (49%) and Zimbabweans (57%),[22] which is a similar frequency to Caucasians (British,

66%)[23] and Asians (Japanese, 67%).[18] The -163 C>A is in strong linkage disequilibrium with other SNPs giving rise to different *CYP1A2* haplotypes.[21,23]

CYP1A2 phenotyping, using caffeine as a probe, indicated that Zimbabwean rural black children have significantly lower enzyme activity than Caucasian and black urban children from Zimbabwe or from Canada.[24] Similarly lower CYP1A2 enzyme activity in African Americans compared to Caucasians is reported.[25] In Ethiopian population, geographical difference between those residing in Ethiopia and Sweden didn't result in any significant difference on CYP1A2 enzyme activity.[21] This might imply contribution of genetic factor in determining enzyme activity. A novel haplotype, *CYP1A2*1K*, with 3 linked SNPs (-730T, -740G and -164A) causing reduced enzyme activity was identified in Ethiopians with a frequency of 3.0%.[21] The *CYP1A2*1K* is equally frequent among Saudi Arabians (3.6%) but is very rare in Caucasians (Spaniards; 0.5%) and is absent in Asians.[18] Similar to *CYP2D6* gene duplication, the *CYP1A2*1K* appears to be localized in East Africa and Middle East. The frequency of *CYP1A2*1C* and **1E* in Egyptians is significantly lower than that in Japanese, while similar frequencies were observed for *CYP1A2*1D* and **1F*.[20] *CYP1A2*1C, *1D* and **1K* remain to be studied in other African populations.

Clinical Significance of CYP1A2 Polymorphism in Africans

Hepatocellular carcinoma is a common neoplasm, especially in sub-Saharan Africa and is to a great extent caused by chronic infection by the hepatitis B virus and intake of dietary aflatoxin.[26,27] Aflatoxins are fungal metabolites that frequently contaminate staple foods in much of sub-Saharan Africa, especially after long-term crop storage because of excessive heat, humidity, and are associated with increased risk of liver cancer.[26] CYP1A2 plays a more important role than CYP3A4 in the bioactivation of aflatoxin at low concentrations in human liver microsomes and in human lung cells expressing CYP1A2.[28,29] Subjects with higher CYP1A2 activity and exposed to dietary aflatoxin B1 might thus be at a higher risk to develop hepatocellular carcinoma. Individual differences in CYP1A2 activity may thus influence individual therapeutic effects of some drugs and susceptibility to liver cancer especially in Sub-Saharan Africa.

CYP2C8

CYP2C8 is involved in the metabolism of several clinically important drugs such as the anticancer drug (paclitaxel), anti-diabetic drug (roziglitazone) all-trans retinoic acid, cerivastatin, and anti-malarial drug (amodiaquine). Clinical studies have demonstrated marked interindividual variation in the response of cancer patients to paclitaxel.[30] Amodiaquine (AQ) is a 4-aminoquinoline derivative that has been widely used for treatment of malaria over the past 50 years. It is essentially more active than the other 4-aminoquinoline, chloroquine, against *Plasmodium falciparum* parasites, which are moderately chloroquine resistant. AQ is therefore increasingly being considered as a replacement for chloroquine as a first line drug in Africa because of widespread chloroquine resistance.[31] AQ is mainly metabolized in the liver and CYP2C8 is the major enzyme that catalyses the formation of *N*-desethylaminodiaquine.[32]

The CYP2C8 is located on chromosome 10q24 along with CYP2C9, CYP2C19 and CYP2C18.[33] Besides polymorphisms in the CYP2C8 promoter region, 10 allelic variants of which 4 variants designated as *2C8*2* (I269F), *2C8*3* (R139K, K399R), *2C8*4* (I264M) and *2C8*5* (475delA) have been well investigated in different populations. In vitro studies using human liver microsomes indicated that these variants lead to enzymes with decreased in vitro activity toward the probe drug paclitaxel.[30,34] Distribution of CYP2C8 allelic variants exhibit wide interethnic differences. *CYP2C8*2* is found in African-Americans with allele frequency of 18%, while it is rare in Caucasian population (1.3%).[30] In contrast, *2C8*3*, is more prevalent in Caucasians with frequency of 14%, but 2% in African-Americans.[34] The frequency of *2C8*4* among Caucasians is 7.5% and the allele appears to be rare in African-Americans.[34] None of the *2C8*2, 2C8*3, 2C8*4* exists in Japanese population but a rare variant allele, *CYP2C8*5* (475delA) occurs with a frequency of 0.25%.[35]

Genetic Polymorphism of CYP2C8 in Africans

Although several variant alleles of the human *CYP2C8* gene exist and CYP2C8 substrates show wide interindividual variation in their metabolism, only two studies investigated *CYP2C8* genetic polymorphism in sub-Saharan Africa populations. Determination of the prevalence of the *CYP2C8* main alleles among 165 unrelated malaria patients from Zanzibar indicated that 3.6% of patients were homozygous for the defective alleles with the allele frequencies of *2C8*2*, *2C8*3* and *2C8*4* being 14, 2.1 and 0.6% respectively.[36] A study in 200 children from Northern Ghana reported allele frequency of *CYP2C8*2* to be 17% but *2C8*3* and *2C8*4* were absent.[37] The frequency of *2C8*2* in Ghanaians is very similar to African Americans,[37] but slightly higher than the population of Zanzibar (14%).[37] In general the frequency of *2C8*2* is much higher in black population compared to Caucasians and the allele frequency is relatively higher in West Africans and African Americans than in East Africans. In contrast the *2C8*3* and **4*, which are more prevalent in Caucasian populations, are absent in West Africans and are very rare in African Americans but present in East Africans. This indicates a heterogeneous nature of *CYP2C8* genetic polymorphism in Africans.

Clinical Significance of CYP2C8 Polymorphism in Africans

CYP2C8 metabolizes several clinically important drugs and endogenous substrates such as arachidonic acid to biologically active metabolites that have significant physiologic roles in pathogenesis of cardiovascular diseases, for instance hypertension and acute myocardial infarction.[38] Hypertension is more frequent and more severe in blacks than in other racial groups.[39,40] The frequency of CYP2C8 PM phenotype in Africans is lower than Caucasians but is higher than Asians and it is mainly due to the higher frequency of *CYP2C8*2*. Considering the higher incidence of hypertension, malaria and wider use of amodiaquine for treatment of malaria in Sub Saharan Africa, genetic polymorphism of *CYP2C8* might have some clinical importance in treatment response and susceptibility to hypertension in Africans.

CYP2C9

CYP2C9 is a major cytochrome P450 enzyme that is involved in the metabolic clearance of a wide variety of therapeutic agents, including nonsteroidal anti-inflammatories, oral anticoagulants, and oral hypoglycemics. To date, more than 50 variants in the *CYP2C9* gene have been described (http://www.imm.ki.se/CYPalleles/cyp2c8.htm) of which two single-nucleotide polymorphisms, *2C9*2* (R144C) and *2C9*3* (I359L), that code for defective enzymes with impaired activity towards a number of substrates are very well characterized in different populations. Patients homozygous for *2C9*2* and **3* receiving the anticoagulant drug, warfarin have several fold higher risk of over anticoagulation and bleeding.[41] *CYP2C9*2* and **3* alleles are the dominant variants in Caucasian populations, whereas some other variants are confined to black and Asian populations. The *CYP2C9*4* (Ile359Thr) polymorphism was first identified in a Japanese epileptic patients and reduces enzyme activity both in vitro and in vivo.[41,42] Some *CYP2C9* alleles, such as *2C9*5*(D360E), *2C9*6*(818delA), *2C9*8*(R150H), and *2C9*11*(R335W) were detected in black populations only.[43-45] *CYP2C9*6* was first described in a female African American with toxicity to phenytoin.[46] The *CYP2C9*5* and **6* alleles are associated with decreased enzyme activity in vivo, whereas the *CYP2C9*8* and **11* variants appear to have minor effects.[43-47]

Genetic Polymorphism of CYP2C9 in Africans

The prevalence of *CYP2C9* variant alleles in Africans in comparison to African Americans, Caucasians and Asians is listed in Table 1. African subjects cannot be considered as a homogeneous group with respect to *CYP2C9*2* allele frequencies, since large discrepancies in these overall frequencies were reported among different populations within Africa. *CYP2C9*2* and **3* are predominantly present in Caucasian populations (Italians; 11% and 9%) and occurs in Egyptians at a similar frequency (12% and 6%) respectively.[48,49] Their frequency in Ethiopians

Table 1. **CYP2C9** ***and*** **CYP2C19** ***allele frequencies in percentage in different African populations, compared with other representative studies in African Americans, Caucasians and Asians***

Allele	Egyptian[49]	Ethiopian[45,48,57]	Tanzanian[45,59]	Zimbabwean[58]	South African Venda[100]	Beninese[44]	African American[43]	Belgian Caucasian[44]	Japanese[50,101]
*2C9*2*	12	4				0	2.5	10.0	0
*2C9*3*	6	2				0	1.3	7.4	1.6
*2C9*4*	0	0				0	0		
*2C9*5*		0	0.8			1.8	0.8	0	0
*2C9*6*							0.08	0	0
*2C9*11*						2.7	0.23	0.4	0
*2C19*2*	11	14	17.9	13.1	21.7	13	25	9.1	26.7
*2C19*3*	0.2	2	0.6	0	0	0	0	0	12.8

(4% and 2%), is significantly lower than Egyptians and Caucasians populations.[48] Interestingly, both variants are absent in West Africans (Beninese). The frequency of *2C9*2* allele in the Beninese[44] is different from the 2.5% frequency reported in African-Americans[43] but the *2C9*3* frequency is similar. However Takahashi, et al reported absence of *2C9*2* in 64 African Americans.[50] Variation in 2C9*2 allele frequency in African Americans from different studies might reflect sampling variations. The allele frequencies of *2C9*2* and **3* in Ethiopians is significantly higher than Beninese and African Americans but is significantly lower than Egyptians (Table 1). Similar to Africans, the allele frequencies for *2C9*2* and *2C9*3* differs significantly between individuals from East Asia compared with South Asian subjects.[51]

Clinical Significance of CYP2C9 Polymorphism in Africans

CYP2C9 exhibits marked inter-individual variability in its expression and catalytic activity due to functionally significant genetic variations that can result in either clinically relevant drug toxicity for instance warfarin-induced bleeding complications in some patients who take standard doses of substrate drugs, or inadequate drug efficacy and therapeutic failure in others.[41] *CYP2C9*2* and **3* polymorphisms are relatively uncommon in blacks, but adverse effects of CYP2C9 substrate drugs such as warfarin and glipizide appear to be more common in this population relative to Caucasians.[52,53] Additional allelic variants such as *CYP2C9*5* and **6* present in Africans and African-Americans may, in part, explain these observations and it is, therefore, important to define their effects on drug metabolism in vivo. Future studies will be required to delineate the effect of rare variant alleles that occurs mainly in black populations on the disposition of such drugs, and the clinical consequences.

CYP2C19

CYP2C19 metabolizes numerous drugs, including omeprazole, diazepam and amitriptyline. To date, 19 variant alleles have been described, of which *CYP2C19**2 and *CYP2C19*3* are the most frequently identified defective alleles. *CYP2C19**2 leads to an aberrant splice site, whereas the *CYP2C19*3* allele produces a premature stop codon.[54] The CYP2C19 polymorphism is of particular importance in African, where malaria is endemic, since this enzyme is responsible for the metabolic activation of the pro-drug proguanil, an anti-malarial drug.

Genetic Polymorphism of CYP2C19 in Africans

The genetic polymorphism of CYP2C19 displays striking interethnic variations. The PM frequency ranges from 2 to 7% in Caucasians, 2-5% in Africans, 14-25% in Asians, and 60% in the Vanuatu.[55,56] About 5.2% Ethiopians and 4% of Zimbabweans are CYP2C19 PMs.[57,58] A relatively high prevalence of PMs in Tanzanians is reported (7.5%).[59] Similar to the situation in Asians, *2C19**2 and *2C19**3 account for 100% PMs in Ethiopians.[57,60] However, *CYP2C19**2 and *3 accounted for only 75% or less of the PM in the Bantu populations.[58,59] African-Americans show *CYP2C19* genotypic trends similar to those of the Bantu populations. The Asian specific allele, *CYP2C19*3* is detected in North and East African populations, namely in Ethiopians, Egyptians and Tanzanians but is absent from West and South African population as well as in African Americans (Table 1).

Several studies conducted among African populations indicated the presence of substantial difference in CYP2C19 activity. Generally Africans display lower CYP2C19 activity compared to Caucasians with a certain genotype.[61] A decreased CYP2C19 capacity to metabolize its substrates, such as omeprazole, mephenytoin and chloroguanide, is observed in Tanzanians,[59,62] Ethiopians[57] and Zimbabweans[58] compared to Europeans. Polymorphism in drug metabolising enzymes is governed both by genetic and environmental factors. In our previous study that aimed at identifying the possible cause of a lower activity among Africans we compared CYP2C19 enzyme activity for each genotype between Ethiopians in Ethiopia, Ethiopians in Sweden and Swedes. The result indicated a higher enzyme activity in Swedes compared to the Ethiopian populations but no significant difference between Ethiopians living in Ethiopia and Ethiopians living in Sweden, regardless of the difference in the environment.[63] This might indicate that the observed differences in the rate of drug metabolism between Black Africans and Europeans of the same CYP genotype might be explained either by hitherto unknown genetic differences of importance for enzyme expression or enzyme function.

Clinical Significance of CYP2C19 Polymorphism in Africans

The observed CYP2C19 activity variation is increasingly emphasized due to its clinical relevance in a geographic area like Africa, where malaria is endemic, since CYP2C19 is responsible for the metabolic activation of proguanil, into its active metabolite, cycloguanil. However there are no clinical data to suggest that PMs of CYP2C19 are at a greater risk for failure of malaria prophylaxis or treatment. The clinical relevance of the reduced metabolism of progruanil to cycloguanil is not clear since progruanil itself has anti-malarial activity through an unknown mechanism. The study in the Vanuatu[56] also showed that the CYP2C19 PM had lower cycloguanil levels but had similar and even better malaria cure rates.

CYP2C19 genetic polymorphism affects pharmacokinetics and pharmacodynamics of proton pump inhibitors (PPIs) such as omeprazole with important implication on treatment response.[64] Cure rates for *Helicobacter pylori* in patients receiving omeprazole and amoxicillin were found to be dependent significantly on the CYP2C19 genotype status, being much lower in *CYP2C19*1/*1* subjects indicating the importance of dose adjustment in extensive metabolizers (EMs). On the other hand the poor metabolizers show higher effectiveness in therapy.[65-67]

CYP2D6

CYP2D6 is responsible for the metabolism and clearance of more than 40 clinically important drugs, including β-receptor blockers, neuroleptics, tricyclic antidepressants antiarrythimics and some analgesics.[68] Human *CYP2D6* gene is localized on chromosome 22 and contains 9 exons.[69] Two to three additional highly homologous pseudogenes are located upstream of *CYP2D6* with in the *CYP2D* locus. The locus is highly polymorphic and more than 70 different allelic variants identified so far, differing for single-base changes, short insertions and deletions, or for major rearrangements such as deletion and duplications of the whole gene (www.imm.ki.se/CYPalleles/cyp2d6). Polymorphisms in *CYP2D6* cause enzyme variants with higher, lower or no activity. They may even lead to total absence of the enzyme.

CYP2D6 genetic polymorphism is widely investigated in several populations including Africans. The gene is extremely polymorphic and so far more than 70 CYP2D6 variant alleles have been described. However a few variants, namely *CYP2D6*2x2*, **4*, **5*, **10* and **17* account for most of the population diversity in enzyme activity. Frequency distribution of these variant alleles displays striking inter-ethnic variation (Table 2), the exception being *CYP2D6*5* (gene deletion) that has relatively similar frequency (4-6%) in Caucasians, Africans and Asians indicating that *CYP2D6* deletion has occurred a long time before the divergence of these continental populations.[70] The separation of people becoming Asians and Caucasians from those becoming today's Africans is estimated to occur about 150,000 years ago and divergence between Asians and Caucasians is estimated to have occurred over 50 000 years ago.[70] It now appears that these different populations may express a somewhat different constellation of *CYP2D6* alleles.[70]

Genetic Polymorphism of CYP2D6 in Africans

Approximately 5-10% of Caucasians are poor metabolizers (PMs), completely lacking CYP2D6 enzyme activity because of inheritance of two mutant *CYP2D6* null alleles, mainly *CYP2D6*4* (22%), which accounts for more than 75% of the mutant alleles in this population.[68] The PM frequency in Asians is <1%. The reported percentage of PM phenotype in African Americans varies greatly from 1.9,[71] 2.6[72] up to 7.1[73] and 8.[74] A comparison of debrisoquine metabolic ratio distribution between Africans, Asians and Caucasian population is illustrated in Figure 1. The frequency of CYP2D6 PM phenotype in Africans is <2% and is mainly due to the absence or very low frequency of *CYP2D6*4* (Table 2). The exception is

Table 2. CYP2D6 allele frequencies in percentage among different African populations, compared with other representative studies in African Americans, Caucasians and Asians

Allele	Ethiopian[63,78]	Tanzanian[75,91]	Zimbabwean[79,92]	South African Venda[100]	Ghanaian[102]	Gabonese[76]	African American[73]	German Caucasian[103]	Chinese[104-106]
*2	13.6	20.3	13	18	11	44	19.1	33.6	13.4
*3	0	0	0	0	0		0	1	0
*4	4.1	1	2.5	3.3	6.3		5.4	19.5	0.8
*5	3.3	6.1	3.8	4.6	6	7	6.6	4.1	5.7
*10	8.6	3.8	5.6	0	3.1		3.6	2.1	50.7
*17	11.3	17	34	24	28	24	21.3	0	
*29		20					7.2		
*41	21.6								
*1 or*2x2	13.6	3.3			1.6		1.6	1.6	1.3
*2x3	1.2								
*2x4	0.8								
*2x2	0.4								
*4x2	0	0.9	0.9				0.3		
PM %	1.8[a]	7[a]	2		2.5[ɔ]		7.1[c]	7,7[b]	1[a]

Phenotyping of CYP2D6 using debrisoquine [a], sparteine [b], dextromethorphan [c]

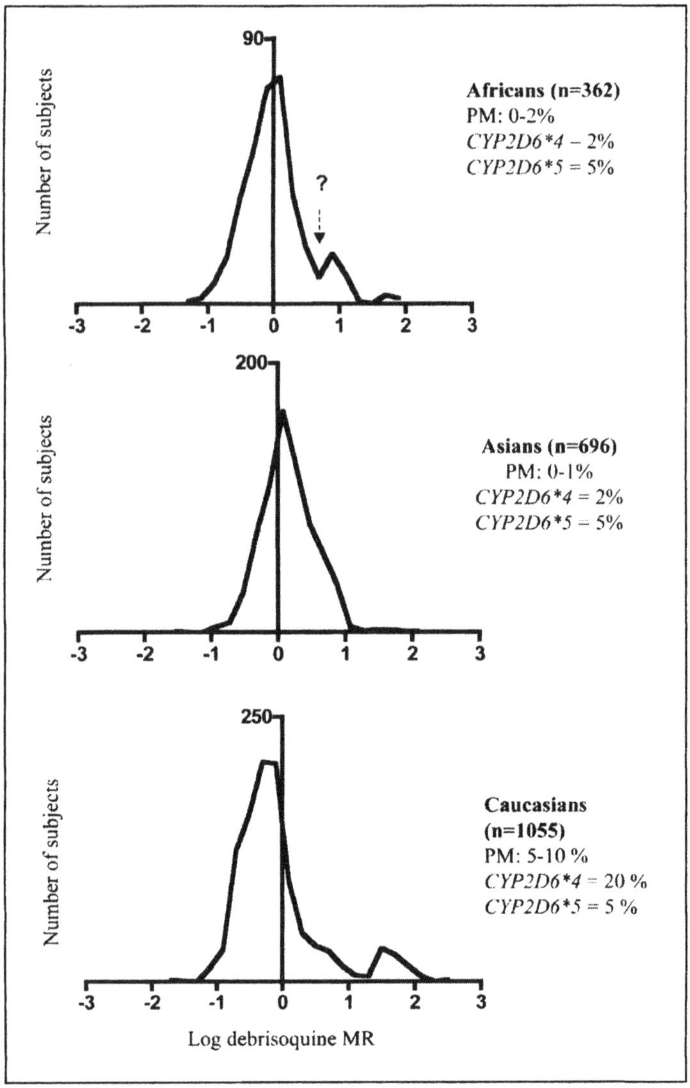

Figure 1. Comparison of debrisoquine metabolic ratio between 362 Africans (180 Ethiopians, 69 Zimbabweans, 113 Nigerians), 696 Asians (282 Han, 128 Mongolia, 156 Wei, 135 Zang) and Caucasians (1055 Swedes).

Tanzanian, where up to 7% of the population is reported to be PMs of debrisoquine.[75] In general, the prevalence of PMs and the frequency of nonfunctional alleles are much lower in Africans and Asians than in Caucasians. Despite lower frequency of defective alleles existing in black population, in general the activity of CYP2D6 is lower for the apparent genotype as compared to Caucasians.[61,75-78] The intermediate metabolizer phenotype is much more prevalent in Africans and Asians compared to Caucasians as indicated by the right shift of debrisoquine MR histogram reflected by higher metabolic ratio (Fig. 2). The right shift has been uniformly observed in several African populations as well as in African Americans.

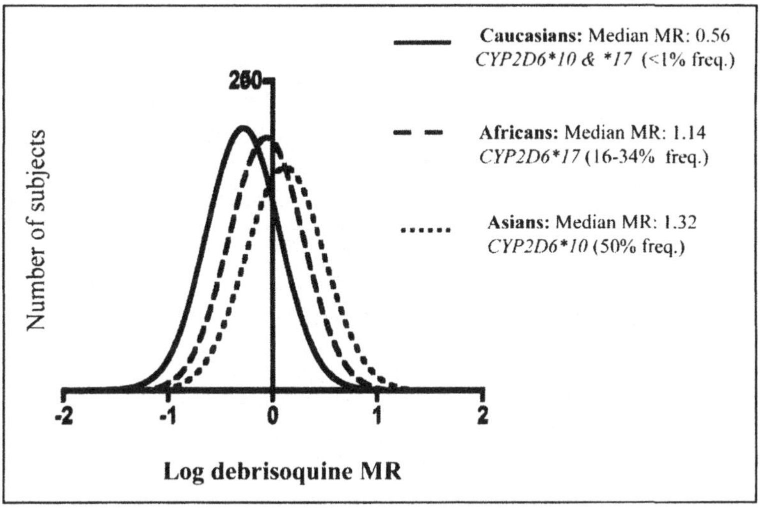

Figure 2. The Gausian distribution curves of metabolic ratio data from Swedes, African (Ethiopians, Yorubas, Zimbabweans) and Asian (Han, Mongolia, Wei, Zang) populations. The larger the log metabolic ratio, the lower the CYP2D6 activity. These distribution curves show that African and Asian populations have generally reduced enzyme activity than Swedes.

We have identified and characterized a specific variant of the CYP2D6 gene (*CYP2D6*17*) that carries three functional mutations (T107I, R296C, and S486T) in Zimbabwean subjects with impaired CYP2D6-dependent hydroxylase activity.[79] In vitro characterization of **17* variant allele indicated that the presence of both the T107I and R296C amino acid substitutions caused 5-fold higher K_m for bufuralol than the wild-type enzyme, whereas the S486T mutation was of little importance.[80] Detailed enzyme kinetic and molecular modeling characterization showed that the *CYP2D6*17* variant results in substrate dependent reduced affinity hence partly explaining the discordant phenotype correlations observed in African populations when different probe drugs were used.[81] The higher prevalence of CYP2D6 intermediate metabolizer phenotype in Africans is due to the presence of a specific variant, *CYP2D6*17* that is associated with reduced enzyme activity. It is also possible that drug biotransformation by proteins associated with reduced activity and contributing to the intermediate metabolizer may affect some CYP2D6 substrates more than others.[82]

The *CYP2D6* gene duplication leads to ultra-rapid metabolism if the duplicated genes are fully active. A pronounced relationship between number of *CYP2D6* gene copies and clearance of nortiptyline in gene dose dependent manner is reported.[83] In Ethiopian population, we described the duplication of *CYP2D6*2* in 2,3,4 or up to 5 copies on the same chromosome resulting in frequent distribution of the ultra rapid metabolizer phenotype.[78] Global distribution of CYP2D6 duplication is indicated in Figure 3. *CYP2D6* duplication is most frequent in Ethiopia and Saudi Arabia whereas the frequency is much lower in other parts of Africa. It is uncommon (1-2%) in Northern Europe and is essentially absent in Asia. It display a north to south gradient in Europe ranging from 1% in Swedes and 4% in Germany up to 10% in southern Europe such as Spain.[84-87]

The number of individuals carrying multiple *CYP2D6* gene copies is highest in Ethiopians (29%) and Saudi Arabia (20%) where up to one third of the population have gene duplication.[78,88] The basis for this event is unknown. The enzyme is not inducible by conventional means, but has a very high affinity for alkaloids.[89] Gene duplications are not stable if they are not beneficial for the organism.[90] A possible adaptation to the environment might have occurred by gene duplication and selection for active CYP2D6 genes in response to specific

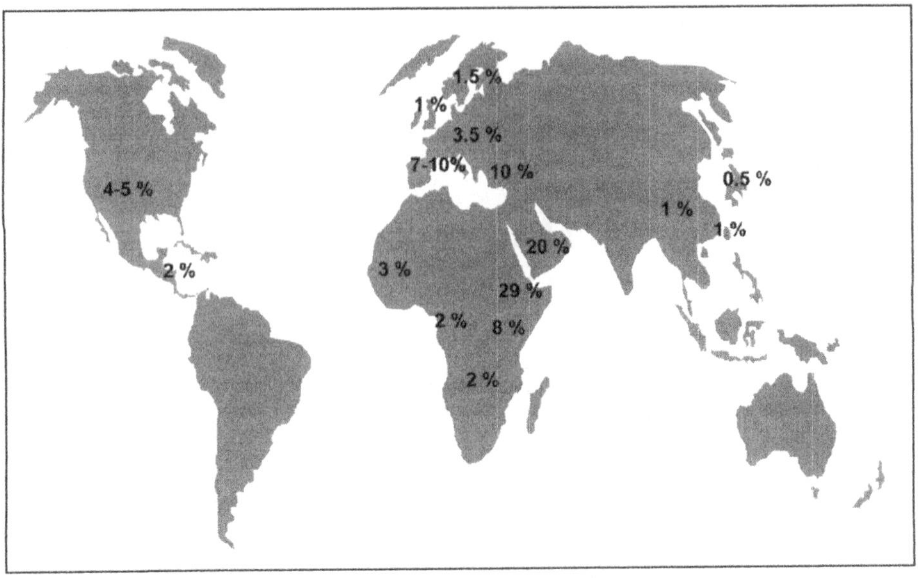

Figure 3. Global frequency distribution of *CYP2D6* gene duplication. (Data from refs. 73, 75, 78, 84-88, 92, 102, 119-122).

diets.[6,63] The presence of varying copy number of *CYP2D6* gene might indicate that the gene duplication has occurred several times during evolution of this locus.

Not only the frequency of CYP2D6 duplication varies within African population, but also the variant allele that is duplicated and number of copies on the same allele. In Ethiopians gene duplication involves a functional allelic variant, *CYP2D6*2*. About 8% of Tanzanians carry duplicated *CYP2D6* gene, the frequency of functional duplicated gene (*CYP2D6*2x2*) and the nonfunctional duplicated gene (*CYP2D6*4x2*) being 3.3% and 1.0% respectively.[91] In Zimbabweans, it is the nonfunctional variant *CYP2D6*4* which is duplicated.[92] Similarly, the frequency of the defective duplicated variant, *CYP2D6*4x2* is more than double that of the functional duplicated variant, *CYP2D6*2x2* in African Americans.[73]

Clinical Significance of CYP2D6 Genetic Polymorphism in Africans

The clinical consequence of CYP2D6 gene duplication is inability to achieve therapeutic drug levels at ordinary dosage.[93] A high level of CYP2D6 activity may be of greater therapeutic concern and challenge for Africans, in particular Ethiopians, than its deficiency. Ethiopians have the highest frequency of CYP2D6 duplication globally and those with duplicated *CYP2D6* genes are the most rapid metabolizer among the Ethiopians. However, subjects with multiple *CYP2D6* genes as well as those homozygous for *CYP2D6*1* generally exhibit a slower rate of debrisoquine metabolism than Caucasians of the same genotype.[63] Having the same genotype, Ethiopians living in Sweden showed higher enzyme activity compared to those living in Ethiopia indicating that the observed reduced enzyme activity is probably due to inhibition by environmental factors such as diet.[63] Thus despite the higher frequency of CYP2D6 duplication in Ethiopia, being an ultra rapid metabolizer phenotype might not be a significant problem for treatment response as the inhibition by environmental factors may counterbalance the effect of gene duplication. However Ethiopians with homozygous gene duplication genotype might suffer from effects of being the ultra rapid metabolizer phenotype when treated with CYP2D6 substrate drugs, upon moving to other parts of the world such as Europe.

Table 3. The distribution of* CYP3A4 *and* CYP3A5 *genetic variants in percentage among different populations

Population	*CYP3A5*3*	*CYP3A5*6*	*CYP3A4*1B*
Caucasian[94,107-111]	91-94	0	4-8
South African (Caucasian)[95]	-	-	43
Chinese[109,111-113]	76	0	0
Malays and Indians[112,113]	60	0	0
Japanese[111,114]	76	0	0
West Africans[7,108,115]	6-12	16	82-87
Pygmies and San[7,95]	7-11	-	-
Bantu (South Africa and Kenya)[7,95,116]	14-19	21	84
Guinea-Bissau, Senegalese and Ghanaians[7,94,117]	31	-	72-81
African American[94,109,111,118]	27	8	55-59
Hispanic American[111]	-	-	9
South Africa (Colored)[116]	48	14	-

CYP3A4/3A5

The CYP3A locus encodes mono-oxygenases that catalyze many reactions involved in drug and carcinogen metabolism, and synthesis of steroids. The *CYP3A* locus consists of four genes, *CYP3A43*, *CYP3A4*, *CYP3A7* and *CYP3A5* spanning over 200-kb. CYP3A4 is the dominant form in the adult liver and intestine while CYP3A7 expression is more pronounced in fetal livers, decreasing rapidly after birth. CYP3A enzymes metabolize more than 50% of the drugs in use today including midazolam, cyclosporine A, and erythromycin. *CYP3A* expression varies between different ethnic groups up to 40-fold and this variation has been attributed to genetic polymorphism and environmental factors. Several SNPs have been observed in *CYP3A4* (**1B-T, *2-*20*) (www.imm.ki.se/CYPalleles/cyp3a4.htm) and in *CYP3A5* (**1A-E, *2-*11*) (www.imm.ki.se/CYPalleles/cyp3a5.htm). The most common CYP3A variants are *CYP3A4*1B* and *CYP3A5*3* of which the former is associated with increased CYP3A4 induction while the latter is associated with severely decreased CYP3A5 activity. Analysis of *CYP3A4* genetic polymorphism shows the *CYP3A4*1B* variant which is associated with increased expression varies from a frequency of 72% among people from Guinea-Bissau, 78% among the Senegalese, 81% among Ghanaians[94] to as high as 84% among Zulu South Africans[95] while that of *CYP3A5*3* is on average <20% in most African populations.[96,97] Table 3 shows the distribution of *CYP3A5* and *CYP3A4* genetic variants in different African populations in comparison to the other populations in the world. Generally, *CYP3A5*3* is highest among Caucasians at over 80%, lower in the Orientals, ranging between 60 and 80% while lowest among black Africans, as low as 6%. The figures for the African Americans are lower than those of both the Caucasians and Orientals but higher than those of the black Africans. *CYP3A4*1B* variant follows a reverse distribution as compared to that of *CYP3A5*3*. It is also interesting to note that another *CYP3A5* variant, *CYP3A5*6* has only been reported among Africans, African Americans and peoples of Mixed Ancestry (known as Colored in South Africa).

Clinical Significance of Polymorphism in CYP3A Loci in Africans

The variation in the levels of *CYP3A* expression in different populations is very important because of the involvement of *CYP3A4/3A5* in the metabolism of cholesterol-lowering drugs (e.g., statins; lovastatin, simvastatin and atorvastatin) anticancer drugs (e.g., anthracyclines), antituberculosis (e.g., rifampicin), antiretrovirals (e.g., nonnucleoside reverse transcriptase inhibitors and protease inhibitors) and other commonly used drugs such as acetaminophen.

Sub-Sahara Africa is home to the highest numbers of HIV/AIDS patients with a prevalence of more than 20% in some populations. In addition to HIV/AIDS, tuberculosis and malaria are two other major killer diseases in this region; thus, there is a high level of use of drugs that are substrates and/or inhibitors of CYP3A In a study by Mirghani et al[96] it was observed that Tanzanians homozygous for low *CYP3A5* gene expression had reduced capacity to convert quinine (an antimalarial) to its major metabolite, 3-hydroxyquinine. Thus, variations in the expression and activity of *CYP3A4/3A5* might have a major impact on pharmacokinetics and fate of drugs, including the occurrence of adverse drug reactions.[98] *CYP3A4* is also involved in the metabolism of irinotecan to inactive metabolites.[99] The observation of increased risk for prostate cancer among carriers of the *CYP3A4*1B* allele, may predispose African males because of the high frequency of this variant among Africans. In a study investigating susceptibility to oesophageal cancer,[97] it was observed that *CYP3A5*1* was associated with increased risk. Thus Africans might be more likely develop cancer when exposed to compounds that need bioactivation activation by *CYP3A5* to reactive intermediates.

Molecular Evolutionary Studies

Starting from the emerging potential of using the genotypic differences in drug metabolising enzymes to optimise treatment in different individuals or ethnic groups, Masimirembwa, proposed the use of the allele frequencies in human molecular evolutionary studies (PhD thesis, 1995). Bertilsson et al also explored such relationships.[68] With more data becoming available on the allele frequencies of many genes coding for drug metabolizing enzymes, the secondary utility of these data in molecular evolutionary studies can be revisited again. This would add to the current genetic data on other genes, which are used to understand the interrelatedness of various populations and towards understanding their migratory patterns in time and space. We have now reanalyzed their potential utility using principal component analysis (PCA) (SIMCA, Umetrics, Umea).

PCA is a multivariate analysis method which can handle complex data which has covariance in its structure, by being able through a series of principal component analysis to extract the information which best describes the dataset. It reduces the dimensionality of multidimentional data while explaining as much of the variation as possible. The first component therefore shows the descriptors (in this case the allelic variants), which differentiate the objects (ethnic groups or populations in this case). In this analysis, the following populations where used, Ethiopians, Tanzanians, Zimbabweans, Venda, African American, Ghanaian, Germans, Chinese, Orientals, Swedes, Americans. The second component then captures information not captured in the first component and is second in importance towards differentiating the populations of interest (Asian, Caucasian and African populations). The results are output as two smaller matrices, the score matrix and the loading matrix. The score matrix is a plane generated by the first two components on which the objects are projected (score plot, Fig. 4B). Objects that are close in the score plot have comparable variance and are similar. The loading matrix relates each PC to the original variables (loading plot, Fig. 4A). From the score plot one can deduce the relative importance of some alleles while the loading plot shows how the score plots relate (affect) population structure/distribution. The score plot and the loading plot are related and variables that are positively correlated to the observation are positioned in the same place in the loading plot as the observations in the score plot. Successful interpretation therefore requires that the loading and score plots be superimposed on each other. Figure 4 shows the score and the loading plots adjacent to each other. The first and second components are shown for the scatter plot of *CYP2D6* (*CYP2D6*2, *3, *4, *5, *10,* and **17*) and *CYP2C19* (*CYP2C19*2, *3*) major allelic variants.

These data show a clear clustering in which the Caucasian and Asian populations are positioned at opposite ends of the first component. This clustering is driven by the ethnic specific existence or distribution of some specific alleles. The African populations cluster between these two extremes. What distinguishes the Asians from the Caucasians are the presence of the *CYP2D6*10,* and *CYP2C19*3* in the Asian and their absence in the Caucasians. This also

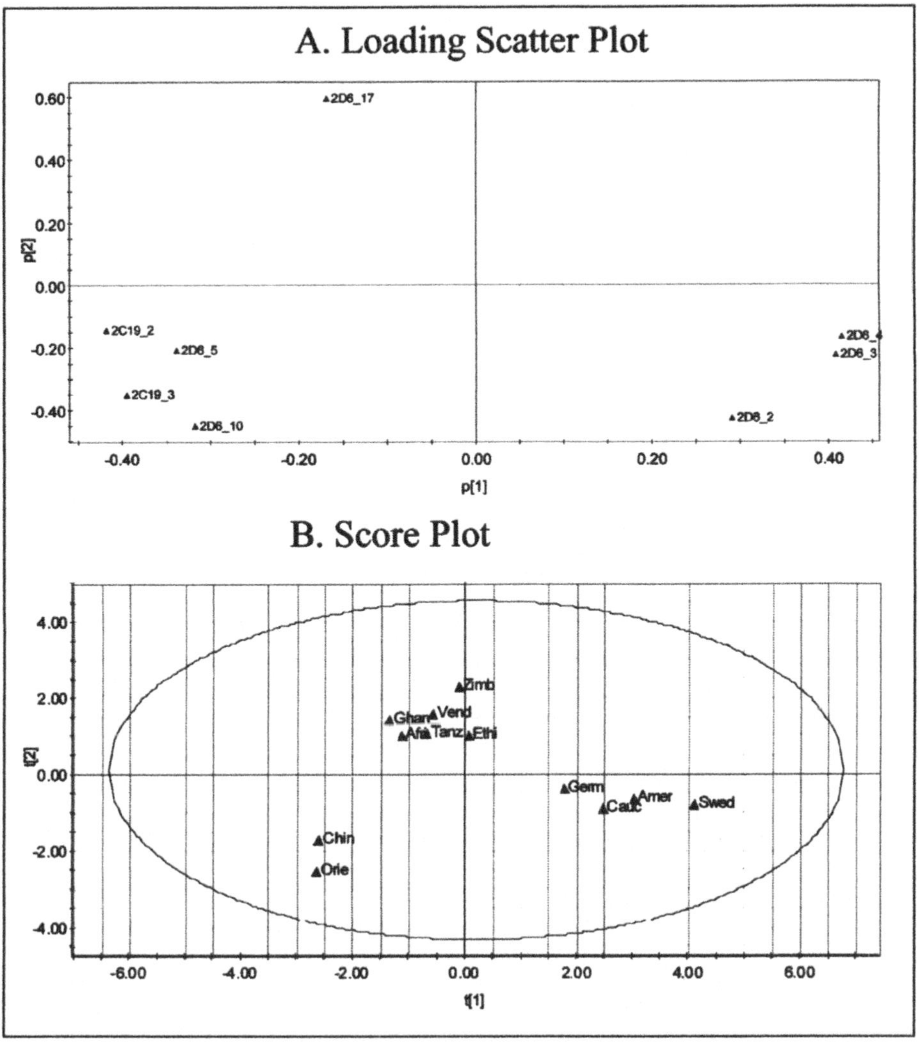

Figure 4. These are principal component analysis (PCA) plots for the first two components (p1 and p2). A) The loading scatter plot of the CYP2D6 and CYP2C19 major alleles measured in the various populations. B) The score plot of the various populations as projected from the alleles analyzed is shown. In the loading plots, the alleles positioned at the far ends of each component scores (t) are the most important in differentiating populations shown in the plot.

explains the difference with the African populations. The Caucasian populations on the other hand are different from the Asians by having a high frequency of *CYP2D6*3* and **4*, which also explains the difference with the African populations. The frequencies of *CYP2C19*2* and *CYP2D6*5* also play a role in clustering these populations. When one considers the second component, it is dominated by the specific presence of the *CYP2D6*17* in African populations which is not detected in the other two populations. In summary, the distribution of the allelic variants clusters the populations in a way that agrees with current knowledge of how these populations have evolved. It also allows us to postulate as to when certain allelic variants occurred as inferred from this evolutionary tree (Fig. 5).

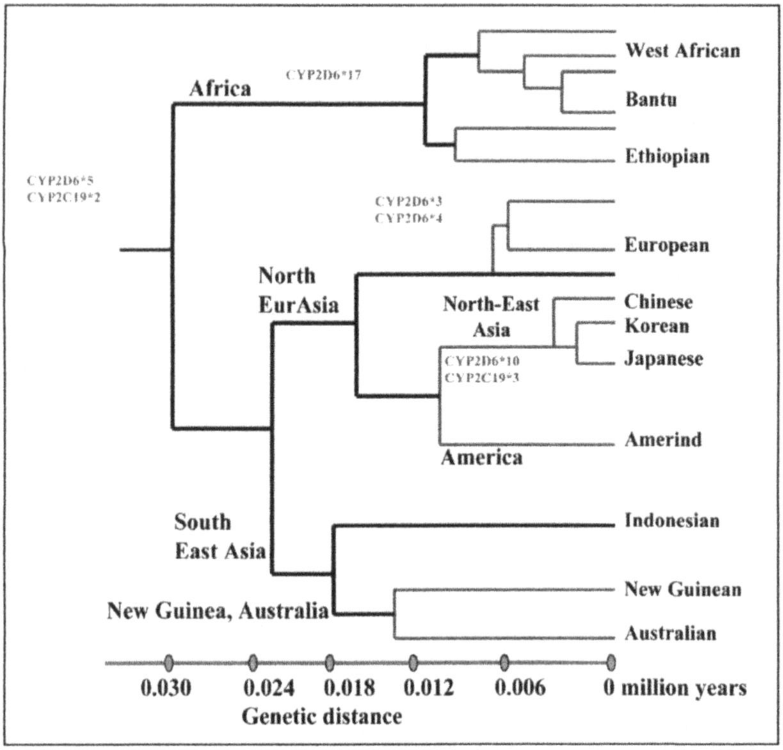

Figure 5. The allocation of major *CYP2D6* and *CYP2C19* variant alleles to the different ethnic groups based on their prevalence analyzed by principal component analysis (PCA) and overlaid on the currently accepted evolutionary tree of world populations from reference 3.

These are very preliminary data that demonstrate that PCA of allelic variants of drug metabolising enzyme genes can also be used for molecular evolutionary studies. As we get more data on more enzymes and on more populations, we will apply this technique in exploring the details of the African population cluster for interethnic structural features that can give us hints of how the Africa populations are interrelated. This information can contribute to other data that seek to address such evolutionary questions, the migratory patterns and genetic admixture of African populations in time and space.

Conclusion

Though few phenotyping and genotyping studies have been done in African populations, the available data already give an insight into the need to carefully consider genetic variability among African populations with respect to response to drugs. Recent developments like the formation of the African Society of Human Genetics (http://www.afshg.org/) and its contributions to the HapMap show how Africa is actively participating in the era of pharmacogenomics/genetics. The formation of the Consortium for the study of pharmacogenetics in African populations (CoPhA) headed by the African Institute of Biomedical Science and Technology (www.aibst.com) is also a significant development. The aim of this consortium is to make a BioBank and Pharmacogenetics database of African populations. Currently the Biobank has over 2000 samples form 12 ethnic groups from 5 different countries (Nigeria, Kenya, Tanzania, Zimbabwe and South Africa). Ongoing genotyping data for genes of drug targets and genes for proteins important in drug pharmacokinetics in these populations will be of both clinical and evolutionary importance with respect to African populations. Work on the potential role of polymorphism of CYP2B6, 3A5, NAT-2, Pgp is of particular interest towards the optimal use of anti-retroviral and anti-TB drugs.

Acknowledgement

Dr. Collen Masimirembwa's funding for this work is from the AiBST Research Board Grant 003. The studies at Leif Bertilsson's laboratory, division of clinical pharmacology, Karolinska Institutet were supported by the Swedish Research Council 3902, NIH(R01 GM60548) and from European and developing countries clinical trials partnerships (EDCTP). Eleni Aklillu's funding for this work is from the Swedish International Development Cooperation Agency, Sida/SAREC.

References

1. Kirchheiner J, Brosen K, Dahl ML et al. CYP2D6 and CYP2C19 genotype-based dose recommendations for antidepressants: A first step towards subpopulation-specific dosages. Acta Psychiatr Scand 2001; 104(3):173-192.
2. Rodriguez-Novoa S, Barreiro P, Jimenez-Nacher I et al. Overview of the pharmacogenetics of HIV therapy. Pharmacogenomics J 2006; 6(4):234-245.
3. Cavalli-Sforza LLMP, Piazza A. The History and Geography of Human Genes. New Jersy: Princeton University Press, 1994.
4. Brooks AS, Helgren DM, Cramer JS et al. Dating and context of three middle stone age sites with bone points in the Upper Semliki Valley, Zaire. Science 1995; 268(5210):548-553.
5. Relethford JH, Jorde LB. Genetic evidence for larger African population size during recent human evolution. Am J Phys Anthropol 1999; 108(3):251-260.
6. Ingelman-Sundberg M. Genetic polymorphisms of cytochrome P450 2D6 (CYP2D6): Clinical consequences, evolutionary aspects and functional diversity. Pharmacogenomics J 2005; 5(1):6-13.
7. Thompson EE, Kuttab-Boulos H, Witonsky D et al. CYP3A variation and the evolution of salt-sensitivity variants. Am J Hum Genet 2004; 75(6):1059-1069.
8. Wooding SP, Watkins WS, Bamshad MJ et al. DNA sequence variation in a 3.7-kb noncoding sequence 5' of the CYP1A2 Gene: Implications for human population history and natural selection. Am J Hum Genet 2002; 71(3):528-542.
9. Gonzalez FJ, Nebert DW. Evolution of the P450 gene superfamily: Animal-plant 'warfare', molecular drive and human genetic differences in drug oxidation. Trends Genet 1990; 6(6):182-186.
10. Nelson DR, Koymans L, Kamataki T et al. P450 superfamily: Update on new sequences, gene mapping, accession numbers and nomenclature. Pharmacogenetics 1996; 6(1):1-42.
11. Ingelman-Sundberg M. Polymorphism of cytochrome P450 and xenobiotic toxicity. Toxicology 2002; 181-182:447-452.
12. McManus ME, Burgess WM, Veronese ME et al. Metabolism of 2-acetylaminofluorene and benzo(a)pyrene and activation of food-derived heterocyclic amine mutagens by human cytochromes P-450. Cancer Res 1990; 50(11):3367-3376.
13. Bertilsson L, Carrillo JA, Dahl ML et al. Clozapine disposition covaries with CYP1A2 activity determined by a caffeine test. Br J Clin Pharmacol 1994; 38(5):471-473.
14. Ikeya K, Jaiswal AK, Owens RA et al. Human CYP1A2: Sequence, gene structure, comparison with the mouse and rat orthologous gene, and differences in liver 1A2 mRNA expression. Mol Endocrinol 1989; 3(9):1399-1408.
15. Eaton DL, Gallagher EP, Bammler TK et al. Role of cytochrome P4501A2 in chemical carcinogenesis: Implications for human variability in expression and enzyme activity. Pharmacogenetics 1995; 5(5):259-274.
16. Jiang Z, Dragin N, Jorge-Nebert LF et al. Search for an association between the human CYP1A2 genotype and CYP1A2 metabolic phenotype. Pharmacogenet Genomics 2006; 16(5):359-367.
17. Sachse C, Brockmoller J, Bauer S et al. Functional significance of a C—>A polymorphism in intron 1 of the cytochrome P450 CYP1A2 gene tested with caffeine. Br J Clin Pharmacol 1999; 47(4):445-449.
18. Takata K, Saruwatari J, Nakada N et al. Phenotype-genotype analysis of CYP1A2 in Japanese patients receiving oral theophylline therapy. Eur J Clin Pharmacol 2006; 62(1):23-28.
19. Nordmark A, Lundgren S, Ask B et al. The effect of the CYP1A2 *1F mutation on CYP1A2 inducibility in pregnant women. Br J Clin Pharmacol 2002; 54(5):504-510.
20. Hamdy SI, Hiratsuka M, Narahara K et al. Genotyping of four genetic polymorphisms in the CYP1A2 gene in the Egyptian population. Br J Clin Pharmacol 2003; 55(3):321-324.

21. Aklillu E, Carrillo JA, Makonnen E et al. Genetic polymorphism of CYP1A2 in Ethiopians affecting induction and expression: Characterization of novel haplotypes with single-nucleotide polymorphisms in intron 1. Mol Pharmacol 2003; 64(3):659-669.
22. Dandara C, Basvi PT, Bapiro TE et al. Frequency of -163 C>A and 63 C>G single nucleotide polymorphism of cytochrome P450 1A2 in two African populations. Clin Chem Lab Med 2004; 42(8):939-941.
23. Sachse C, Bhambra U, Smith G et al. Polymorphisms in the cytochrome P450 CYP1A2 gene (CYP1A2) in colorectal cancer patients and controls: Allele frequencies, linkage disequilibrium and influence on caffeine metabolism. Br J Clin Pharmacol 2003; 55(1):68-76.
24. Masimirembwa CM, Beke M, Hasler JA et al. Low CYP1A2 activity in rural Shona children of Zimbabwe. Clin Pharmacol Ther 1995; 57(1):25-31.
25. Relling MV, Lin JS, Ayers GD et al. Racial and gender differences in N-acetyltransferase, xanthine oxidase, and CYP1A2 activities. Clin Pharmacol Ther 1992; 52(6):643-658.
26. Motola-Kuba D, Zamora-Valdes D, Uribe M et al. Hepatocellular carcinoma. An overview. Ann Hepatol 2006; 5(1):16-24.
27. Uwaifo AO, Bababunmi EA. Liver carcinogenesis in tropical Africa. IARC Sci Publ 1984; 63:59-88.
28. Gallagher EP, Kunze KL, Stapleton PL et al. The kinetics of aflatoxin B1 oxidation by human cDNA-expressed and human liver microsomal cytochromes P450 1A2 and 3A4. Toxicol Appl Pharmacol 1996; 141(2):595-606.
29. Van Vleet TR, Mace K, Coulombe Jr RA. Comparative aflatoxin B(1) activation and cytotoxicity in human bronchial cells expressing cytochromes P450 1A2 and 3A4. Cancer Res 2002; 62(1):105-112.
30. Dai D, Zeldin DC, Blaisdell JA et al. Polymorphisms in human CYP2C8 decrease metabolism of the anticancer drug paclitaxel and arachidonic acid. Pharmacogenetics 2001; 11(7):597-607.
31. Molta NB, Oguche S, Pam SD et al. Amodiaquine treatment of uncomplicated malaria in children, in an area of chloroquine-resistant Plasmodium falciparum in north-central Nigeria. Ann Trop Med Parasitol 2003; 97(7):663-669.
32. Li XQ, Bjorkman A, Andersson TB et al. Identification of human cytochrome P(450)s that metabolise anti-parasitic drugs and predictions of in vivo drug hepatic clearance from in vitro data. Eur J Clin Pharmacol 2003; 59(5-6):429-442.
33. Finta C, Zaphiropoulos PG. The human CYP2C locus: A prototype for intergenic and exon repetition splicing events. Genomics 2000; 63(3):433-438.
34. Bahadur N, Leathart JB, Mutch E et al. CYP2C8 polymorphisms in Caucasians and their relationship with paclitaxel 6alpha-hydroxylase activity in human liver microsomes. Biochem Pharmacol 2002; 64(11):1579-1589.
35. Nakajima M, Fujiki Y, Noda K et al. Genetic polymorphisms of CYP2C8 in Japanese population. Drug Metab Dispos 2003; 31(6):687-690.
36. Cavaco I, Stromberg-Norklit J, Kaneko A et al. CYP2C8 polymorphism frequencies among malaria patients in Zanzibar. Eur J Clin Pharmacol 2005; 61(1):15-18.
37. Rower S, Bienzle U, Weise A et al. Short communication: High prevalence of the cytochrome P450 2C8*2 mutation in Northern Ghana. Trop Med Int Health 2005; 10(12):1271-1273.
38. Totah RA, Rettie AE. Cytochrome P450 2C8: Substrates, inhibitors, pharmacogenetics, and clinical relevance. Clin Pharmacol Ther 2005; 77(5):341-352.
39. Opie LH, Seedat YK. Hypertension in sub-Saharan African populations. Circulation 2005; 112(23):3562-3568.
40. Nesbitt SD. Hypertension in black patients: Special issues and considerations. Curr Hypertens Rep 2005; 7(4):244-248.
41. Lindh JD, Lundgren S, Holm L et al. Several-fold increase in risk of overanticoagulation by CYP2C9 mutations. Clin Pharmacol Ther 2005; 78(5):540-550.
42. Imai J, Ieiri I, Mamiya K et al. Polymorphism of the cytochrome P450 (CYP) 2C9 gene in Japanese epileptic patients: Genetic analysis of the CYP2C9 locus. Pharmacogenetics 2000; 10(1):85-89.
43. Dickmann LJ, Rettie AE, Kneller MB et al. Identification and functional characterization of a new CYP2C9 variant (CYP2C9*5) expressed among African Americans. Mol Pharmacol 2001; 60(2):382-387.
44. Allabi AC, Gala JL, Desager JP et al. Genetic polymorphisms of CYP2C9 and CYP2C19 in the Beninese and Belgian populations. Br J Clin Pharmacol 2003; 56(6):653-657.
45. Yasar U, Aklillu E, Canaparo R et al. Analysis of CYP2C9*5 in Caucasian, Oriental and black-African populations. Eur J Clin Pharmacol 2002; 58(8):555-558.
46. Kidd RS, Curry TB, Gallagher S et al. Identification of a null allele of CYP2C9 in an African-American exhibiting toxicity to phenytoin. Pharmacogenetics 2001; 11(9):803-808.

47. Allabi AC, Gala JL, Horsmans Y et al. Functional impact of CYP2C95, CYP2C96, CYP2C98, and CYP2C911 in vivo among black Africans. Clin Pharmacol Ther 2004; 76(2):113-118.
48. Scordo MG, Aklillu E, Yasar U et al. Genetic polymorphism of cytochrome P450 2C9 in a Caucasian and a black African population. Br J Clin Pharmacol 2001; 52(4):447-450.
49. Hamdy SI, Hiratsuka M, Narahara K et al. Allele and genotype frequencies of polymorphic cytochromes P450 (CYP2C9, CYP2C19, CYP2E1) and dihydropyrimidine dehydrogenase (DPYD) in the Egyptian population. Br J Clin Pharmacol 2002; 53(6):596-603.
50. Takahashi H, Wilkinson GR, Nutescu EA et al. Different contributions of polymorphisms in VKORC1 and CYP2C9 to intra- and inter-population differences in maintenance dose of warfarin in Japanese, Caucasians and African-Americans. Pharmacogenet Genomics 2006; 16(2):101-110.
51. Garcia-Martin E, Martinez C, Ladero JM et al. Interethnic and intraethnic variability of CYP2C8 and CYP2C9 polymorphisms in healthy individuals. Mol Diagn Ther 2006; 10(1):29-40.
52. White RH, Beyth RJ, Zhou H et al. Major bleeding after hospitalization for deep-venous thrombosis. Am J Med 1999; 107(5):414-424.
53. Shorr RI, Ray WA, Daugherty JR et al. Individual sulfonylureas and serious hypoglycemia in older people. J Am Geriatr Soc 1996; 44(7):751-755.
54. Desta Z, Zhao X, Shin JG et al. Clinical significance of the cytochrome P450 2C19 genetic polymorphism. Clin Pharmacokinet 2002; 41(12):913-958.
55. Goldstein JA, Ishizaki T, Chiba K et al. Frequencies of the defective CYP2C19 alleles responsible for the mephenytoin poor metabolizer phenotype in various Oriental, Caucasian, Saudi Arabian and American black populations. Pharmacogenetics 1997; 7(1):59-64.
56. Kaneko A, Lum JK, Yaviong L et al. High and variable frequencies of CYP2C19 mutations: Medical consequences of poor drug metabolism in Vanuatu and other Pacific islands. Pharmacogenetics 1999; 9(5):581-590.
57. Persson I, Aklillu E, Rodrigues F et al. S-mephenytoin hydroxylation phenotype and CYP2C19 genotype among Ethiopians. Pharmacogenetics 1996; 6(6):521-526.
58. Masimirembwa C, Bertilsson L, Johansson I et al. Phenotyping and genotyping of S-mephenytoin hydroxylase (cytochrome P450 2C19) in a Shona population of Zimbabwe. Clin Pharmacol Ther 1995; 57(6):656-661.
59. Herrlin K, Massele AY, Jande M et al. Bantu Tanzanians have a decreased capacity to metabolize omeprazole and mephenytoin in relation to their CYP2C19 genotype. Clin Pharmacol Ther 1998; 64(4):391-401.
60. Roh HK, Dahl ML, Tybring G et al. CYP2C19 genotype and phenotype determined by omeprazole in a Korean population. Pharmacogenetics 1996; 6(6):547-551.
61. Bathum L, Skjelbo E, Mutabingwa TK et al. Phenotypes and genotypes for CYP2D6 and CYP2C19 in a black Tanzanian population. Br J Clin Pharmacol 1999; 48(3):395-401.
62. Herrlin K, Massele AY, Rimoy G et al. Slow chloroguanide metabolism in Tanzanians compared with white subjects and Asian subjects confirms a decreased CYP2C19 activity in relation to genotype. Clin Pharmacol Ther 2000; 68(2):189-198.
63. Aklillu E, Herrlin K, Gustafsson LL et al. Evidence for environmental influence on CYP2D6-catalysed debrisoquine hydroxylation as demonstrated by phenotyping and genotyping of Ethiopians living in Ethiopia or in Sweden. Pharmacogenetics 2002; 12(5):375-383.
64. Klotz U. Clinical impact of CYP2C19 polymorphism on the action of proton pump inhibitors: A review of a special problem. Int J Clin Pharmacol Ther 2006; 44(7):297-302.
65. Furuta T, Shirai N, Watanabe F et al. Effect of cytochrome P4502C19 genotypic differences on cure rates for gastroesophageal reflux disease by lansoprazole. Clin Pharmacol Ther 2002; 72(4):453-460.
66. Tanigawara Y, Aoyama N, Kita T et al. CYP2C19 genotype-related efficacy of omeprazole for the treatment of infection caused by Helicobacter pylori. Clin Pharmacol Ther 1999; 66(5):528-534.
67. Kita T, Tanigawara Y, Aoyama N et al. CYP2C19 genotype related effect of omeprazole on intragastric pH and antimicrobial stability. Pharm Res 2001; 18(5):615-621.
68. Bertilsson L, Dahl ML, Dalen P et al. Molecular genetics of CYP2D6: Clinical relevance with focus on psychotropic drugs. Br J Clin Pharmacol 2002; 53(2):111-122.
69. Gough AC, Smith CA, Howell SM et al. Localization of the CYP2D gene locus to human chromosome 22q13.1 by polymerase chain reaction, in situ hybridization, and linkage analysis. Genomics 1993; 15(2):430-432.
70. Bertilsson L. Geographical/interracial differences in polymorphic drug oxidation. Current state of knowledge of cytochromes P450 (CYP) 2D6 and 2C19. Clin Pharmacokinet 1995; 29(3):192-209.
71. Relling MV, Cherrie J, Schell MJ et al. Lower prevalence of the debrisoquin oxidative poor metabolizer phenotype in American black versus white subjects. Clin Pharmacol Ther 1991; 50(3):308-313.
72. Wan YJ, Poland RE, Han G et al. Analysis of the CYP2D6 gene polymorphism and enzyme activity in African-Americans in southern California. Pharmacogenetics 2001; 11(6):489-499.

73. Gaedigk A, Bradford LD, Marcucci KA et al. Unique CYP2D6 activity distribution and genotype-phenotype discordance in black Americans. Clin Pharmacol Ther 2002; 72(1):76-89.
74. Leathart JB, London SJ, Steward A et al. CYP2D6 phenotype-genotype relationships in African-Americans and Caucasians in Los Angeles. Pharmacogenetics 1998; 8(6):529-541.
75. Wennerholm A, Johansson I, Massele AY et al. Decreased capacity for debrisoquine metabolism among black Tanzanians: Analyses of the CYP2D6 genotype and phenotype. Pharmacogenetics 1999; 9(6):707-714.
76. Panserat S, Sica L, Gerard N et al. CYP2D6 polymorphism in a Gabonese population: Contribution of the CYP2D6*2 and CYP2D6*17 alleles to the high prevalence of the intermediate metabolic phenotype. Br J Clin Pharmacol 1999; 47(1):121-124.
77. Masimirembwa CM, Johansson I, Hasler JA et al. Genetic polymorphism of cytochrome P450 CYP2D6 in Zimbabwean population. Pharmacogenetics 1993; 3(6):275-280.
78. Aklillu E, Persson I, Bertilsson L et al. Frequent distribution of ultrarapid metabolizers of debrisoquine in an ethiopian population carrying duplicated and multiduplicated functional CYP2D6 alleles. J Pharmacol Exp Ther 1996; 278(1):441-446.
79. Masimirembwa C, Persson I, Bertilsson L et al. A novel mutant variant of the CYP2D6 gene (CYP2D6*17) common in a black African population: Association with diminished debrisoquine hydroxylase activity. Br J Clin Pharmacol 1996; 42(6):713-719.
80. Oscarson M, Hidestrand M, Johansson I et al. A combination of mutations in the CYP2D6*17 (CYP2D6Z) allele causes alterations in enzyme function. Mol Pharmacol 1997; 52(6):1034-1040.
81. Bapiro TE, Hasler JA, Ridderstrom M et al. The molecular and enzyme kinetic basis for the diminished activity of the cytochrome P450 2D6.17 (CYP2D6.17) variant. Potential implications for CYP2D6 phenotyping studies and the clinical use of CYP2D6 substrate drugs in some African populations. Biochem Pharmacol 2002; 64(9):1387-1398.
82. Wennerholm A, Dandara C, Sayi J et al. The African-specific CYP2D617 allele encodes an enzyme with changed substrate specificity. Clin Pharmacol Ther 2002; 71(1):77-88.
83. Dalen P, Dahl ML, Ruiz ML et al. 10-Hydroxylation of nortriptyline in white persons with 0, 1, 2, 3, and 13 functional CYP2D6 genes. Clin Pharmacol Ther 1998; 63(4):444-452.
84. Bathum L, Johansson I, Ingelman-Sundberg M et al. Ultrarapid metabolism of sparteine: Frequency of alleles with duplicated CYP2D6 genes in a Danish population as determined by restriction fragment length polymorphism and long polymerase chain reaction. Pharmacogenetics 1998; 8(2):119-123.
85. Aynacioglu AS, Sachse C, Bozkurt A et al. Low frequency of defective alleles of cytochrome P450 enzymes 2C19 and 2D6 in the Turkish population. Clin Pharmacol Ther 1999; 66(2):185-192.
86. Scordo MG, Caputi AP, D'Arrigo C et al. Allele and genotype frequencies of CYP2C9, CYP2C19 and CYP2D6 in an Italian population. Pharmacol Res 2004; 50(2):195-200.
87. Bernal ML, Sinues B, Johansson I et al. Ten percent of North Spanish individuals carry duplicated or triplicated CYP2D6 genes associated with ultrarapid metabolism of debrisoquine. Pharmacogenetics 1999; 9(5):657-660.
88. McLellan RA, Oscarson M, Seidegard J et al. Frequent occurrence of CYP2D6 gene duplication in Saudi Arabians. Pharmacogenetics 1997; 7(3):187-191.
89. Fonne-Pfister R, Meyer UA. Xenobiotic and endobiotic inhibitors of cytochrome P-450dbl function, the target of the debrisoquine/sparteine type polymorphism. Biochem Pharmacol 1988; 37(20):3829-3835.
90. Clark AG. Invasion and maintenance of a gene duplication. Proc Natl Acad Sci USA 1994; 91(8):2950-2954.
91. Wennerholm A, Johansson I, Hidestrand M et al. Characterization of the CYP2D6*29 allele commonly present in a black Tanzanian population causing reduced catalytic activity. Pharmacogenetics 2001; 11(5):417-427.
92. Masimirembwa C, Hasler J, Bertilssons L et al. Phenotype and genotype analysis of debrisoquine hydroxylase (CYP2D6) in a black Zimbabwean population. Reduced enzyme activity and evaluation of metabolic correlation of CYP2D6 probe drugs. Eur J Clin Pharmacol 1996; 51(2):117-122.
93. Bertilsson L, Dahl ML, Sjoqvist F et al. Molecular basis for rational megaprescribing in ultrarapid hydroxylators of debrisoquine. Lancet 1993; 341(8836):63.
94. Zeigler-Johnson CM, Walker AH, Mancke B et al. Ethnic differences in the frequency of prostate cancer susceptibility alleles at SRD5A2 and CYP3A4. Hum Hered 2002; 54(1):13-21.
95. Chelule PK, Gordon M, Palanee T et al. MDR1 and CYP3A4 polymorphisms among African, Indian, and white populations in KwaZulu-Natal, South Africa. Clin Pharmacol Ther 2003; 74(2):195-196.
96. Mirghani RA, Sayi J, Aklillu E et al. CYP3A5 genotype has significant effect on quinine 3-hydroxylation in Tanzanians, who have lower total CYP3A activity than a Swedish population. Pharmacogenet Genomics 2006; 16(9):637-645.

97. Dandara C, Li DP, Walther G et al. Gene-environment interaction: The role of SULT1A1 and CYP3A5 polymorphisms as risk modifiers for squamous cell carcinoma of the oesophagus. Carcinogenesis 2006; 27(4):791-797.
98. Bertz RJ, Granneman GR. Use of in vitro and in vivo data to estimate the likelihood of metabolic pharmacokinetic interactions. Clin Pharmacokinet 1997; 32(3):210-258.
99. de Jong FA, de Jonge MJ, Verweij J et al. Role of pharmacogenetics in irinotecan therapy. Cancer Lett 2006; 234(1):90-106.
100. Dandara C, Masimirembwa CM, Magimba A et al. Genetic polymorphism of CYP2D6 and CYP2C19 in east- and southern African populations including psychiatric patients. Eur J Clin Pharmacol 2001; 57(1):11-17.
101. Fukushima-Uesaka H, Saito Y, Maekawa K et al. Genetic variations and haplotypes of CYP2C19 in a Japanese population. Drug Metab Pharmacokinet 2005; 20(4):300-307.
102. Griese EU, Asante-Poku S, Ofori-Adjei D et al. Analysis of the CYP2D6 gene mutations and their consequences for enzyme function in a West African population. Pharmacogenetics 1999; 9(6):715-723.
103. Griese EU, Zanger UM, Brudermanns U et al. Assessment of the predictive power of genotypes for the in-vivo catalytic function of CYP2D6 in a German population. Pharmacogenetics 1998; 8(1):15-26.
104. Bertilsson L, Lou YQ, Du YL et al. Pronounced differences between native Chinese and Swedish populations in the polymorphic hydroxylations of debrisoquin and S-mephenytoin. Clin Pharmacol Ther 1992; 51(4):388-397.
105. Johansson I, Oscarson M, Yue QY et al. Genetic analysis of the Chinese cytochrome P4502D locus: Characterization of variant CYP2D6 genes present in subjects with diminished capacity for debrisoquine hydroxylation. Mol Pharmacol 1994; 46(3):452-459.
106. Wang SL, Huang JD, Lai MD et al. Molecular basis of genetic variation in debrisoquin hydroxylation in Chinese subjects: Polymorphism in RFLP and DNA sequence of CYP2D6. Clin Pharmacol Ther 1993; 53(4):410-418.
107. Gervasini G, Vizcaino S, Gasiba C et al. Differences in CYP3A5*3 genotype distribution and combinations with other polymorphisms between Spaniards and Other Caucasian populations. Ther Drug Monit 2005; 27(6):819-821.
108. Garsa AA, McLeod HL, Marsh S. CYP3A4 and CYP3A5 genotyping by Pyrosequencing. BMC Med Genet 2005; 6:19.
109. Hustert E, Haberl M, Burk O et al. The genetic determinants of the CYP3A5 polymorphism. Pharmacogenetics 2001; 11(9):773-779.
110. Roy JN, Lajoie J, Zijenah LS et al. CYP3A5 genetic polymorphisms in different ethnic populations. Drug Metab Dispos 2005; 33(7):884-887.
111. Ball SE, Scatina J, Kao J et al. Population distribution and effects on drug metabolism of a genetic variant in the 5' promoter region of CYP3A4. Clin Pharmacol Ther 1999; 66(3):288-294.
112. Balram C, Zhou Q, Cheung YB et al. CYP3A5*3 and *6 single nucleotide polymorphisms in three distinct Asian populations. Eur J Clin Pharmacol 2003; 59(2):123-126.
113. Chowbay B, Li H, David M et al. Meta-analysis of the influence of MDR1 C3435T polymorphism on digoxin pharmacokinetics and MDR1 gene expression. Br J Clin Pharmacol 2005; 60(2):159-171.
114. Saeki M, Saito Y, Nakamura T et al. Single nucleotide polymorphisms and haplotype frequencies of CYP3A5 in a Japanese population. Hum Mutat 2003; 21(6):653.
115. Kittles RA, Chen W, Panguluri RK et al. CYP3A4-V and prostate cancer in African Americans: Causal or confounding association because of population stratification? Hum Genet 2002; 110(6):553-560.
116. Dandara C, Ballo R, Parker MI. CYP3A5 genotypes and risk of oesophageal cancer in two South African populations. Cancer Lett 2005; 225(2):275-282.
117. Cavaco I, Reis R, Gil JP et al. CYP3A4*1B and NAT2*14 alleles in a native African population. Clin Chem Lab Med 2003; 41(4):606-609.
118. Kuehl P, Zhang J, Lin Y et al. Sequence diversity in CYP3A promoters and characterization of the genetic basis of polymorphic CYP3A5 expression. Nat Genet 2001; 27(4):383-391.
119. Ingelman-Sundberg M. Duplication, multiduplication, and amplification of genes encoding drug-metabolizing enzymes: Evolutionary, toxicological, and clinical pharmacological aspects. Drug Metab Rev 1999; 31(2):449-459.
120. Dahl ML, Johansson I, Bertilsson L et al. Ultrarapid hydroxylation of debrisoquine in a Swedish population. Analysis of the molecular genetic basis. J Pharmacol Exp Ther 1995; 274(1):516-520.
121. Mitsunaga Y, Kubota T, Ishiguro A et al. Frequent occurrence of CYP2D6*10 duplication allele in a Japanese population. Mutat Res 2002; 505(1-2):83-85.
122. Ji L, Pan S, Marti-Jaun J et al. Single-step assays to analyze CYP2D6 gene polymorphisms in Asians: Allele frequencies and a novel *14B allele in mainland Chinese. Clin Chem 2002; 48(7):983-988.

CHAPTER 8

Pharmacogenomics in the Indian Population

M. Ravindra Kumar and C. Adithan*

Abstract

India is one of the most diverse countries in the world. There are more than 200 languages spoken in India and 22 of them are considered official language. The population is segregated into endogamous groups based on religion (Hindus, Muslims, Christians, Buddhists, Sikhs etc.), caste and ethnicity. Studies are under way in various aspects of pharmacogenomics in the subcontinent, especially aimed at determining interindividual variability in drug response with regard to genetic polymorphisms in drug metabolizing enzymes, drug transporters and targets. Studies are also focusing on identification of disease susceptibility markers. Many of these studies are SNP (single nucleotide polymorphisms) based approaches. These are used to identify the frequency distribution of various polymorphisms in different ethnic groups of India and thereby predict a subset of the population who would benefit most from therapy. In this review, we focus on studies related to polymorphisms in genes encoding enzymes involved in Phase I *(CYP2C9, CYP2C19, CYP2D6* and *CYP2E1)* and Phase II (*GST* and *NAT)* drug metabolism, and the extensively studied 3435C>T polymorphism in the *ABCBI* gene which codes for the multidrug resistant (MDR-1) or P-glycoprotein transporter. Not much work has been done on pharmacological targets of drugs (receptors and proteins) but the available information has been included in the review.

Introduction

The world population is genetically about 99.9% identical and the difference of 0.1% in genetic sequence accounts for several phenotypic variations leading to altered drug response. It is known that drugs do not produce the same effect in all patients and this inter-individual variation in drug response has been attributed to many causes such as age, sex, concomitant medication, disease status, organ function etc. Another important source of variation is the mutations/polymorphisms in the genes that a person carries. The genome-wide screening for the mutations or polymorphisms responsible for this inter-individual response to drugs is known as pharmacogenomics. Frequencies of variant alleles of these polymorphisms and their expression may vary from one ethnic group to another. Several studies have been published investigating these polymorphisms and their clinical importance but most of them have studied the Caucasians and the East Asians (Chinese and Japanese).[1-4] Very few studies have been done in Indian populations. This chapter reviews available literature in Indians living in India, the second most populous country in the world.

*Corresponding Author: C. Adithan—Pharmacogenomics Laboratory, Department of Pharmacology, JIPMER, Pondicherry, India. Email: jipgene@jipmer.edu

Pharmacogenomics in Admixed Populations, edited by Guilherme Suarez-Kurtz.

Indian Population

The population of India exceeds 1 billion and the subcontinent comprises groups of people with different anthropological, genetic, cultural and linguistic characteristics which are largely due to the varied topography of the country. India has the most genetic diversity (except Africa) in sharp contrast to other global regions.[5]

Anthropologically Indian population is grouped into four major ethnic categories viz., Australoid, Indo-Caucasoid, Indo-Mongoloid and Negrito populations. On linguistic basis and the likely major migrations into India, the Indian population can be broadly assigned into one of the four language families viz., Indo-European, Dravidian, Austro-Asiatic and Sino-Tibetan speakers. Of these migrations Austric speakers are believed to be the earliest, followed by Dravidian speakers from middle-east Asia and Sino-Tibetan speakers from China and the last migrants were Indo-European speakers from west Asia.[6]

All communities of Austric speakers are exclusively tribals and are thought to be the oldest inhabitants of India. Sino-Tibetan speakers of India include many tribal groups that are concentrated along the Himalayas. The major portions of the mainland Indian population are of Dravidian and Indo-European speakers. Among them Dravidians were earlier settlers and were suggested to have been pushed to south India by the later migrants, the Indo-European speakers.[6]

On cultural basis Indian population can be broadly stratified into tribals and nontribals. Tribals are considered to be the original inhabitants and they constitute about 8% of the total population.[7] The nontribal populations of India belong to many religious groups of which the major one is Hindus (80.5%) and the others are Muslims (13.4%), Christians (2.3%), Sikhs (1.9%), Buddhists (0.8%) and Jains (0.4%) as per 2001 Indian census data.

The Hindu religious people were segregated into four categories based on their occupations viz., Brahmin (priestly class), Kshatriya (warrior class), Vysya (business class) and Sudra (menial labour class). These were hierarchically placed with Brahmin, Kshatriya and Vysya considered at higher status and the Sudras at the lower status. There are multiple sub-castes in each one of them. In general the nontribal populations predominantly belong to the Indo-European and Dravidian language families.[7]

The segregation, isolation and the strict practice of endogamy across all social ranks has resulted in a remarkable number of around 4635 recognized ethnic communities in India. This explains the immense and interesting diversity of the Indian population in the form of groups with specific social traditions and linguistic dialects.

Most of the pharmacogenomic studies undertaken in India have chosen their subjects based on their geographical location. They are mainly from the north Indian (Indo-European) and south Indian (Dravidians) populations. It is well accepted that these two geographic groups differ linguistically and in socio-cultural habits. No pharmacogenomic studies have been reported from northeastern (Mongoloid) and native Andaman (Negrito) populations.

Indian Government Initiative in Pharmacogenomic Research

Realizing the importance and the potential benefits of pharmacogenomic research, the Government of India has initiated several programmes. The Ministry of Health has allotted about £13.3m ($20m) for medical genomics research for five years starting from 2001.[8] Government funding agencies like Department of Biotechnology (DBT), Department of Science and Technology (DST) and Indian Council of Medical Research (ICMR) have listed pharmacogenomics as a priority area for research support. In 2005, the Department of Biotechnology convened a meeting of experts at Mumbai to prepare the road map for pharmacogenomics research in India. In a process to set up National Biotechnology Regulatory Authority the Ministry of Science and Technology, Government of India had put on the website a draft of "National Biotechnology Development Strategy" open for public debate in April 2005. In this draft under the head of "Diagnostics for Emerging Medical Paradigm" one of the strategic actions mentioned states *"Establish a cell for Diagnostic Biotechnology to encour-*

age and support studies into the clinical application of pharmacogenomics". Based on the comments received a draft cabinet note was prepared and circulated to the concerned ministers/ departments in August 2006.

Pharmacogenomic Research in India

The majority of pharmacogenomic studies in India have aimed at establishing the frequency of variant alleles of polymorphic genes encoding drug metabolizing enzymes, drug transporters and drug targets (receptors) in healthy individuals. Only a few studies have attempted to establish genotype-phenotype relationships and their clinical implications. This chapter reviews the currently available information on various polymorphisms in India and their clinical implications.

Polymorphisms in Drug Metabolism

Phase I Enzymes

Metabolism of drugs is mainly mediated by Phase I and Phase II enzymes which render them polar and favour their elimination. Many clinically relevant drugs are metabolized by cytochrome (CYP) P450s which are the main Phase I drug metabolizing enzymes involved in the oxidative degradation of drugs. Extensive variation is seen in CYP activity due to genetic polymorphisms, which can lead to potential pharmacokinetic consequences such as lack of prodrug activation, increased effective dose, metabolism by alternative pathways and clinical consequences such as extended pharmacological effect, adverse drug reactions, drug toxicity, exacerbated drug-drug interactions etc.

The CYP super family consists of more than 30 families of drug metabolizing enzymes. Three of them, viz., CYP1, CYP2 and CYP3, contribute to the metabolism of the majority of drugs and xenobiotics whose elimination depends primarily on metabolic route. There is clear evidence of ethnic variation in the frequency distribution of genetic polymorphisms in drug metabolizing enzymes.

Indian Studies on *CYP2C9*

The CYP2C subfamily includes CYP2C8, CYP2C9, CYP2C18 and CYP2C19, which constitutes about 18% of the CYP protein content in human liver and metabolizes approximately 20% of currently prescribed drugs like phenytoin, warfarin, NSAIDs, and oral hypoglycemic agents.[9] The gene encoding CYP2C9 has at least 30 polymorphic forms (http:// www.cypalleles.ki.se// accessed in August 2006). Most of them encode for an enzyme with decreased activity which leads to increased substrate concentration and might predispose to toxicity in patients who are carriers of variant alleles. Of these the most frequently studied are *CYP2C9*2* and **3*. *CYP2C9*2* is due to a 430C>T substitution on exon 3 which leads to Arg144Cys conversion resulting in the formation of less active enzyme.[10] Further reduction in enzyme activity is seen with *CY2PC9*3* allele which is due to a 1075C>T polymorphism on exon 7 which results in an altered protein with an Ile359Leu substitution.[11]

The first study of CYP2C9 in Indian populations was done in 135 Tamilian subjects of South India. It revealed that the frequency of *CYP2C9*3* (6.7%) was higher than that of **2* (2.6%).[12] A subsequent study was performed in 346 healthy subjects belonging to three other south Indian states viz., Karnataka, Andhra Pradesh and Kerala.[13] Collectively, these two studies revealed that *CYP2C9*3* occurred at 8% in south Indian individuals. The average frequency of *CYP2C9*2* in south Indians was 4%, with the Kerala population having the lowest (2%), and the Karnataka population the highest (6%) frequency (Table 1). Studies done in Indians residing in Malaysia revealed similar frequency of *CYP2C9*2* and **3* (4.4 and 8.2% respectively).[14] This could be explained by the fact that the majority of the Malaysian Indians are migrants from south India.

The **3* mutant allele frequency in south Indians (8%) was similar to that of Caucasians (8%) but higher than that of Chinese (3%). The **2* allele frequency (4%) was found to be between that of Chinese (0%) and Caucasians (12%).[15]

Table 1. Allele frequencies of* CYP2C9 *gene polymorphisms in Indian populations

Population	N	Frequency of *CYP2C9* Variant Alleles (%)		References
		*2	*3	
Tamil Nadu population	135	2.6	6.7	12
Karnataka population	110	6	8	13
Andhra Pradesh population	116	4	9	13
Kerala population	120	2	8	13
South Indian (pooled)	481	4	8	12,13
Malaysian Indians	92	4.4	8.2	14

The hydroxylation activity of the CYP2C9 enzyme in subjects carrying *CYP2C9*2* and *3 alleles was studied in 27 south Indian subjects, and a significant correlation was found between phenytoin hydroxylation and the *CYP2C9* genotype.[16] There was only one subject with the *3/*3 genotype whose metabolic ratio was nearly three times (109.9) higher compared to the wild genotype (34.8 ± 21.8). This could be the result of the mutant genotype or possibly due to large variation seen in the mean value of metabolic ratio.

Indian Studies on *CYP2C19*

CYP2C19 metabolizes many clinically important drugs such as omeprazole, lansoprazole, proguanil, diazepam, S-mephenytoin, amitriptyline etc. S-mephenytoin and omeprazole are frequently used as the probe drugs to assess CYP2C19 activity. Based on the parent drug/metabolite ratio (a pharmacokinetic phenotype), individuals can be categorized as poor metabolizers (PM), who have decreased enzyme activity or extensive metabolizers (EM) who have normal enzyme activity.

Many studies have investigated the frequency of CYP2C19 pharmacokinetic phenotypes and have correlated this with *CYP2C19* genotype in the Indian population. The first such study was reported from Bombay (western part of India) using S-mephenytoin as a probe drug.[17] In this study the PM frequency was 20%. A subsequent study using omeprazole as probe drug in 100 north Indian subjects residing in Chandigarh and adjoining area, reported an antimode of 50.1 and a PM frequency of 11%.[18] The same authors conducted a further study to investigate the phenotype-genotype correlation.[19] Another 100 unrelated healthy north Indians were phenotyped for CYP2C19 using omeprazole. Then, 100 EM and 21 PM subjects were genotyped for *CYP2C19* *2 and *3 mutant alleles. The frequency of *2 allele was 30% and no individual carried the *3 allele in this population. Out of the 21 PMs, only 9 subjects had the *CYP2C19*2/*2* genotype which correlated with their phenotype. The remaining 12 PM subjects were genotyped as either *CYP2C19*1/*1* or *CYP2C19*1/*2*. The PM status of these subjects could be due to other polymorphisms in the *CYP2C19* gene which was not investigated in the study. An in vitro study using liver microsomes of 15 north Indians showed concordance between the in vitro activity of omeprazole hydroxylase and *CYP2C19* genotype.[20]

A *CYP2C19* genotype study was done in 453 south Indians belonging to Tamil Nadu, Kerala, Andhra Pradesh and Karnataka. The frequencies of *CYP2C19*2* and *3 were 35% and 1% respectively.[13,21] An interesting finding is that *CYP2C19*3* was present in all the south Indian states except in subjects from Andhra Pradesh.[13] Even though the *3 allele was present in the south Indians, the genotypes were either *CYP2C19*1/*3* or *CYP2C19*2/*3*. None of them were homozygous for the mutant *3 allele. The frequency of the *2 allele was similar to that of north Indians (Table 2). In a subsequent study the association between *CYP2C19* genotype and phenotype was investigated in 300 south Indians using omeprazole as probe drug.

Table 2. Allele frequencies of* CYP2C19 *gene polymorphisms in Indian populations

Population	N	Frequency of *CYP2C19* Variant Alleles (%) *2	*3	References
Tamil Nadu population	112	37.9	2.2	21
Karnataka population	108	39	1	13
Andhra Pradesh population	115	33	0	13
Kerala population	118	31	1	13
South Indian (Pooled)	453	35	1	13,21
North Indian	121	30	0	19

The estimated antimode was 14.4 and the frequency of PM was 14% which is similar to the results reported for north Indians.[22] There was no correlation between genotype and phenotype in 23 out of 300 subjects. The PM outliers were genotyped for **2B*, **4*, **5* and **6* alleles, but none was detected.

Indian Studies on *CYP2D6*

CYP2D6 phenotype studies done in Indian population showed the PM frequency to be between 1.8 and 4.8%.[23] The first phenotype study done in Bombay population using debrisoquine as probe drug reported the PM frequency of 2%.[24] Another study conducted at Chandigarh (North India) using dextromethorphan as probe drug found 3% of the subjects to be PM.[25] Two studies done in the population of Andhra Pradesh, disclosed PM frequencies of 1.8% in Kakinada (a coastal city) and 3.2% in Hyderabad (cosmopolitan city).[26,27] The PM frequencies in other three south Indian states viz., Karnataka, Tamil Nadu and Kerala were 4%, 3.6% and 4.8% respectively.[26,28,29] The reported frequencies of PMs in Indians (1.8–4.8%) were lower than in Caucasians (5–10%) and higher than in Chinese (<1.6%) and Zimbabwean and Tanzanian population (0–2%).[23]

The PM phenotype could be due to the polymorphic forms of the *CYP2D6* gene. Among *CYP2D6* allelic forms, the frequency of **2* (normal enzyme activity), **3*, **4*, **5*, **14* (absent enzyme activity), **10* and **17* (reduced enzyme activity) were studied in Indians. The frequency of *CYP2D6*4* varies from 7.3% in south Indians to 11.9% in north Indians. Among the south Indians studied lowest frequency was found in the population of Karnataka (4.8%). Other alleles were studied only in south Indian population and in Indians residing in Malaysia. The **2* allele is the most frequent variant allele found in south Indians (34.8%). The frequency of the low activity allele **10* is similar between south Indians and the Malaysian Indians. The **17* allele was not found in south Indians whereas it was present in Malaysian Indians (1%) (Table 3).

The frequency of **10* allele in Indians (10.2%) is higher than Caucasians (1–2%), Black Africans (6%), Ethiopians and Saudi Arabians (3–9%). The frequency of *4 allele in Indians (7.3–11.9%) is lower than that of Caucasians (12–21%) and higher than that seen in Black Africans (2%), Ethiopians and Saudi Arabians (1–4%). The *17 allele is found predominantly in Black Africans (20-35%).[30]

Indian Studies on *CYP2E1*

CYP2E1 is involved in the metabolism of substances having toxicological and carcinogenic importance, such as aromatic and halogenated hydrocarbons. Many of the halogenated anesthetics (halothane, enflurane etc.), alcohols (ethanol, methanol etc.), nitrosamines (N,N-dimethylnitrosamine), anti-epileptics (phenobarbital) are also substrates for this enzyme.[35] Of the many allelic variants found in the *CYP2E1* gene, **1B*, **6* and **5B* alleles have been studied in Indian population.[36,37] *CYP2E1*1B* allele or Taq1 polymorphism is due to base

Table 3. Allele frequencies of* CYP2D6 *gene polymorphisms in Indian populations

Population	N	Frequency of *CYP2D6* Polymorphisms (%)							Refs.
		*2	*2xN	*3	*4	*5	*10	*17	
South Indians (Pooled)	447	34.8	-	0	7.3	1.9	10.2	0	31
Western & Central India	881	-	-	-	9.25	-	-	-	32
North Indians	250	-	-	-	11.9	-	-	-	33
Malaysian Indians	86	-	2	0	8	1	15	1	34

change 9896C>T in intron 7 of *CYP2E1* gene. A person is said to have *A1* allele if there is no cutting site and *A2* allele if there is a cutting site for the Taq1 enzyme. The study on 123 unrelated healthy volunteers of Tamil Nadu (South India) revealed a frequency of 21% of *A1* allele of *CYP2E1*1B* or Taq1 polymorphism, 0.4% of *c2* allele of *CYP2E1*5B* polymorphism and 16% of *C* allele of *CYP2E1*6* polymorphism.[37] In a north Indian study evaluating the role of *CYP2E1* polymorphisms in susceptibility to leukoplakia, 227 controls were genotyped for *CYP2E1*5B* and **6* polymorphism.[36] The frequency of *c2* allele of *CYP2E1*5B* was 0.8% and *C* allele of *CYP2E1*6* was 19% which was similar to that seen in south Indians.

Phase II Enzymes

Phase II reactions are mainly involved in the conjugation of drugs with endogenous compounds and are catalyzed by enzymes such as glutathione S-transferases, N-acetyl transferases, UDP-glucuronosyl transferases etc. The genes encoding these enzymes have been shown to exhibit polymorphisms and some of them have been studied in the Indian population.

Indian Studies on GSTs

Glutathione S-transferase is a major group of detoxifying enzymes comprising of at least five distantly related gene families (designated class Alpha, Mu, Pi, Sigma, and Theta *GST*). Among these isoenzymes, Mu (M), Theta (T) and Pi (P) have been frequently studied. Individuals with *GSTM1*0/*0* and *GSTT1*0/*0* (homozygous gene deletions) lack GST enzymes and these null genotypes have been extensively studied. *GSTM1-null* and *GSTT1-null* frequencies were studied in populations from Orissa (a north-eastern state), western and central India, south India and north India.[38-42] There was heterogeneity in the distribution of *GSTM1-null* and *GSTT1-null* genotypes (Table 4). The frequency of *GSTM1-null* was lowest in Orissa population (24%) and highest in the population of north Indians (33%) while the frequency was almost similar in the population of south and north India. People of Orissa are of Indo-European lineage and most of them are settled in rural areas. The tribals called adivasis constitute nearly 22% of the Orissa population. The Oriyan population forms a separate subgroup explaining a different frequency. The range of frequency of the *GSTM1-null* genotype in Indians (24–33%) is lower than that of Caucasians (42–60%), Japanese (47.6%) and Koreans (52.1%).[43]

The distribution of *GSTT1*0/*0* (*null* genotype) in Indians ranges from 13.0 to 18.4%. The highest frequency was observed in north Indians (18.4%) and the lowest being in populations from western and central India (Table 4). The Indian data on *GSTT1-null* genotype is closer to that obtained from Caucasians (13–26%) but lower than that of Japanese (35.3%) and Koreans (51.5%).[43]

The frequency of *GSTP1* wild type homozygous genotype (Ile/Ile) is similar in south Indians and north Indians (43.6 and 44.3% respectively) but the homozygous mutant genotype (Val/Val) is more common in south Indians (9%) than north Indians (5.4%). The frequency of homozygous mutants was higher in Indians (5.4–9%) in comparison to Chinese (0.8%), Japanese (3.1%) and Koreans (2.5%).[44]

Table 4. Allele and genotype frequencies of* GST *polymorphisms in Indian populations

Population	N	GSTM1 *0/*0	GSTT1 *0/*0	GSTM1 *0/*0 & GSTT1 *0/*0	GSTP1 Ile/Ile	GSTP1 Ile/Val	GSTP1 Val/Val	Refs.
South Indians (Pooled)	517	30.4	16.8	4.6		-		41
Tamilians	133				43.6	47.4	9.0	37
North Indians	370	33.0	18.4	7.0	44.3	50.3	5.4	40
Western and Central India	883	26.6	13.0	-		-		38
Orissa (North-eastern India)	72	24.0	-	-		-		42
Singapore Indians	139 (for GSTM1) 152 (for GSTT1 and double null)	32.0	16.0	5.0		-		39,45

GST gene polymorphisms have been correlated with adverse effects of drugs as this family of enzymes is involved in detoxifying many chemicals and drugs like busulfan.[46] Subjects with polymorphisms that encode defective enzyme activity are more likely to experience adverse effects. A case control study was conducted to correlate GST polymorphisms with the hepatotoxicity induced by anti-tuberculous drug therapy. Sixty-six patients were recruited, of whom 33 developed hepatotoxicity (cases) and 33 did not (controls). The subjects were genotyped for *GSTM1*, *GSTT1* and *NAT2* mutations. The frequency of *GSTM1*0/*0* was found to be significantly higher among cases than controls with a relative risk of 2.13 (95% CI 1.25-3.1, $p<0.05$) suggesting the role of this polymorphism in anti-tubercular drugs induced hepatotoxicity.[47]

Indian Studies on NAT

N-Acetyltransferases (NATs) are mainly involved in the bioconversion of heterocyclic arylamines into carcinogens and in the metabolism of clinically used drugs such as hydralazine, dapsone, procainamide, nitrazepam, and caffeine.[48] Two functional NATs are present in humans NAT1 and NAT2; both are polymorphic and the encoding genes are located on chromosome 8.[49,50] In the *NAT1* gene the **3*, **4*, **10* and **11* polymorphisms were studied in three ethnic groups viz., Indians, Chinese and Malays, residing in Singapore.[51] The frequency of *NAT1*4* was found to be higher in Indians (51%) when compared to Chinese (35%) and Malays (30%) while that of *NAT1*10* was lower in Indians (17%) than in Chinese (30%) or Malays (39%). NAT1 polymorphism has been studied for its association with colorectal cancer, pancreatic cancer and breast cancer but didn't show a clear cut association. In *NAT2* gene, the wild-type allele is represented as *NAT2*4*, and 34 mutated alleles have so far been described (http://www.louisville.edu/medschool/pharmacology/ NAT.html accessed in August 2006). These mutants are characterized by the possession of combinations of 1 to 4 nucleotide substitutions that occur at defined positions within the 870 bp coding sequence. Based on the NAT-2 phenotype, an individual may be classified as a rapid or a slow acetylator.[52]

In south Indian populations, *NAT2* was genotyped for seven SNPs at positions 191, 282, 341, 481, 590, 803 and 857 in 166 subjects belonging to eight tribal ethnic communities viz., Adiya, Kani, Kattunaikkar, Kurichiya, Kuruma, Malapandram, Paniya, Pulaya of Dravidian origin and the modern Dravidian populations are believed to have evolved from these communities. The predominant mutation was 282C>T (44%) and the least frequent was 481C>T. Slow acetylators were found to be predominant in all these populations, with a frequency of 74%.[53] The **6A* allele was the most frequently observed (23.6%) and **5B/*6A* was the most

Table 5. Allele frequency of NAT2 gene polymorphisms in Indian populations

Population	N	Allele Frequency (%)				Reference
		NAT2*5	NAT2*6	NAT2*7	NAT2*14	
South Indians	166	22	37	25	0	53
North Indians	140	50	30	25	0	54

common genotype (16.3%). The **4/*6A* genotype was found to be the most frequent rapid acetylator genotype (8.2%). The study also reported a novel deletion mutation at position 859 of the BamHI site, which was present in 30% of the individuals of the Malapandaram tribe.

The percentage of slow acetylators in north Indians (44%) was lower than south Indians (74%).[54] The frequency of different alleles of *NAT2* in north Indians and south Indians are almost similar except for *5 which is higher in north Indians (50%) than in south Indians (22%) (Table 5). The slow acetylator phenotype is determined by the presence of two variant alleles (*NAT*5, NAT*6* or *NAT*7*). In south Indians **12* and **13* alleles were also studied and the genotypes having these variant alleles were more common in south Indian population explaining the higher frequency of slow acetylators.

Indian Studies on Butyrylcholinesterase

Butyrylcholinesterase (BChE) is involved in the metabolism of succinylcholine and mivacurium. A study done in 226 people of Vysya community in Coimbatore (south Indian city in Tamil Nadu) identified a novel mutation in exon 2 of the *BCHE* gene resulting in substitution of leucine 307 by proline. This mutation can lead to prolonged apnea after a normal dose of succinylcholine or mivacurium. The frequency of this mutation in Vysya community was 4.16% which is 4000 times higher (genotypes leading to low BChE activity is 1 in 3500 or ~0.03%) when compared to other populations.[55,56]

Drug Transporters

Multidrug transporters belonging to ATP-binding cassette (ABC) family facilitate the transport of diverse drugs across cell membranes. The most extensively studied is the MDR1 (or ABCB1) multidrug transporter, an efflux pump for drugs like protease inhibitors, antidepressants, antipsychotics etc. To date, 48 SNPs have been described for the *ABCB1* gene, of which the 3435C>T on exon 26 has been the most extensively studied.

The MDR1 gene expression was studied in Indian patients with leukemia, oral and gastric cancer.[57-60] Few studies examined the 3435C>T polymorphism in the Indian population. The frequency of the T allele was 54% among 185 unrelated healthy volunteers of Tamil Nadu (South India)[61] and 62% among 93 unrelated healthy Indians residing in Malaysia.[62] The frequency of this allele in Indians is closer to that of British Caucasians (52%) and Chinese population (47%) but higher than that of African American (16%) and Kenyan population (17%).[63]

Drug Targets

Genetic polymorphisms not only alter drug pharmacokinetic but also modulate the pharmacodynamic response, by affecting the responsiveness of the receptor or enzyme or protein upon which the drug acts. One study evaluated the influence of polymorphisms in beta 2 adrenergic receptor (*ADRB2*) gene in the response to salbutamol.[64] Bronchial asthma patients were genotyped for 10 SNPs in *ADRB2* gene and those who were homozygous for Arg16 responded poorly to salbutamol. No other associations were disclosed between the genetic polymorphisms and drug response. Similar association was found between the Arg16Gly polymorphism of *ADRB2* gene and response to albuterol (salbutamol) in Korean, Japanese and Puerto Rican population but not in the Mexicans.[65-67]

Figure 1. Map of India showing different states and some cities that are mentioned in this review.

Susceptibility to ADR

Pharmacovigilance is a national health program in India initiated to monitor the safety of drugs and to increase awareness relating to adverse drug reactions. It is expected that knowledge and application of pharmacogenomics in clinical practice may help to reduce the adverse drug reactions. For example, tardive dyskinesia (TD) is a distressing adverse drug effect seen with long term treatment with typical antipsychotics, characterized by choreoathetotic movement. Four studies were reported from a laboratory in India which attempted to identify the genetic cause(s), if any, underlying the development of TD in schizophrenic patients.[33,68-70] Out of 335 patients of north India, 96 (29%) developed TD. Among them, 28 had been on typical antipsychotics, 23 on atypical antipsychotics and 45 were on both drugs in the course of their illness. The genotype-phenotype association studies showed that polymorphisms in the *DRD4* (120 bp duplication, 1.2 kb upstream from initiation codon) and *COMT* genes (*COMT 408C>G* and *COMT 472G>A*) were significantly associated with TD.[69] There was no association between polymorphisms in the drug metabolizing enzyme genes *CYP1A2* (alleles **1C, *1F, *2, *4, *5* and **6*), *CYP3A4* (allele **1B*) and *CYP2D6* (allele **4*).[33,70] Similarly, no association was found between TD and the *5HT2A* or *5HT2C* receptor genes.[68] The SNPs in *DRD4* and *COMT* genes may serve as markers to identify the susceptible individuals to TD induced by antipsychotic drugs.

Similar to these findings significant association was found between *DRD4* gene polymorphisms and TD in Italian schizophrenic patients but not in Israeli patients.[71,72] Contrast to the findings in Indian patients no association between *COMT* gene polymorphisms and TD was found in Chinese and Japanese.[73,74] Most consistent association was found between Ser9Gly polymorphism of *DRD3* gene and TD in schizophrenia patients of China, Israel and South Korea.[75-77]

Summary

India is the second most populous country next only to China. The Indian population is diverse in social, cultural, lingual and dietary habits. For the past four to five years there has been an increase in the number of pharmacogenetic studies in the subcontinent. It is also in the priority areas for funding by agencies like Department of Biotechnology (DBT) and Indian Council of Medical Research (ICMR).

Most pharmacogenetic studies reported from the Indian subcontinent are concerned with the drug metabolizing CYP enzymes, especially in relation to SNPs in the *CYP2C9, CYP2C19* and *CYP2D6* genes. The frequencies of the variant alleles of all these genes in the Indian population are distinct from those seen in subjects from other ethnic groups (Caucasian, Chinese, Japanese, African Americans). Also there are differences in frequency between the north Indian and the south Indian populations. As far as drug metabolizing enzymes are concerned, not only do the people of India form a group distinct from nonIndian groups but there are also differences within India. There is a pressing need for pharmacogenetic studies involving polymorphisms of pharmacological targets in Indian population to get an overall picture of genotype-based drug response.

With so much emphasis given to pharmacogenomic studies, there are important issues which warrant attention in this research area in India. The first is concerned with the evidence that response to drugs is not solely determined by genes, but is also influenced by environmental factors and even more complicated gene-environment interaction. The second is the ethical concern involved in pharmacogenomic research which might lead to patient stratification based on educational, racial or socioeconomic parameters. Funding agencies such as ICMR and DBT have proposed ethical guidelines for conduction of biomedical research including genetic screening in human subjects and they have undertaken a joint exercise to review the current guidelines which is very much essential with the rapid progress in genomic research. With continuing progress in this field in India, people of this country will also be able to enjoy the fruits of safer and better therapy.

References

1. Constable S, Johnson MR, Pirmohamed M. Pharmacogenetics in clinical practice: Considerations for testing. Expert Rev Mol Diagn 2006; 6:193-205.
2. Eichelbaum M, Ingelman-Sundberg M, Evans WE. Pharmacogenomics and individualized drug therapy. Annu Rev Med 2006; 57:119-137.
3. Andersson T, Flockhart DA, Goldstein DB et al. Drug-metabolizing enzymes: Evidence for clinical utility of pharmacogenomic tests. Clin Pharmacol Ther 2005; 78:559-581.
4. Shastry BS. Pharmacogenetics and the concept of individualized medicine. Pharmacogenomics J 2006; 6:16-21.
5. Majumder PP. People of India: Biological diversity and affinities. Evol Anthrop 1998; 6:100-110.
6. Gadgil M, Joshi NV, Shambu Prasad UV et al. Peopling of India. In: Balasubramanian D, Appaji Rao N, eds. The Indian Human Heritage. Hyderabad: Universities Press, 1997:100-29.
7. Majumder PP. Ethnic populations of India as seen from an evolutionary perspective. J Biosci 2001; 26:533-545.
8. Ganapati Mudur. India invests heavily in genomics research. Br Med J 2001; 322:576.
9. Lee CR, Goldstein JA, Pieper JA. Cytochrome P450 2C9 polymorphisms. A comprehensive review of the in-vitro and human data. Pharmacogenetics 2002; 12:251-263.
10. Rettie AE, Wienkers LC, Gonzalez FJ et al. Impaired (S)-warfarin metabolism catalysed by the R144C allelic variant of CYP2C9. Pharmacogenetics 1994; 4:39-42.
11. Haining RL, Hunter AP, Veronese ME et al. Allelic variants of human cytochrome P450 2C9: Baculovirus-mediated expression, purification, structural characterization, substrate stereoselectivity, and prochiral selectivity of the wild-type and I359L mutant forms. Arch Biochem Biophys 1996; 333:447-458.

12. Adithan C, Gerard N, Vasu S et al. Allele and genotype frequency of CYP2C9 in Tamilnadu population. Eur J Clin Pharmacol 2003; 59:707-709.
13. Jose R, Chandrasekaran A, Sam SS et al. CYP2C9 and CYP2C19 genetic polymorphisms: Frequencies in the south Indian population. Fundam Clin Pharmacol 2005; 19:101-105.
14. Seng KC, Gin GG, Sangkar JV et al. Frequency of cytochrome P450 2C9 CYP2C9 alleles in three ethnic groups in Malaysia. As Pac J Mol Biol Biotechnol 2003; 11:83-91.
15. Xie HG, Prasad HC, Kim RB et al. CYP2C9 allelic variants: Ethnic distribution and functional significance. Adv Drug Deliv Rev 2002; 54:1257-1270.
16. Rosemary J, Surendiran A, Rajan S et al. Influence of the CYP2C9 and CYP2C19 polymorphisms on phenytoin hydroxylation in healthy individuals from south India. Indian J Med Res 2006; 123:665-670.
17. Doshi BS, Kulkarni RD, Chauhan BL et al. Frequency of impaired mephenytoin 4'-hydroxylation in an Indian population. Br J Clin Pharmacol 1990; 30:779-780.
18. Lamba JK, Dhiman RK, Kohli KK. Genetic polymorphism of the hepatic cytochrome P450 2C19 in north Indian subjects. Clin Pharmacol Ther 1998; 63:422-427.
19. Lamba JK, Dhiman RK, Kohli KK. CYP2C19 genetic mutations in North Indians. Clin Pharmacol Ther 2000; 68:328-335.
20. Lamba JK, Dhiman RK, Singh R et al. Correlation between omeprazole hydroxylase and CYP2C19 genotype in North Indians. Eur J Clin Pharmacol 2001; 57:649-652.
21. Adithan C, Gerard N, Vasu S et al. Allele and genotype frequency of CYP2C19 in a Tamilian population. Br J Clin Pharmacol 2003; 56:331-333.
22. Rosemary J, Adithan C, Padmaja N et al. The effect of the CYP2C19 genotype on the hydroxylation index of omeprazole in South Indians. Eur J Clin Pharmacol 2005; 61:19-23.
23. Abraham BK, Adithan C. Genetic polymorphism of CYP2D6. Ind J Pharmacol 2001; 33:147-169.
24. Idle JR, Smith RL. The debrisoquine hydroxylation gene: A gene of multiple consequences. In: Lemberger L, Reidenberg MW, Washington DC, eds. Am Soc Pharmacol Exp Ther. Proceedings of the Second World Conference on Clinical Pharmacology and Therapeutics, 1984; 148-164, (Ref Type: Conference Proceeding).
25. Lamba V, Lamba JK, Dilawari JB et al. Genetic polymorphism of CYP2D6 in North Indian subjects. Eur J Clin Pharmacol 1998; 54:787-791.
26. Abraham BK, Adithan C, Kiran PU et al. Genetic polymorphism of CYP2D6 in Karnataka and Andhra Pradesh population in India. Acta Pharmacol Sin 2000; 21:494-498.
27. Marandi T, Dahl ML, Rago L et al. Debrisoquine and S-mephenytoin hydroxylation polymorphisms in a Russian population living in Estonia. Eur J Clin Pharmacol 1997; 53:257-260.
28. Abraham BK, Adithan C, Shashindran CH et al. Genetic polymorphism of CYP2D6 in a keralite South India population. Br J Clin Pharmacol 2000; 49:285-286.
29. Abraham BK, Adithan C, Mohanasundaram J et al. Genetic polymorphism of CYP2D6 in Tamil population. Eur J Clin Pharmacol 2001; 56:849-850.
30. Ingelman-Sundberg M. Genetic polymorphisms of cytochrome P450 2D6 CYP2D6): Clinical consequences, evolutionary aspects and functional diversity. Pharmacogenomics J 2005; 5:6-13.
31. Naveen AT, Adithan C, Soya SS et al. CYP2D6 genetic polymorphism in South Indian populations. Biol Pharm Bull 2006; 29:1655-1658.
32. Buch S, Kotekar A, Kawle D et al. CYP and GST gene loci. Prevalence in the Indian population. Eur J Clin Pharmacol 2001; 57:553-555.
33. Tiwari AK, Deshpande SN, Rao AR et al. Genetic susceptibility to tardive dyskinesia in chronic schizophrenia subjects: III. Lack of association of CYP3A4 and CYP2D6 gene polymorphisms. Schizophr Res 2005; 75:21-26.
34. Ismail R, Teh LK. Genetic polymorphism of CYP2D6: Malaysian Indians have the highest frequency for CYP2D6*4 in Asia. Eur J Clin Pharmacol 2001; 57:617-618.
35. Tanaka E, Terada M, Misawa S. Cytochrome P450 2E1: Its clinical and toxicological role. J Clin Pharm Ther 2000; 25:165-175.
36. Sikdar N, Mahmud SA, Paul RR et al. Polymorphism in CYP1A1 and CYP2E1 genes and susceptibility to leukoplakia in Indian tobacco users. Cancer Lett 2003; 195:33-42.
37. Soya SS, Padmaja N, Adithan C. Genetic polymorphisms of CYP2E1 and GSTP1 in a South Indian population—comparison with North Indians, Caucasians and Chinese. Asian Pac J Cancer Prev 2005; 6:315-319.
38. Buch S, Kotekar A, Kawle D et al. Polymorphisms at CYP and GST gene loci. Prevalence in the Indian population. Eur J Clin Pharmacol 2001; 57:553-555.
39. Lee EJ, Wong JY, Yeoh PN et al. Glutathione S transferase-theta GSTT1) genetic polymorphism among Chinese, Malays and Indians in Singapore. Pharmacogenetics 1995; 5:332-334.

40. Mishra DK, Kumar A, Srivastava DS et al. Allelic variation of GSTT1, GSTM1 and GSTP1 genes in North Indian population. Asian Pac J Cancer Prev 2004; 5:362-365.
41. Naveen AT, Adithan C, Padmaja N et al. Glutathione S-transferase M1 and T1 null genotype distribution in South Indians. Eur J Clin Pharmacol 2004; 60:403-406.
42. Roy B, Dey B, Chakraborty M et al. Frequency of homozygous null mutation at the glutathione-s-transferase M1 locus in some populations of Orissa, India. Anthropol Anz 1998; 56:43-47.
43. Garte S, Gaspari L, Alexandrie AK et al. Metabolic gene polymorphism frequencies in control populations. Cancer Epidemiol Biomarkers Prev 2001; 10:1239-1248.
44. Cho HJ, Lee SY, Ki CS et al. GSTM1, GSTT1 and GSTP1 polymorphisms in the Korean population. J Korean Med Sci 2005; 20:1089-1092.
45. Zhao B, Lee EJ, Wong JY et al. Frequency of mutant CYP1A1, NAT2 and GSTM1 alleles in normal Indians and Malays. Pharmacogenetics 1995; 5:275-280.
46. Czerwinski M, Gibbs JP, Slattery JT. Busulfan conjugation by glutathione S-transferases alpha, mu, and pi. Drug Metab Dispos 1996; 24:1015-1019.
47. Roy B, Chowdhury A, Kundu S et al. Increased risk of antituberculosis drug-induced hepatotoxicity in individuals with glutathione S-transferase M1 'null' mutation. J Gastroenterol Hepatol 2001; 16:1033-1037.
48. Harmer D, Evans DA, Eze LC et al. The relationship between the acetylator and the sparteine hydroxylation polymorphisms. J Med Genet 1986; 23:155-156.
49. Grant DM, Blum M, Demierre A et al. Nucleotide sequence of an intronless gene for a human arylamine N-acetyltransferase related to polymorphic drug acetylation. Nucleic Acids Res 1989; 17:3978.
50. Hickman D, Risch A, Buckle V et al. Chromosomal localization of human genes for arylamine N-acetyltransferase. Biochem J 1994; 297(Pt 3):441-445.
51. Zhao B, Lee EJ, Yeoh PN et al. Detection of mutations and polymorphism of N-acetyltransferase 1 gene in Indian, Malay and Chinese populations. Pharmacogenetics 1998; 8:299-304.
52. Vatsis KP, Weber WW, Bell DA et al. Nomenclature for N-acetyltransferases. Pharmacogenetics 1995; 5:1-17.
53. Anitha A, Banerjee M. Arylamine N-acetyltransferase 2 polymorphism in the ethnic populations of South India. Int J Mol Med 2003; 11:125-131.
54. Srivastava DS, Mittal RD. Genetic polymorphism of the N-acetyltransferase 2 gene, and susceptibility to prostate cancer: A pilot study in north Indian population. BMC Urol 2005; 5:12.
55. Gardiner SJ, Begg EJ. Pharmacogenetics, drug-metabolizing enzymes, and clinical practice. Pharmacol Rev 2006; 58:521-590.
56. Manoharan I, Wieseler S, Layer PG et al. Naturally occurring mutation Leu307Pro of human butyrylcholinesterase in the Vysya community of India. Pharmacogenet Genomics 2006; 16:461-468.
57. Ramesh S, Shanthi P, Krishnan KB et al. Multidrug resistance 1 gene expression in Indian patients with gastric carcinoma. Indian J Gastroenterol 2003; 22:19-21.
58. Gurbuxani S, Singh AL, Raina V et al. Significance of MDR1, MRP1, GSTpi and GSTmu mRNA expression in acute lymphoblastic leukemia in Indian patients. Cancer Lett 2001; 167:73-83.
59. Ralhan R, Swain RK, Agarwal S et al. P-glycoprotein is positively correlated with p53 in human oral premalignant and malignant lesions and is associated with poor prognosis. Int J Cancer 1999; 19(84):80-85.
60. Gurbuxani S, Zhou D, Simonin G et al. Expression of genes implicated in multidrug resistance in acute lymphoblastic leukemia in India. Ann Hematol 1998; 76:195-200.
61. Kesavan R, Soya SS, Adithan C. Allele and genotype frequency of MDR1 C3435T in Tamilian population. Drug Metab Pharmacokinet 2006; (accepted for publication).
62. Balram C, Sharma A, Sivathasan C et al. Frequency of C3435T single nucleotide MDR1 genetic polymorphism in an Asian population: Phenotypic-genotypic correlates. Br J Clin Pharmacol 2003; 56:78-83.
63. Ameyaw MM, Regateiro F, Li T et al. MDR1 pharmacogenetics: Frequency of the C3435T mutation in exon 26 is significantly influenced by ethnicity. Pharmacogenetics 2001; 11:217-221.
64. Kukreti R, Bhatnagar P, Rao C et al. Beta(2)-adrenergic receptor polymorphisms and response to salbutamol among Indian asthmatics*. Pharmacogenomics 2005; 6:399-410.
65. Choudhry S, Ung N, Avila PC et al. Pharmacogenetic differences in response to albuterol between Puerto Ricans and Mexicans with asthma. Am J Respir Crit Care Med 2005; 171:563-570.
66. Kotani Y, Nishimura Y, Maeda H et al. Beta2-adrenergic receptor polymorphisms affect airway responsiveness to salbutamol in asthmatics. J Asthma 1999; 36:583-590.

67. Cho SH, Oh SY, Bahn JW et al. Association between bronchodilating response to short-acting beta-agonist and nonsynonymous single-nucleotide polymorphisms of beta-adrenoceptor gene. Clin Exp Allergy 2005; 35:1162-1167.
68. Deshpande SN, Varma PG, Semwal P et al. II. Serotonin receptor gene polymorphisms and their association with tardive dyskinesia among schizophrenia patients from North India. Psychiatr Genet 2005; 15:157-158.
69. Srivastava V, Varma PG, Prasad S et al. Genetic susceptibility to tardive dyskinesia among schizophrenia subjects: IV. Role of dopaminergic pathway gene polymorphisms. Pharmacogenet Genomics 2006; 16:111-117.
70. Tiwari AK, Deshpande SN, Rao AR et al. Genetic susceptibility to tardive dyskinesia in chronic schizophrenia subjects: I. Association of CYP1A2 gene polymorphism. Pharmacogenomics J 2005; 5:60-69.
71. Lattuada E, Cavallaro R, Serretti A et al. Tardive dyskinesia and DRD2, DRD3, DRD4, 5-HT2A variants in schizophrenia: An association study with repeated assessment. Int J Neuropsychopharmacol 2004; 7:489-493.
72. Segman RH, Goltser T, Heresco-Levy U et al. Association of dopaminergic and serotonergic genes with tardive dyskinesia in patients with chronic schizophrenia. Pharmacogenomics J 2003; 3:277-283.
73. Lai IC, Wang YC, Lin CC et al. Negative association between catechol-O-methyltransferase (COMT) gene Val158Met polymorphism and persistent tardive dyskinesia in schizophrenia. J Neural Transm 2005; 112:1107-1113.
74. Matsumoto C, Shinkai T, Hori H et al. Polymorphisms of dopamine degradation enzyme (COMT and MAO) genes and tardive dyskinesia in patients with schizophrenia. Psychiatry Res 2004; 127:1-7.
75. Chong SA, Tan EC, Tan CH et al. Polymorphisms of dopamine receptors and tardive dyskinesia among Chinese patients with schizophrenia. Am J Med Genet B Neuropsychiatr Genet 2003; 116:51-54.
76. Woo SI, Kim JW, Rha E et al. Association of the Ser9Gly polymorphism in the dopamine D3 receptor gene with tardive dyskinesia in Korean schizophrenics. Psychiatry Clin Neurosci 2002; 56:469-474.
77. Segman R, Neeman T, Heresco-Levy U et al. Genotypic association between the dopamine D3 receptor and tardive dyskinesia in chronic schizophrenia. Mol Psychiatry 1999; 4:247-253.

CHAPTER 9

Pharmacogenetics and Ethnicity:
An Asian Perspective

Su Pin Choo, Suman Lal and Balram Chowbay*

Abstract

The improved understanding and merging of the areas of genomic research and therapeutics have resulted in a rapid acceleration in the area of pharmacogenomics. The clinical implications of this translational research discipline supported by the insights gained so far has further emphasized the importance of ethnicity in determining predictive end points in drug therapy. Singapore provides a unique opportunity to study three ethnic groups, namely the Chinese, Malays and Indians. This chapter aims to review the current understanding of functional genetic polymorphisms in the drug metabolizing enzymes and drug efflux proteins in the multi-ethnic population of Singapore. The roles of particular genetic polymorphisms which are likely to be responsible for interindividual and interethnic pharmacokinetic variations and dose limiting toxicities are also discussed.

Introduction

Pharmacogenetics uses genetic information to predict an individual's drug response[1] and has potential implications on drug development and drug therapy. Fuelled by the information available from the Human Genome Project, the fields of pharmacogenomics and pharmacogenetic research have gained enormous momentum in recent years.

It has become increasingly clear that drug metabolism can significantly differ between ethnic groups and the recommended drug dosage regimens for one population hence cannot be directly extrapolated to other ethnic groups. These differences pose potential problems in clinical practice such as unexpected adverse drug reactions or exacerbations in drug toxicity, lack of predictable response to normal drug dosage leading to therapeutic failure and undesirable drug-drug interactions.[2] There is accumulating evidence that single nucleotide polymorphisms (SNPs) in drug metabolizing enzymes, drug transporters, receptors and other drug targets[3] account for the interindividual and interethnic variations in response to many therapeutic agents by causing functional alterations in the encoded proteins. The multi-racial population in Singapore where 76.8% are Chinese, 13.9% are Malays and 8% are Indians (http://www.singstat.gov.sg/keystats), presents an excellent opportunity for comparative inter ethnic pharmacogenetic studies. In this chapter, we look at how the frequencies of functional SNPs vary between the different ethnic groups in Singapore and examine their clinical implications.

*Corresponding Author: Balram Chowbay—Laboratory of Clinical Pharmacology, Division of Medical Sciences, National Cancer Centre, 11 Hospital Drive, 169610 Singapore. Email: ctebal@nccs.com.sg

Pharmacogenomics in Admixed Populations, edited by Guilherme Suarez-Kurtz.

Phase 1 Drug Metabolising Enzymes

The cytochrome P450 monooxygenase enzymes (CYPs) forms the major enzyme system for phase I drug metabolism and are involved in approximately 80% of oxidative drug metabolism and elimination of approximately 50% of commonly used drugs.[4] There are two major functional classes of the mammalian CYP families: those that are involved in the biosynthesis of steroids, bile acids and fatty acids and those that are involved in the metabolism of xenobiotics. At least 57 CYPs have been identified in humans, but only a relatively small number of the encoded proteins, mainly in CYP1, CYP2 and CYP3 families appear to play significant roles in drug metabolism.[5]

CYP1A1

CYP1A1 and CYP1A2 constitute the two main members in the CYP1A subfamily, the former expressed mainly in lung, placenta and lymphocytes while the latter is a major CYP1A isoform in human liver and constitutes almost 13% of the total hepatic CYP content.[5] The 2.6kb *CYP1A1* gene is located on chromosome 15 near the MPI locus at 15q22-24[6] and contains seven exons. To date, at least fifteen allelic variants have been described in both the upstream and downstream regions of the CYP1A1 gene (http://www.imm.ki.se/CYPalleles/cyp1a1.htm). The **2A* and **2B* variant alleles have a 3801T>C base change in intron 6 and have been associated with a highly inducible phenotype of the enzyme and increased lung cancer risk.[7] The 2455A>G base changes at codon 462 in exon 7 results in altered amino acid (Ile462Val) near the heme binding region and increases CYP1A1 activity two-fold.[8] The 2455A>G polymorphism (*CYP1A1*2B* and **2C*) have been studied in three distinct ethnic groups in Singapore (Table 1).[9] The frequencies of the variant G allele in Chinese (28%) and Malays (31%) were similar to those reported in Japanese subjects (22-24%).[10] Interestingly, the frequency of G allele in the Indian population was only 18% and similar to the frequencies reported in Caucasian populations.[11]

As an aryl hydrocarbon hydroxylase, CYP1A1 can convert polycyclic hydrocarbons to reactive electrophiles that can cause DNA damage, resulting in carcinogenic transformation of cells.[12] CYP1A1 is also involved in bioactivation of the teratogen thalidomide[13] and is presumed to play a critical role in etiology of breast and prostate malignancies.[14] Several studies have examined the relationship between *CYP1A1* variants in relation to the inducibility of the enzyme[15] and risks of developing lung, oesophageal, colon and breast cancers.[16,17] Ethnicity based studies have also shown that CYP1A1 inducibility is higher among Asians and Caucasians when compared to African-Americans.[18]

The *CYP1A1 Msp*I and 2455A>G (Ile462Val) polymorphisms were genotyped and tested for association with lung cancer risk in a Singapore study comprising a total of 126 incident lung cancer cases (of which 87.7% were pathologically confirmed) and 162 age-matched hospital controls.[19] An elevated risk of lung cancer was observed among individuals with the *Msp*I CC (OR = 1.7, 95% CI = 0.9-3.3) and GG genotypes (OR = 2.8, 95%CI = 1.1-7.6). After stratifying by environmental tobacco smoke (ETS) exposure, the risk of lung cancer associated with both polymorphisms was higher among individuals with lower exposure to ETS, compared with those who reported at least weekly exposure. Individuals with the *Msp*I CC genotype

Table 1. Distribution of CYP1A1 genotype and variant alleles, N(%)[9]

Ethnic Groups	*AA*	*AG*	*GG*	*A*	*G*
Chinese	100 (54)	70 (37)	17 (9)	270 (72)	104 (28)
Malays	79 (54)	44 (30)	23 (16)	202 (69)	90 (31)
Indians	95 (68)	37 (27)	7 (5)	227 (82)	51 (18)

showed a two-fold higher risk of lung cancer if they were also null for either *GSTM1* or *T1* (OR = 2.3, 95% CI =1 .0-5.0 and OR = 2.7, 95% CI = 1.1-6.9, respectively, compared to other genotype combinations combined). These findings suggested that *CYP1A1* is a susceptibility gene for lung cancer among nonsmoking Asian women and this association can be influenced by ETS exposure and genetic variation at *GST* genes.

The role of polymorphic variants of the *CYP1A1* gene in affecting the disposition of drugs is not clear due to lack of related studies in Asian subjects. As an extra-hepatic enzyme, CYP1A1 is unlikely to have a major role in the disposition of a variety of clinically important drugs and thus polymorphic variants of this gene may only have a minimal role in influencing the metabolism of these drug substrates.

CYP2C9

The *CYP2C9* isoform is a major component of the CYP2C subfamily and has been specifically mapped to chromosome 10q24.2, spans approximately 55kb in length and encodes a protein consisting of 490 amino acids.[20] Several population based studies[21,22] have been done on polymorphisms in the 5'-flanking region and the promoter sequences of *CYP2C9* gene which contains several transcription factor binding sites including consensus sequences for the glucocorticoid response elements (GREs) and putative binding sites such as TATA box, Barbie box, CAAT box and hepatic nuclear factor-1 (HNF-1).[20]

Polymorphisms in the coding region of *CYP2C9* are well described (http://www.imm.ki.se/CYPalleles/cyp2c9.htm) and exhibit marked interethnic variation in genotype and allelic frequencies.[23] Disruption of CYP2C9 activity by metabolic inhibition or pharmacogenetic variability underlies many of the adverse drug reactions that are associated with the enzyme.[24] The newly discovered genetic variations in the Vitamin K epoxide reductase 1 gene (*VKORC1*) and the gamma glutamyl carboxylase (*GGCX*) gene is now being intensively studied along with the polymorphisms in *CYP2C9* to determine the dosage requirements of warfarin in different ethnic groups.[25-27] In a study conducted among the multi ethnic population of Singapore, *CYP2C9*2* and *VKORC1* coding region variants were rare (<2%), whereas *CYP2C9*3* associated with lower warfarin requirements was less common in Chinese and Malays (7% and 9%, respectively) than in Indians (18%) and could not account for their lower warfarin requirements. *VKORC1 H1* and *H7/H8/H9* haplotypes were associated with lower and higher warfarin requirements, respectively, *VKORC1 H1* haplotype (requiring low warfarin doses) was common in Chinese (87%) and Malays (65%) but uncommon in Indians (12%), whereas H7, H8, and H9 haplotypes (requiring high warfarin doses) were rare in Chinese (9%), intermediate in Malays (30%), and common in Indians (82%). However, the interethnic differences in warfarin requirements became nonsignificant when adjusted for *VKORC1* haplotype.[28]

CYP2D6

The CYP2D6 isoenzyme plays a role in the metabolism of a wide range of drugs which includes psychotropic and cardiovascular drugs. The *CYP2D6* gene is located on chromosome 22.[29] Pharmacogenetic variations results in four major phenotypes: the poor, intermediate, rapid and the ultra-rapid metabolisers. Individuals who have functionally defective *CYP2D6* alleles are at increased risk of developing greater toxicity and interactions when administered CYPD6 drug substrates.

Polymorphic variants of *CYP2D6* are rare in the Asian population. Table 2 summarizes the prevalence of the different *CYP2D6* alleles in Singapore ethnic groups. Among the Chinese[30] *CYP2D6*3* alleles were not detected while *CYP2D6*4, CYP2D6*5* and the replicated allele *CYP2D6*rep* were low (1%, 7%, and 4% respectively) compared to Caucasians.[31] The *CYP2D6*10* allele which causes functionally diminished enzyme activity was more frequent in the Chinese and Malay ethnic groups (48% vs 37%, respectively) and lower in the Indian population (22%). The allele frequency of *CYP2D6*10* in the Chinese and Malay populations were comparable to that of Japanese (45%), Korean (51%) and other Chinese (47%) populations.

Table 2. **CYP2D6** ***allelic frequencies, N (%)***[30]

Ethnic Groups	*CYP2D6*rep*	*CYP2D6*3*	*CYP2D6*4*	*CYP2D6*5*	*CYP2D6*10*
Chinese	7 (4)	0 (0)	2 (1)	14 (7)	92 (48)
Malays	6 (8)	-	-	14 (12)	44 (37)
Indians	24 (25)	-	-	8 (8)	20 (22)

The *CYP2D6*10* probably represents a high penetrant allele in the Asian population and may partially account for the lower metabolic capacity of Asian patients to metabolize certain pharmacologic agents such as the anti-depressant and neuroleptic CYP2D6 substrates and hence the need for lower doses in these ethnic groups. The frequency of *CYP2D6*rep* allele in the Indian Singaporean group (25%) was significantly different compared to the 4% and 8% frequencies in Chinese and Malays, respectively. This allele confers ultra-rapid metabolism and often results in therapeutic failures in these subjects as only suboptimal plasma concentrations of drugs can be achieved at normal doses.

CYP3A4

CYP3A subfamily of P450 enzymes, which are the most abundant CYPs in intestinal epithelium and the liver,[32] catalyzes the oxidative-reductive reactions of more than 60% of currently known therapeutic agents. There are four main CYP3A subfamilies and include CYP3A4, CYP3A5, CYP3A7[33] and CYP3A43[34] enzymes.

The *CYP3A4* gene is located on chromosome 7q22.1 and is 27kb long consisting of 13 exons and 12 introns.[35] To date, more than 35 allelic variants of *CYP3A4* have been reported (http://www.imm.ki.se/CYPalleles). In the Singaporean population, four *CYP3A4* allelic variants (*CYP3A4*1B, *4, *5* and **6*) have been investigated among Chinese, Malay and Indian healthy subjects.[36] The **1B* (-392 A>G) polymorphic variant in the 5' regulatory region and **5* (653C>G; Pro218Arg) variant in exon 5 were absent in all three racial groups, similar to Japanese[37] and Chinese subjects.[38] *CYP3A4*4* (352A>G; Ile118Val) variant in exon 4 was found in 2 out of 110 Chinese subjects whereas the *CYP3A4*6* (830-831insA) was detected in 1 of 104 Malays and 1 of 101 Indians.[36] The genetic polymorphisms in *CYP3A4* seem to be more prevalent in Caucasian populations and are unlikely to be the cause of variability in drug disposition in Asians.[39]

CYP3A5

CYP3A5 is the primary extra hepatic CYP3A enzyme that is expressed in a variety of tissues including kidney, lung and leukocytes and accounts for about 7-8% of CYP3A content.[34] The *CYP3A5* gene is located in a cluster on chromosome 7q21-q22.1 and consists of 13 exons.[40] So far, at least 23 allelic variants and identified haplotypes of *CYP3A5* have been reported (http://www.imm.ki.se/CYPalleles). There are 23 additional *CYP3A5* variants with unidentified haplotypes.[41] The reference genotype is *CYP3A5*1*1* which is linked to high levels of CYP3A5 expression[34,42] unlike the *CYP3A5*3* and *CYP3A5*6* alleles. The *CYP3A5*2* (27289C>A) allele contains a point mutation resulting in a Thr398Asn change in 5-10% of Caucasians.[43] The *CYP3A5*3* alleles contain 10 alleles designated as **3A* to **3J* with an 6986A>G mutation in intron 3[34] or other changes.[41] The *CYP3A5*3* allele contains a splice variant and encodes a truncated nonfunctional protein. Homozygosity for the *CYP3A5*3* allele is common in several ethnic populations including the Caucasians,[34] African-Americans,[34] Japanese[44] and Chinese.[45] The *CYP3A5*4* allele contains a mutation of 14763A>G and Gln200Arg change and the *CYP3A5*5* allele has a 12952T>C change at intron 5 splicing donor site.[45] Studies have shown that the *CYP3A5*6* and *CYP3A5*7* alleles are infrequent among populations.

***Table 3. Distribution of* CYP3A5 *genotypes and variant alleles (%)*[47]**

Ethnic Groups	*1/*1	*1/*3	*1/*6	*3/*3	*6/*6	*1	*3	*6
Chinese	8.3	32.4	0	59.3	0	25	76	0
Malays	10.2	57.1	0	32.7	0	39	61	0
Indians	12.2	56.7	0	31.1	0	41	59	0

In a pharmacogenetic study of healthy Chinese (N = 108), Malays (N = 98) and Indians (N = 90) in Singapore[47] (Table 3), the frequency of the *CYP3A5* reference genotype was similar in each of the three ethnic groups (8.3% to 12.2%). However, the frequency of *CYP3A5*3/*3* genotype was approximately 60% in the Chinese healthy population, almost two-fold higher compared to the Malay (32.7%) and the Indian (31.1%) populations. There was a statistically significant ($p < 0.001$) difference between the frequencies of *CYP3A5*1* and *CYP3A*3* alleles in the Chinese compared with the Malays and Indians. The *CYP3A5*6* allele which is present in approximately 7.5% of African-Americans, was absent in the Chinese, Malay and Indian populations and similar to published reports in Caucasian[34] and Japanese[48] populations. The higher frequency of the *CYP3A5*1* reference allele in the Malay and Indian populations, which is similar to African-Americans, suggest that they are likely to have a higher expression of the CYP3A5 protein than the Chinese, Japanese and Caucasians.

Phase II Drug Metabolising Enzymes

Phase II reactions are conjugation reactions involved in further detoxification of metabolites formed in Phase I reactions. The major Phase II enzymes in humans are glutathione S-transferases (GSTs),[49] N-acetyltransferases (NATs),[50] uridine diphosphate glucuronosyltransferases (UGTs), sulfotransferases[51] and methyltransferases[52] and are abundant in both hepatic and extrahepatic tissues. Apart from catalyzing the formation of reactive metabolites of drugs, resulting in drug toxicity, there is evidence that the induction of Phase II detoxifying enzymes can modulate the threshold for chemical carcinogenesis and increase cellular resistance to carcinogen exposure.[53] Allelic variants in genes encoding Phase II metabolizing enzymes can cause differences in enzyme activities, and lead to partial or total absence of an enzyme. There are considerable variations in genetic polymorphisms of these enzymes between different ethnic groups.[54]

Glutathione S-Transferase

Glutathione S-transferases (GSTs) are involved in conjugation of a large number of hydrophobic and electrophilic compounds by catalyzing a variety of reduced glutathione-dependent reactions. The cytosolic GSTs in man are divided into 4 main classes: alpha (GSTA1, GSTA2), mu (GST1, GSTM1), pi (GST3, GSTP) and theta (GSTT).[49] Five mu (GSTM1-5) isozymes have been mapped on chromosome 1p13.[55] GSTM1 is abundantly expressed in hepatocytes, stomach, brain and other tissues. *GSTM1*0* represents an allele with a deleted mutation at exon 5. Subjects who are homozygous for this allele (*GSTM1* null genotype) lack expression of GSTM1 protein.[56] *GSTM1*A* and *GSTM1*B* alleles differ by a single base at exon 7 and encode monomers with similar catalytic activities that form active homo and heterodimeric enzymes.[55] Two liver specific theta class enzymes, GSTT1-1 and GSTT2-2 exists and the common null polymorphisms have been identified at the GSTT1 locus. The genotype of *GSTM1* and *GSTT1* are not linked as individuals with *GSTM1-null* are not necessarily *GSTT1-null* and vice versa.[57]

Previous studies have shown that between 40% and 54% of Caucasians[58,59] and 46 and 47% of Japanese[60] harbour the *GSTM1*0* genotype. In a Singapore study, only 33.1% of

Table 4. Distribution of GSTM1 *genotypes and allelic frequencies (%)*[61]

Ethnic groups	*0/*0 (null)	*0/A or *0/B	*0	*A or *B	No: of Alleles
Chinese	63.1	36.9	79	21	236
Malays	61.6	38.4	79	21	180
Indians	33.1	66.9	57	43	92

Indians had the homozygous *GSTM1*0* genotype compared to Malays (61.6%) and Chinese (63.1%), ($p < 0.001$) (Table 4)[61] and was similar to the results reported among Hong Kong Chinese (67%).[62] The reference allele *GSTM1*A* was present in at least 37% of Chinese, while 10% carried at least one *GSTM1*B* allele. Among the Malays and Indians, 7.5% and 7.2% respectively carried at least one *GSTM1*B* allele similar to results from Caucasians.[63] The estimated allelic frequencies of *GSTM1*0, GSTM1*A* and *GSTM1*B* alleles in Chinese, Malay and Indian subjects were 79%, 16% and 5%; 79%, 17% and 4%; and 57%, 39% and 4%, respectively. The Chinese and Malays had similar *GSTM1* genotype frequencies that were different from the Indians ($p < 0.001$).

Several studies have linked *GSTM1* polymorphisms to the risk of developing lung cancer.[16,64,65] In a study conducted among Chinese women in Singapore, increased intake of dietary isothiocyanates reduced lung cancer risk in individuals with *GSTM1*0* genotype but not in those with detectable *GSTM1*.[66] A similar association was found in individuals with *GSTM1* and *T1* null genotypes and risk of colorectal cancer.[67] In an unpublished study by the same group, *GSTM1* and *GSTT1* did not seem to play a major role in nasopharyngeal carcinoma among Singaporean patients. These results were consistent with those of a Taiwanese group[68] which did not find any association between *GSTM1*, *GSTT1* and *GSTP1* polymorphisms with NPC risks. *GST* polymorphisms have also been associated with chemotherapy drug toxicity, efficacy and resistance[49] and thus increased risk of cancer relapse.[69]

*N-Acetyltransferases (*NAT1 *and* NAT2*)*

N-acetylation by hepatic arylamine N-acetyltransferases (NAT) is an important route of biotransformation for a large number of hydrazine and arylamine drugs including several chemical carcinogens.[70] The *NAT1* and *NAT2* genes responsible for N-acetyltranferase activity are located on chromosome 8p22,[71] a region often recognized for loss of heterozygosity in human tumors. The *NAT3* gene has multiple deleterious mutations and does not encode a functional NAT protein and most likely represents a pseudogene.[72] *NAT1* derives its entire transcript from a single exon while *NAT2* mRNA is derived from both the protein-coding exon and a second noncoding exon of 100bp located about 8kb upstream of the translation start site.[72] Both *NAT1* and *NAT2* polymorphisms have been associated with bladder,[73] breast,[74] gastric[75] and prostate[76] malignancies, the *NAT1*10* allele being particularly associated with advanced gastric cancer.[75] The metabolic activation of many N-hydroxy heterocyclic amine carcinogens is catalyzed to a greater extent by human NAT2 than NAT1, but tissue-specific expression is also likely to be important in determining the effect of the NAT1 and NAT2 acetylation polymorphisms in relation to risk of carcinogenesis. Differences in *NAT* polymorphisms have also been implicated in various noncancer conditions like Parkinsons disease,[77] spina bifida,[78] and allergies.[79]

The *NAT2*4* allele is considered as the reference allele for *NAT2* since missense and silent substitutions are absent in this group. Phenotypically, individuals can be divided into rapid, intermediate or slow acetylators for NAT2. Those alleles with the 191G>A, 341T>C, 434A>C, 590G>A and/or 857G>A missense substitutions are associated with slow acetylator phenotype(s).[70] Human *NAT2*5, *6, *14* and **17* yield variable reductions in catalytic activity

***Table 5. Distribution of* NAT1 *alleles (%)*[83]**

Ethnic Groups	*3	*4	*10	*11
Chinese	33	35	30	2
Malays	29	30	39	2
Indians	30	51	17	2

associated with low acetylator phenotype, whereas human *NAT2*12* and **13* clusters catalyze N-, O- and NO-acetyltranferase activities at levels associated with rapid acetylator *NAT2*4*[70] The *NAT2*5A*, **6A* and **7A* alleles account for more than 95% of slow acetylators in Caucasians.[80] *NAT1*4* is the reference allele for NAT1 as it is reported as the most frequent in most ethnic groups studied.[81] The genotypes containing *NAT1*4*, *NAT1*3* and *NAT1*11* alleles are slow acetylators while those with *NAT1*10* allele are considered rapid acetylators.[82] A comprehensive consensus nomenclature of the currently known *NAT1* and *NAT2* alleles along with the corresponding nucleotide and amino acid changes are available at http://www.louisville.edu/medschool/pharmacology/NAT.html.

The proportion of rapid and slow acetylators has been reported to vary between ethnic groups. Unlike Caucasian populations[82] where the frequency of the reference *NAT1*4* allele is high (77%), Asians reportedly have a much lower frequency at 51% for the Indians, 30% for the Malays and 35% for the Chinese (Table 5). *NAT1*10* allele which is considered a fast acetylator, is more common among Chinese (30%) and Malays (39%) compared with Indians (17%)[83] who had frequencies similar to that reported in Caucasians (15%).[82] The frequency of slow acetylator alleles *NAT2*4*, **5A*, **6A* and **7A* among Singaporean Chinese was reported to be similar to that of Hong Kong Chinese[84] at 51%, 7.5%, 32% and 10%.[61] Among the Malays and Indians studied, the observed frequencies of the reference type *NAT2*4*, **5A*, **6A* and **7A* alleles were 41%, 12%, 38% and 9% and 44%, 20%, 32% and 4%, respectively.[9] The frequency of *NAT2*7A* allele (often referred to as the "oriental mutation") in Indians (4.7%) was lower than both Malay (9.3%) and Chinese (10.2%) ethnic groups but higher that that in Caucasians (0.7%).[84]

There is a predominance of rapid acetylators over slow acetylators among Singaporeans, similar to Thailand[85] and Japanese[86] but unlike Caucasian,[87] Iranian,[88] Turkish,[89] Arab[90] and Russian[91] populations. The slow-acetylator allele and genotype frequencies are significantly higher among Indians (38%) and Malays (43%) compared to only 23% slow acetylators in the Chinese population. These frequencies are lower than in Caucasians (58-65%),[84] while more similar to other Asian populations like the Japanese (9-20%). The Singaporean Indians had a lower frequency of slow acetylators compared to their Indian counterparts in India who had a frequency of 74%.[92] The frequency of slow acetylators among Singaporean Chinese was similar to that reported in Chinese from mainland China (16.7%).[93]

N-acetylation reactions are important in the activation and de-activation of numerous arylamine carcinogens and hydrazine drugs which can affect individual development of cancer and other diseases.[70] Exposure to environmental tobacco smoke was associated with an increased risk for colorectal cancer among *NAT2* fast acetylators. Frequent red meat consumption significantly increased colorectal cancer risk for all *NAT2* fast acetylators or carriers of the *NAT1*10* allele but not among those with "slow" *NAT1* and *NAT2* genotypes.[94] In a study done in Singapore[95] (Table 6) on 216 colorectal carcinoma patients and 187 normal individuals of Chinese descent, there was no difference in the distribution of genotypes associated with rapid acetylation when comparing colorectal carcinoma patients with control individuals. However, the frequencies of the slow acetylators *NAT2*4*, **5A*, **6A* and **7A* alleles (51%, 7%, 32% and 10%, respectively) in control individuals were significantly different from those in

***Table 6. Distribution of* NAT2 *allelic frequencies (%)*[95]**

Ethnic Groups	*4 (reference)	*5	*6	*7	No. of Alleles
Chinese	51	7	32	10	374
Malays	41	12	38	9	292
Indians	44	12	38	9	278

patients (49%, 6%, 26% and 19%, respectively, $p < 0.01$). A significant increase in the frequency of colorectal cancer in patients who were compound heterozygotes for *NAT2*7A* and a variant non*NAT2*7A* allele has been observed.

There have been conflicting results on the association between acetylator status and lung cancer risks. *NAT2* rapid acetylator genotypes, especially the *NAT2*4/*4* genotype, have been linked to increased risk of lung cancer[96] especially among nonsmokers.[97] However, some studies have shown increased lung cancer risk with slow acetylation[98] while others have shown a lack of association.[99] Among 294 Singaporean Chinese women (153 with lung cancer and 141 age-matched controls), of which 217 were nonsmokers, the risk for lung cancer was higher in nonsmokers with slow acetylator genotype and this was independent of age.[100] When divided into histological subtypes, there was a 2-fold increase in risk of lung adenocarcinoma among slow acetylators. The frequency of *NAT2*7A* allele was higher among cancer cases at 26.5% compared with 19.1% of controls. There was no association between acetylator status and lung cancer among current or ex-smokers.

*Uridine Diphosphate Glucuronosyltransferase 1A1 (*UGT1A1*)*

Mammalian UGTs are part of a gene superfamily consisting of enzymes that catalyse the addition of UDP-glucuronic acid to substrates such as steroids, bile acids, bilirubin, dietary constituents and xenobiotics, which include drugs, environmental toxicants and carcinogens.[101] Except for bilirubin which is solely metabolized by UGT1A1 isoform, the UGT isoforms generally have nonspecific substrate specificities for both exogenous and endogenous compounds. The UGT family in humans is divided into 2 major classes, UGT1 and UGT2.[102] *UGT1A* is located on chromosome 2q.37 and encodes UGT1A1, UGT1A3, UGT1A4, UGT1A5, UGT1A6, UGT1A7, UGT1A8, UGT1A9 and UGT1A10 proteins. The *UGT1A* locus spans 160kb and contains exon 1 complex and common exons 2-5.[103] The exon 1 complex comprises at least 13 unique exon 1's each preceded by its own promoter region and encoding a unique UGT isoform. The mRNA encoding each UGT isoform is formed by the fusion of specific exon 1 to the common exons 2-5.[102] Thus, gene mutations in the exon 1 complex or promoter region may only affect the unique isoform involved, while mutations in exons 2-5 can result in changes in activity and/or expression of additional isoforms. The UGT1 isozymes are expressed in the liver as well as extra-hepatic sites.[104]

To date, thirty-three mutant *UGT1A1* alleles have been reported, of which 9 occur in exon 1 complex and 22 in exons 2 to 5.[102] Genetic polymorphisms in the promoter region of the *UGT1A1* gene are due to TA (thymine adenine) repeats in the TATA-box upstream of *UGT1A1* gene. The reference allele, *UGT1A1*1* has six TA repeats. The presence of seven TA repeats (*UGT1A1*28*) is associated with reduced UGT1A1 expression and is responsible for higher plasma levels of unconjugated bilirubin.[105,106] Asians have been reported to have lower frequencies (10-16.8%)[107] of *UGT1A1*28* allele compared to Caucasian and African-American populations (38.7-42.6%). Our study of 266 individuals among the Singaporean ethnic groups found that *UGT1A1*28* variant frequency was highest among Indians (35.1%) compared to only 16.2% in Chinese and 18.8% in Malays (Table 7).[108] These findings are in contrast to that reported by other investigators in Asian populations. In a Taiwanese study involving 290

***Table 7. Distribution of* UGT1A1 *genotype and allele frequencies (%)*[108]**

Ethnic Groups	6/6	6/7	7/7	6s	7s
Chinese	71	27	2	84	16
Malays	67	29	4	81	19
Indians	43	44	13	65	35

subjects, no *UGT1A1*28* variant was found.[109] The *UGT1A1*28* variant was not found among Asians in a study by Lampe et al[110]and was rare among Japanese.[111] The frequency of *UGT1A1*28* variant among the Malay and Chinese populations were similar to that reported in black Africans (10-25%),[106] while the frequency of *UGT1A1*28* genotype in Singaporean Indian population was very similar to those reported for Caucasians.[106] The mean steady state bilirubin level was highest among individuals with *UGT1A1*28* genotype and the correlation between promoter length polymorphism and serum bilirubin levels was consistent in all three ethnic groups.[108] The frequencies of 5 (*UGT1A1*36*) or 8 (*UGT1A1*37*) TA repeats are very rare in Orientals. Mutations in the coding region of *UGT1A1* has been postulated to be responsible for hyperbilirubinemia as seen in the Taiwanese[109] population and Japanese patients with Gilbert's syndrome.[112]

The UGTs play an important role in the detoxification of drugs and its metabolites and hence modulate the pharmacodynamic effects of several drugs. Irinotecan is a topoisomerase I inhibitor widely used in the treatment of metastatic colorectal cancer and other malignancies and both hepatic (UGT1A1 and UGT1A9) and extrahepatic (UGT1A7) enzymes are involved in the glucuronidation of its active metabolite, SN-38. Among the UGT enzymes involved in the glucuronidation of SN-38, UGT1A1 has the highest enzymatic capacity.. Genetic polymorphisms in the *UGT1A1* gene have been associated with modulatory effects on disposition and toxicity of this drug.[113] UGT1A1 inactivates SN-38 into the more polar SN-38 glucuronide, which is then eliminated in bile and urine. The presence of $(TA)_7$ repeats (*UGT1A1*28*) in the TATA box of the *UGT1A1* gene is associated with decreased glucuronidation capacity and cancer patients homozygous for the *UGT1A1*28* allele have been shown to be at increased risk of irinotecan-induced severe diarrhoea and neutropenia. Owing to the small frequency of occurrence of the *UGT1A1*28* allele in Chinese and Malay populations, the incidence of irinotecan-induced diarrhoea is rare in these populations.[108,113] Recent reports have highlighted the importance of the *UGT1A1*6* allele as a genotypic biomarker for severe neutropenia in Asians. The frequency of the *UGT1A1*6* allele is reported to be higher in Asian patients of Japanese and Korean origin (~20%).[114,115] Unpublished results from our laboratory in cancer patients receiving irinotecan has revealed that the homozygous *UGT1A1*6* variant was present in approximately 1% of our cancer patients who were mainly of Chinese origin. These results suggest that frequencies of marker SNPs may differ between populations of similar ethnic origin but residing in different geographic locations.

*Thiopurine Methyltransferase (*TPMT*)*

Drugs such as azathiopurine, 6-mercaptopurine and 6-thioguanine, are inactive prodrugs that requires multistep metabolic activation to active thioguanine nucleotides (TGN) that exert cytotoxic effects required in the treatment of leukemia, autoimmune diseases and organ transplants. Thiopurine methyltransferase (TPMT) catalyzes the S-methylation of these drugs to inactive metabolites,[116] thus reducing the intracellular concentration of TGN.[117] TPMT activity is inherited as an autosomal codominant trait and the *TPMT* gene is localized on chromosome 6p22.3 and is encoded by a 34kb gene consisting of 10 exons and nine introns which is highly polymorphic between different ethnic groups.[118] TPMT activity in Caucasians

Table 8. Distribution of TPMT *allelic frequencies (%)*[126]

Ethnic Groups	*3A	*3C	*6
Chinese	0	3	0
Malays	0	2.3	0.3
Indians	0.5	0.8	0

exhibit a trimodal distribution (89-94% with high, 6-11% with intermediate and 0.33% with low) enzyme activity[119] while the Chinese have been found to exhibit an unimodal distribution.[120] *TPMT*1* is the reference allele and the common mutant alleles are *TPMT*2*(238G>C),[121] *TPMT*3A*(460G>A,719A>G),[122] *TPMT*3B*(460G>A) and *TPMT*3C*(719A>G).[123] The most prevalent variant allele among Caucasians is *TPMT*3A*[124] while *TPMT*3C* is predominant in Chinese and Africans.[124,125]

In a study on cord blood of 600 newborn babies in Singapore (200 Chinese, 200 Malays and 200 Indians),[126] the *TPMT*3C* variant was the most common variant allele, found in all 3 ethnic groups with frequencies of 3%, 2.3% and 0.8% in Chinese, Malays and Indians respectively (Table 8). The *TPMT*3A* allele was detected only in the Indian group at a low frequency of 0.5% and the *TPMT*6* variant was found in a Malay subject. Among 100 children with acute lymphoblastic leukemia, one Chinese patient was heterozygous for the *TPMT*3A* variant and showed sensitivity to 6-mercaptopurine during maintenance therapy.[126] Three Chinese patients and one Malay patient were heterozygous for the *TPMT*3C* variant but mercaptopurine sensitivity could be validated in only one. Acute lymphoblastic leukemia patients with at least one mutant *TPMT* allele tend to have an improved response to mercaptopurine therapy and better chances of treatment efficacy, compared with patients who were homozygous for the reference *TPMT* alleles[127] but this may also mean an increased risk of developing a thiopurine-related second tumor because of reduced *TPMT* expression.

Drug Transporters

Drug transport is often the result of the concerted action of different pumps located in the basolateral and apical membranes of epithelial cells and constitute key determinants of drug disposition and response.[128] The expression pattern of different drug transporters vary in different anatomic regions of the human body and the differential uptake and efflux activities of these proteins account for the preferential transfer of drugs between different compartments in the body. The drug transporters may be classified as either primary or secondary active transporters. The primary transporters include mainly the ATP-binding cassette (ABC) transporters that utilize the ATP hydrolysis as the driving force for solute transport. The secondary transporters are driven by an exchange of intracellular and extracellular ions and include the solute carrier family of transporters (SLC).

Efflux Carrier Systems

Multidrug Resistance Gene (*MDR1;ABCB1*)

The *MDR1* gene product, P-glycoprotein (ABCB1) is a membrane protein which functions as an ATP-dependent exporter of xenobiotics from cells. Its expression level varies in different tissues, the intestinal levels of P-glycoprotein have been shown to vary significantly between individuals and results in wide inter-individual variation in pharmacokinetics of several orally administered drugs.[129] MDR1 is encoded by the *ABCB1* gene, which extends over more than 100kb on human chromosome 7q21 containing a cDNA sequence of 4.7 kb with 28 exons that codes for 1280 amino acids.[130] To date, at least 105 *ABCB1* gene variants have been reported, the majority of which being either intronic or noncoding.[131] Most studies on

***Table 9. Distribution of* MDR1 *allele and genotype frequencies (%)*[36]**

Exon	Ethnic Groups	Allele Frequency			Genotype Frequency					
12		*C*	*T*		*CC*	*CT*	*TT*			
	Chinese	28.1	71.9		8.3	39.6	52.1			
	Malays	34.2	65.8		12.0	44.6	43.5			
	Indians	32.8	67.2		13.8	37.9	48.3			
21		*A*	*G*	*T*	*AA*	*AG*	*AT*	*GG*	*GT*	*TT*
	Chinese	12.5	37.5	50.0	1.0	8.3	14.6	16.7	33.3	26.0
	Malays	3.3	52.7	44.0	0.0	2.2	4.4	28.3	46.7	18.5
	Indians	6.9	33.3	59.8	0.0	8.1	5.8	13.8	31.0	41.4
26		*C*	*T*	*CC*	*CT*	*TT*				
	Chinese	46.9	53.1	25.0	43.8	31.3				
	Malays	48.9	51.1	27.2	43.5	29.4				
	Indians	36.8	63.2	18.4	36.8	44.8				

Adapted from: Chowbay B et al. Pharmacogenetics 2003; 13(2):89-95, with permission from Lippincott Williams & Wilkins.[36]

genotype-related function of *ABCB1* have been performed on three high frequency variants in exons 12 (1236C>T), 21 (2677G>T/A) and 26 (3435C>T) which have been genotyped in Singaporean Asians (Table 9).[36] The frequencies of the homozygous CC and TT variants at exon 12 (1236C>T) were similar in the Indian, Chinese and Malay populations. The distributions of the reference C allele in exon 12 for the three populations were similar to that of Japanese (38.5%)[132] and Caucasian (41%)[133] populations but higher than that of African-Americans (15%).[133] When compared to other ethnic groups, the frequency of the variant TT genotype at exon 12 was higher in Singaporean Asians (43.5% to 52.1%) and the Japanese (61.5%) populations compared with those found in the Caucasians (13.3%).[134]

The triallelic SNP at exon 21 (2677G>T/A) was also highly polymorphic in the Asian population.[36] Of the three allelic variants, the A allele was more common in the Chinese, the G allele in the Malays and the T allele in the Indians. Although the A allele was present at low frequency among Asians, it seemed to be more common in the Singaporean population (3.3-12.5%) compared with the Caucasian and African populations in whom the allele was not detected.[135] The frequency of G and T alleles were similar to that in the Caucasians (46%) but different from African-Americans[133] where their frequencies were 93.5% and 6.5%, respectively. The frequency of TT was much lower in the Malays (19%) than in the Indians (41%).[36,135] With regards to the wobble SNP in exon 26 (3435C>T), there were no statistically significant differences in the distribution of CC and TT genotypes among the Chinese, Malays and Indians.[135]

Linkage analysis revealed strong association between the SNPs at the three studied loci in the Asian population. In each ethnic group, the allele C at exon 12 was positively associated with alleles A and G at exon 21 ($p < 0.01$). Similarly, alleles A and G at exon 21 were positively associated with allele C at exon 26 ($p < 0.01$). Twelve haplotypes were observed in the Chinese, 8 in the Malays and 10 in the Indians. T-T-T was the most common haplotype in all groups (50% in Indians, 41% in Chinese and 37% in Malays). The other three major haplotypes were C-A-C, C-G-C and T-G-C which accounted for 80.3%, 79.7% and 77.3% respectively of the haplotypes in the Chinese, Malays and Indians. The C-G-C and T-G-C haplotypes were predominant in the Malay population at 39% compared to 29% in the Chinese and 21.5% in the Indians. Two

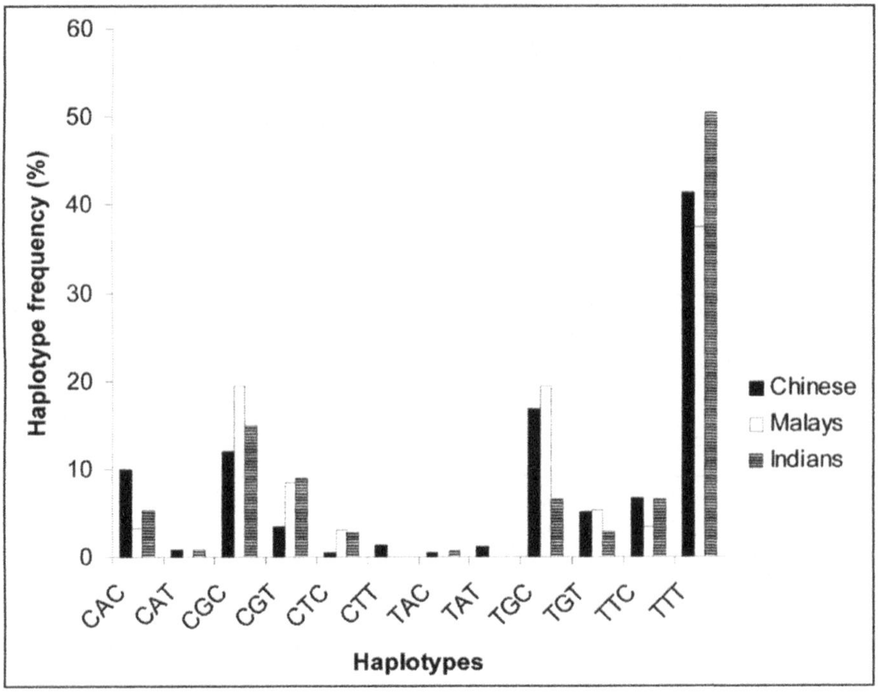

Figure 1. *ABCB1* haplotypes based on high frequency SNPs in exons 12, 21 and 26 in Asian population. Adapted from: Chowbay B et al. Pharmacogenetics 2003; 13(2):89-95, with permission from Lippincott Williams & Wilkins.[36]

low frequency haplotypes were also reported. They were C-A-T haplotype in the Chinese and Indian populations and the T-A-T haplotype exclusively in the Chinese population. The frequency of the homozygous mutant genotype at all three loci (TT-TT-TT) was highest in the Indians (31%) compared with 19% in the Chinese and 15% in the Malays. The frequencies of haplotypes based on high frequency SNPs in exons 12, 21 and 26 are shown in Figure 1.

Recent studies have highlighted the importance of *ABCB1* haplotypes in affecting the disposition of MDR1 substrates. Among Singaporean heart transplant patients, the haplotypic association of these SNPs was found to influence cyclosporineA exposure levels (Table 10). Patients homozygous for the CC-GG-CC genotypes had lower peak and trough cyclosporine concentrations as well as $AUC_{0\text{-}4h}$ and $AUC_{0\text{-}12h}$, while patients carrying genotypes with mutations at all three points (TT-TT-TT) and the T-T-T haplotypes had higher values. Patients with heterozygote genotype (CT-GT-CT) had intermediate values. This phenotypic outcome was most significant in the Indian population suggesting that this ethnic group may require dosages of cyclosporine that are different to that of Chinese and Malay populations. There was no influence of *CYP3A4* SNPS on cyclosporine disposition in this study.[135]

Uptake (Influx) Carrier Systems

Influx transporters are expressed on the basolateral domains of hepatocytes and facilitate uptake of drugs intracellularly prior to elimination into bile, which is mediated by efflux transporters.[136] Among the influx transporter family, members of the organic anion-transporting polypeptides (OATPs) are able to transport a large array of structurally divergent drugs.[136] Amongst them, OATP2B1,[137] OATP1B1[137] and OATP1B3[138] are the major OATPs expressed at the basolateral membrane of human hepatocytes.

Table 10. ABCB1 haplotypes and normal pharmacokinetics (mean ± SD) of cyclosporine in Singaporean heart transplant patients[36]

1236 C>T	2677G> T/A	3435 C>T	N	$AUC_{0\text{-}4h}$ (ug*h/L)	$AUC_{0\text{-}12h}$ (ug*h/L)	C_{max}	C_{min}
CC	*GG*	*CC*	3	2021.8 (421.4)	3799.9 (749.2)	826.7 (270.5)	127.3 (53.8)
CT	*GT*	*CT*	3	2303.3 (692.5)	3995.3 (1397.2)	952.3 (209.1)	141.7 (73.0)
CC	*TT*	*TT*	3	4922.5 (511.0)	4922.5 (511.0)	1218.3 (234.1)	159.3 (13.6)

Adapted from: Chowbay B et al. Pharmacogenetics 2003; 13(2):89-95, with permission from Lippincott Williams & Wilkins.[36]

Organic Anion-Transporting Polypeptide (*OATP1B1*)

OATP1B1 (gene *SLCO1B1*) which is also known as liver specific transporter (LST-1) or OATP2, mediates the uptake of a wide array of chemically divergent compounds and drugs.[139] Variant polymorphisms in the *SLCO1B1* gene among different ethnic groups have been shown to affect the pharmacokinetics of different substrates[140] including pravastatin[139] and irinotecan.[141]

Recent studies using HEK293 cells stably transfected with the reference *SLCO1B1*1a* allele have shown that SN-38, but not irinotecan and SN-38G, to be a substrate of OATP1B1.[141] Inhibitory uptake studies in *Xenopus* oocytes injected with cRNA of *SLCO1B1*1a* in the presence of oestrone-3-sulphate, however, showed that both irinotecan and SN-38 were transported by OATP1B1, although the affinity was greater for SN-38 compared with irinotecan. The authors also showed that the uptake of SN-38 was significantly decreased in *Xenopus* oocytes expressing the **15* variant compared with cells expressing the reference *SLCO1B1* gene. This *in vitro* study provided the first evidence of the involvement of OATP1B1 and its associated genetic variants in the hepatic disposition of irinotecan and SN-38.

In a study on 284 healthy and 71 cancer patients from Singapore,[142] the frequencies of *SLCO1B1 *1a, *1b, *5* and **15* alleles were 24%, 69%, 1% and 7% in the healthy subjects and this was similar to that in the patient group (Table 11). The **1b* alleles was the most common SNP variation among the Chinese, Indians and Malays (56%-79%), similar to reports in African-Americans[140]and Japanese.[144] The frequencies of **5* (0%-2%) and **15* alleles (2%-11%) among the Singaporean ethnic groups were also similar to that in Japanese subjects (0-0.7% and 10-15% respectively). This differed significantly from studies conducted in Caucasian populations where the frequencies of **5* allele and **15* alleles were higher at 8%[140,145] and 14% respectively.[145] There was no difference in haplotype frequency distributions between healthy subjects and cancer patients who were ethnic Chinese and Malays.[143] The **1a/*5* diplotype was rare in the Singaporean subjects, being present only in healthy Indian subjects.

Genotypic-phenotypic correlative studies in cancer patients receiving irinotecan showed that patients with the **15* haplotype had significantly decreased clearance of irinotecan and increased exposure levels of SN-38. The total clearance of irinotecan was approximately 3-fold lower in cancer patients habouring at least one **15* allele compared with patients harboring the reference genotype *(*1a/*1a)* or those with at least one **1b* allele. These in vivo findings are in concordance with previously reported in vitro findings[141] These findings may have therapeutic implications in Asian cancer patients especially among the Chinese and Malay groups who harbor higher frequencies of the **15* haplotype (7% and 13% respectively) compared to the Indians.

Table 11. Distribution of* SLCO1B1 **1a, *1b, *5 and *15 genotypic and allelic frequencies in healthy subjects and cancer patients (%)[142]

Ethnic Groups	No:	*1a/ *1a	*1b/ *1b	*1a/ *1b	*1a /*5	*1b/* 15	*15/* 15	*1a	*1b	*5	*15
Healthy Subjects											
Chinese	94	5.3	47.9	26.6	0	19.1	1.1	19	71	0	0.11
Malays	97	2.1	60.8	19.6	0	16.5	1.0	12	79	0	0.09
Indians	93	15.1	30.1	48.4	3.2	3.2	0	41	56	0.02	0.02
Pooled Healthy Asians	284	7.4	46.5	31.3	1.1	13.0	0.7	24	69	0.01	0.07
Cancer Patients											
Chinese	61	9.8	42.6	32.8	0	14.8	0	26	66	0	7
Malays	8	0	62.5	25	0	0	12.5	13	75	0	13
Indians	1	0	0	100	0	0	0	50	50	0	0
Pooled Cancer Patients	71	8.5	43.7	32.4	0	14.1	1.4		67	0	9

Adapted from: Xiang XQ et al. Pharmacogenet Genomics 16(9):683-691; ©2006 with permission from Lippincott Willams & Wilkins.[142]

Conclusion

Ethnicity specific genotypic differences in drug response have been attributed to sequence variations in the relevant candidate genes of drug metabolism and transport. Haplotype identification and characterization looking at linkage is likely to prove vital in determining the final predictive value and functional significance of these polymorphisms. With increasing knowledge of functional polymorphisms of drug metabolizing enzyme and drug transporter genes, dosage individualization based on a patient's genotype status is becoming a reality. Current recommendations for a variety of drug dosing regimens are primarily optimized for the Caucasian population; the reasons could be partly attributed to the subject characteristics taken into consideration during the developmental pipeline. Our experience over the years has shown that pharmacogenetics of the concerned genes in Asian population differs from that of other ethnic groups, in particular the Caucasian and African populations. In the context of a large amount of research leading to the idea of testing in different clinical populations, the observed variations in pharmacogenetics and its pharmacokinetic correlates are currently evolving as an important influence in determining dose schedules and drug dosing regimens among different ethnic groups. These insights and new findings are also proving valuable in tailoring drug dosage regimens to an individual to maximize therapeutic efficacy and minimize adverse drug reactions observed in these populations.

Acknowledgements

This work was supported in part by grants from the Singapore Cancer Syndicate (PS0023) and NMRC grants 0814 and 0885.

References

1. Ulrich CM, Robien K, McLeod HL. Cancer pharmacogenetics: Polymorphisms, pathways and beyond. Nat Rev Cancer 2003; 3(12):912-920.
2. Wood AJ. Ethnic differences in drug disposition and response. Ther Drug Monit 1998; 20(5):525-526.
3. Evans WE, Johnson JA. Pharmacogenomics: The inherited basis for interindividual differences in drug response. Annu Rev Genomics Hum Genet 2001; 2:9-39.

4. Wilkinson GR. Drug metabolism and variability among patients in drug response. N Engl J Med 2005; 352(21):2211-2221.
5. Rendic S. Summary of information on human CYP enzymes: Human P450 metabolism data. Drug Metab Rev 2002; 34(1-2):83-448.
6. Hildebrand CE, Gonzalez FJ, McBride OW et al. Assignment of the human 2,3,7,8 -tetrachlorodibenzo-p-dioxin-inducible cytochrome P1-450 gene to chromosome 15. Nucleic Acids Res 1985; 13(6):2009-2016.
7. Hayashi SI, Watanabe J, Nakachi K et al. PCR detection of an A/G polymorphism within exon 7 of the CYP1A1 gene. Nucleic Acids Res 1991; 19(17):4797.
8. Crofts F, Cosma GN, Currie D et al. A novel CYP1A1 gene polymorphism in African-Americans. Carcinogenesis 1993; 14(9):1729-1731.
9. Zhao B, Lee EJ, Wong JY et al. Frequency of mutant CYP1A1, NAT2 and GSTM1 alleles in normal Indians and Malays. Pharmacogenetics 1995; 5(5):275-280.
10. Miyoshi Y, Ando A, Hasegawa S et al. Association of genetic polymorphisms in CYP19 and CYP1A1 with the oestrogen receptor-positive breast cancer risk. Eur J Cancer 2003; 39(17):2531-2537.
11 Chang BL, Zheng SL, Isaacs SD et al. Polymorphisms in the CYP1A1 gene are associated with prostate cancer risk. Int J Cancer 2003; 106(3):375-378.
12. Shimada T, Yun CH, Yamazaki H et al. Characterization of human lung microsomal cytochrome P-450 1A1 and its role in the oxidation of chemical carcinogens. Mol Pharmacol 1992; 41(5):856-864.
13. Miyata M, Tamura E, Motoki K et al. Thalidomide-induced suppression of embryo fibroblast proliferation requires CYP1A1-mediated activation. Drug Metab Dispos 2003; 31(4):469-475.
14. Silig Y, Pinarbasi H, Gunes S et al. Polymorphisms of CYP1A1, GSTM1, GSTT1, and prostate cancer risk in Turkish population. Cancer Invest 2006; 24(1):41-45.
15 Smart J, Daly AK. Variation in induced CYP1A1 levels: Relationship to CYP1A1, Ah receptor and GSTM1 polymorphisms. Pharmacogenetics 2000; 10(1):11-24.
16 Chen S, Xue K, Xu L et al. Polymorphisms of the CYP1A1 and GSTM1 genes in relation to individual susceptibility to lung carcinoma in Chinese population. Mutat Res 2001; 458(1-2):41-47.
17. Zhang Y, Wise JP, Holford TR et al. Serum polychlorinated biphenyls, cytochrome P-450 1A1 polymorphisms, and risk of breast cancer in Connecticut women. Am J Epidemiol 2004; 160(12):1177-1183.
18. Cosma G, Crofts F, Currie D et al. Racial differences in restriction fragment length polymorphisms and messenger RNA inducibility of the human CYP1A1 gene. Cancer Epidemiol Biomarkers Prev 1993; 2(1):53-57.
19. Ng DP, Tan KW, Zhao B et al. CYP1A1 polymorphisms and risk of lung cancer in nonsmoking Chinese women: Influence of environmental tobacco smoke exposure and GSTM1/T1 genetic variation. Cancer Causes Control 2005; 16(4):399-405.
20. de Morais SM, Schweikl H, Blaisdell J et al. Gene structure and upstream regulatory regions of human CYP2C9 and CYP2C18. Biochem Biophys Res Commun 1993; 194(1):194-201.
21. Shintani M, Ieiri I, Inoue K et al. Genetic polymorphisms and functional characterization of the 5'-flanking region of the human CYP2C9 gene: In vitro and in vivo studies. Clin Pharmacol Ther 2001; 70(2):175-182.
22. Takahashi H, Ieiri I, Wilkinson GR et al. 5'-Flanking region polymorphisms of CYP2C9 and their relationship to S-warfarin metabolism in white and Japanese patients. Blood 2004; 103(8):3055-3057.
23. Lee CR, Goldstein JA, Pieper JA. Cytochrome P450 2C9 polymorphisms: A comprehensive review of the in-vitro and human data. Pharmacogenetics 2002; 12(3):251-263.
24. Rettie AE, Jones JP. Clinical and toxicological relevance of CYP2C9:drug-drug interactions and pharmacogenetics. Annu Rev Pharmacol Toxicol 2005; 45:477-494.
25. Takahashi H, Wilkinson GR, Nutescu EA et al. Different contributions of polymorphisms in VKORC1 and CYP2C9 to intra- and inter- population differences in maintenance dose of warfarin in Japanese, Caucasians and African-Americans. Pharmacogenet Genomics 2006; 16(2):101-110.
26. Herman D, Peternel P, Stegnar M et al. The influence of sequence variations in factor VII, gamma-glutamyl carboxylase and vitamin K epoxide reductase complex genes on warfarin dose requirement. thromb Haemost 2006; 95(5):782-787.
27. Schalekamp T, Brasse BP, Roijers JF et al. VKORC1 and CYP2C9 genotypes and acenocoumarol antocoagulation status: Interaction between both genotypes affects overanticoagulation. Clin Pharmacol Ther 2006; 80(1):13-22.
28. Lee SC, Ng SS, Oldenburg J et al. Interethnic variability of warfarin maintenance requirement is explained by VKORC1 genotype in an Asian population. Clin Pharmacol Ther 2006; 79(3):197-205.
29. Gough AC, Miles JS, Spurr NK et al. Identification of the primary gene defect at the cytochrome P450 CYP2D locus. Nature 1990; 347(6295):773-776.

30. Wong JY, Seah ES, Lee EJ. Pharmacogenetics: The molecular genetics of CYP2D6 dependent drug metabolism. Ann Acad Med Singapore 2000; 29(3):401-406.
31. Mortimer O, Lindstrom B, Laurell H et al. Dextromethorphan: Polymorphic serum pattern of the O-demethylated and didemethylated metabolites in man. Br J Clin Pharmacol 1989; 27(2):223-227.
32. Shimada T, Yamazaki H, Mimura M et al. Interindividual variations in human liver cytochrome P-450 enzymes involved in the oxidation of drugs, carcinogens and toxic chemicals: Studies with liver microsomes of 30 Japanese and 30 Caucasians. J Pharmacol Exp Ther 1994; 270(1):414-423.
33. Rodriguez-Antona C, Jande M, Rane A et al. Identification and phenotype characterization of two CYP3A haplotypes causing different enzymatic capacity in fetal livers. Clin Pharmacol Ther 2005; 77(4):259-270.
34. Kuehl P, Zhang J, Lin Y et al. Sequence diversity in CYP3A promoters and characterization of the genetic basis of polymorphic CYP3A5 expression. Nat Genet 2001; 27(4):383-391.
35. Hashimoto H, Toide K, Kitamura R et al. Gene structure of CYP3A4, an adult-specific form of cytochrome P450 in human livers, and its transcriptional control. Eur J Biochem 1993; 218(2):585-595.
36. Chowbay B, Cumaraswamy S, Cheung YB et al. Genetic polymorphisms in MDR1 and CYP3A4 genes in Asians and the influence of MDR1 haplotypes on cyclosporin disposition in heart transplant recipients. Pharmacogenetics 2003; 13(2):89-95.
37. Naoe T, Takeyama K, Yokozawa T et al. Analysis of genetic polymorphism in NQO1, GST-M1, GST-T1, and CYP3A4 in 469 Japanese patients with therapy-related leukemia/ myelodysplastic syndrome and de novo acute myeloid leukemia. Clin Cancer Res 2000; 6(10):4091-4095.
38. Hsieh KP, Lin YY, Cheng CL et al. Novel mutations of CYP3A4 in Chinese. Drug Metab Dispos 2001; 29(3):268-273.
39. Dai D, Tang J, Rose R et al. Identification of variants of CYP3A4 and characterization of their abilities to metabolize testosterone and chlorpyrifos. J Pharmacol Exp Ther 2001; 299(3):825-831.
40. Schuetz EG, Relling MV, Kishi S et al. PharmGKB update: II. CYP3A5, cytochrome P450, family 3, subfamily A, polypeptide 5. Pharmacol Rev 2004; 56(2):159.
41. Saeki M, Saito Y, Nakamura T et al. Single nucleotide polymorphisms and haplotype frequencies of CYP3A5 in a Japanese population. Hum Mutat 2003; 21(6):653.
42. Hustert E, Haberl M, Burk O et al. The genetic determinants of the CYP3A5 polymorphism. Pharmacogenetics 2001; 11(9):773-779.
43. Jounaidi Y, Hyrailles V, Gervot L et al. Detection of CYP3A5 allelic variant: A candidate for the polymorphic expression of the protein? Biochem Biophys Res Commun 1996; 221(2):466-470.
44. Hiratsuka M, Takekuma Y, Endo N et al. Allele and genotype frequencies of CYP2B6 and CYP3A5 in the Japanese population. Eur J Clin Pharmacol 2002; 58(6):417-421.
45. Chou FC, Tzeng SJ, Huang JD. Genetic polymorphism of cytochrome P450 3A5 in Chinese. Drug Metab Dispos 2001; 29(9):1205-1209.
46. Shih PS, Huang JD. Pharmacokinetics of midazolam and 1'-hydroxymidazolam in Chinese with different CYP3A5 genotypes. Drug Metab Dispos 2002; 30(12):1491-1496.
47. Balram C, Zhou Q, Cheung YB et al. CYP3A5*3 and *6 single nucleotide polymorphisms in three distinct Asian populations. Eur J Clin Pharmacol 2003; 59(2):123-126
48. Fukuen S, Fukuda T, Maune H et al. Novel detection assay by PCR-RFLP and frequency of the CYP3A5 SNPs, CYP3A5*3 and *6, in a Japanese population. Pharmacogenetics 2002; 12(4):331-334.
49. Hayes JD, Pulford DJ. The glutathione S-transferase supergene family: Regulation of GST and the contribution of the isoenzymes to cancer chemoprotection and drug resistance. Crit Rev Biochem Mol Biol 1995; 30(6):445-600.
50. Meisel P. Arylamine N-acetyltransferases and drug response. Pharmacogenomics 2002; 3(3):349-366.
51. Glatt H, Meinl W. Pharmacogenetics of soluble sulfotransferases (SULTs). Naunyn Schmiedebergs Arch Pharmacol 2004; 369(1):55-68.
52. Weinshilboum RM. Methylation pharmacogenetics: Thiopurine methyltransferase as a model system. Xenobiotica 1992; 22(9-10):1055-1071.
53. Boysen G, Kenney PM, Upadhyaya P et al. Effects of benzyl isothiocyanate and 2-phenethyl isothiocyanate on benzo[a]pyrene and 4-(methylnitrosamino)-1-(3-pyridyl)-1-butanone metabolism in F-344 rats. Carcinogenesis 2003; 24(3):517-525.
54. Butcher NJ, Boukouvala S, Sim E et al. Pharmacogenetics of the arylamine N-acetyltransferases. Pharmacogenomics J 2002; 2(1):30-42.
55. DeJong JL, Chang CM, Whang-Peng J et al. The human liver glutathione S-transferase gene superfamily: Expression and chromosome mapping of an Hb subunit cDNA. Nucleic Acids Res 1988; 16(17):8541-8554.

56. Seidegard J, DePierre JW, Pero RW. Hereditary interindividual differences in the glutathione transferase activity towards trans-stilbene oxide in resting human mononuclear leukocytes are due to a particular isozyme(s). Carcinogenesis 1985; 6(8):1211-1216.
57. Pemble S, Schroeder KR, Spencer SR et al. Human glutathione S-transferase theta (GSTT1): cDNA cloning and the characterization of a genetic polymorphism. Biochem J 1994; 300(Pt 1):271-276.
58. Fryer AA, Zhao L, Alldersea J et al. Use of site-directed mutagenesis of allele-specific PCR primers to identify the GSTM1 A, GSTM1 B, GSTM1 A,B and GSTM1 null polymorphisms at the glutathione S-transferase, GSTM1 locus. Biochem J 1993; 295(Pt 1):313-315.
59. Zhong S, Wolf CR, Spurr NK. Chromosomal assignment and linkage analysis of the human glutathione S-transferase mu gene (GSTM1) using intron specific polymerase chain reaction. Hum Genet 1992; 90(4):435-439.
60. Harada S, Nakamura T, Misawa S. Polymorphism of pentanucleotide repeats in the 5' flanking region of glutathione S-transferase (GST) pi gene. Hum Genet 1994; 93(2):223-224.
61. Lee EJ, Zhao B, Moochhala SM et al. Frequency of mutant CYPIA1, NAT2 and GSTM1 alleles in a normal Chinese population. Pharmacogenetics 1994; 4(6):355-358.
62. Ge H, Lam WK, Lee J et al. Analysis of L-myc and GSTM1 genotypes in Chinese nonsmall cell lung carcinoma patients. Lung Cancer 1996; 15(3):355-366.
63. Brockmoller J, Kaiser R, Kerb R et al. Polymorphic enzymes of xenobiotic metabolism as modulators of acquired P53 mutations in bladder cancer. Pharmacogenetics 1996; 6(6):535-545.
64. Wenzlaff AS, Cote ML, Bock CH et al. GSTM1, GSTT1 and GSTP1 polymorphisms, environmental tobacco smoke exposure and risk of lung cancer among never smokers: A population-based study. Carcinogenesis 2005; 26(2):395-401.
65. Quinones L, Lucas D, Godoy J et al. CYP1A1, CYP2E1 and GSTM1 genetic polymorphisms. The effect of single and combined genotypes on lung cancer susceptibility in Chilean people. Cancer Lett 2001; 174(1):35-44.
66. Seow A, Vainio H, Yu MC. Effect of glutathione-S-transferase polymorphisms on the cancer preventive potential of isothiocyanates: An epidemiological perspective. Mutat Res 2005; 592(1-2):58-67.
67. Seow A, Yuan JM, Sun CL et al. Dietary isothiocyanates, glutathione S-transferase polymorphisms and colorectal cancer risk in the Singapore Chinese Health Study. Carcinogenesis 2002; 23(12):2055-2061.
68. Cheng YJ, Chien YC, Hildesheim A et al. No association between genetic polymorphisms of CYP1A1, GSTM1, GSTT1, GSTP1, NAT2, and nasopharyngeal carcinoma in Taiwan. Cancer Epidemiol Biomarkers Prev 2003; 12(2):179-180.
69. Stanulla M, Schrappe M, Brechlin AM et al. Polymorphisms within glutathione S-transferase genes (GSTM1, GSTT1, GSTP1) and risk of relapse in childhood B-cell precursor acute lymphoblastic leukemia: A case-control study. Blood 2000; 95(4):1222-1228.
70. Hein DW. N-Acetyltransferase genetics and their role in predisposition to aromatic and heterocyclic amine-induced carcinogenesis. Toxicol Lett 2000; 112-113:349-356.
71. Hickman D, Risch A, Buckle V et al. Chromosomal localization of human genes for arylamine N-acetyltransferase. Biochem J 1994; 297(Pt 3):441-445.
72. Blum M, Grant DM, McBride W et al. Human arylamine N-acetyltransferase genes: Isolation, chromosomal localization, and functional expression. DNA Cell Biol 1990; 9(3):193-203.
73. Garcia-Closas M, Malats N, Silverman D et al. NAT2 slow acetylation, GSTM1 null genotype, and risk of bladder cancer: Results from the Spanish Bladder Cancer Study and meta-analyses. Lancet 2005; 366(9486):649-659
74. van der Hel OL, Peeters PH, Hein DW et al. NAT2 slow acetylation and GSTM1 null genotypes may increase postmenopausal breast cancer risk in long-term smoking women. Pharmacogenetics 2003; 13(7):399-407.
75. Boissy RJ, Watson MA, Umbach DM et al. A pilot study investigating the role of NAT1 and NAT2 polymorphisms in gastric adenocarcinoma. Int J Cancer 2000; 87(4):507-511.
76. Hamasaki T, Inatomi H, Katoh T et al. N-acetyltransferase-2 gene polymorphism as a possible biomarker for prostate cancer in Japanese men. Int J Urol 2003; 10(3):167-173.
77. Bandmann O, Vaughan J, Holmans P et al. Association of slow acetylator genotype for N-acetyltransferase 2 with familial Parkinson's disease. Lancet 1997; 350(9085):1136-1139.
78. Jensen LE, Hoess K, Mitchell LE et al. Loss of function polymorphisms in NAT1 protect against spina bifida. Hum Genet 2006.
79. Zielinska E, Niewiarowski W, Bodalski J. The arylamine N-acetyltransferase (NAT2) polymorphism and the risk of adverse reactions to cotrimoxazole in children. Eur J Clin Pharmacol 1998; 54(9-10):779-785.

80. Lin HJ, Han CY, Lin BK et al. Slow acetylator mutations in the human polymorphic N-acetyltransferase gene in 786 Asians, blacks, Hispanics, and whites: Application to metabolic epidemiology. Am J Hum Genet 1993; 52(4):827-834.
81. Cascorbi I, Brockmoller J, Mrozikiewicz PM et al. Arylamine N-acetyltransferase activity in man. Drug Metab Rev 1999; 31(2):489-502.
82. Bell DA, Badawi AF, Lang NP et al. Polymorphism in the N-acetyltransferase 1 (NAT1) polyadenylation signal: Association of NAT1*10 allele with higher N-acetylation activity in bladder and colon tissue. Cancer Res 1995; 55(22):5226-5229.
83. Lee E, Huang Y, Zhao B et al. Genetic polymorphism of conjugating enzymes and cancer risk: GSTM1, GSTT1, NAT1 and NAT2. J Toxicol Sci 1998; 23(Suppl 2):140-142.
84. Lin HJ, Han CY, Lin BK et al. Ethnic distribution of slow acetylator mutations in the polymorphic N-acetyltransferase (NAT2) gene. Pharmacogenetics 1994; 4(3):125-134.
85. Kukongviriyapan V, Lulitanond V, Areejitranusorn C et al. N-acetyltransferase polymorphism in Thailand. Hum Hered 1984; 34(4):246-249.
86. Koizumi A, Nomiyama T, Tsukada M et al. Evidence on N-acetyltransferase allele-associated metabolism of hydrazine in Japanese workers. J Occup Environ Med 1998; 40(3):217-222.
87. Cascorbi I, Drakoulis N, Brockmoller J et al. Arylamine N-acetyltransferase (NAT2) mutations and their allelic linkage in unrelated Caucasian individuals: Correlation with phenotypic activity. Am J Hum Genet 1995; 57(3):581-592.
88. Bakayev VV, Mohammadi F, Bahadori M et al. Arylamine N-acetyltransferase 2 slow acetylator polymorphisms in unrelated Iranian individuals. Eur J Clin Pharmacol 2004; 60(7):467-471.
89. Aynacioglu AS, Cascorbi I, Mrozikiewicz PM et al. Arylamine N-acetyltransferase (NAT2) genotypes in a Turkish population. Pharmacogenetics 1997; 7(4):327-331.
90. Woolhouse NM, Qureshi MM, Bastaki SM et al. Polymorphic N-acetyltransferase (NAT2) genotyping of Emiratis. Pharmacogenetics 1997; 7(1):73-82.
91. Gaikovitch EA, Cascorbi I, Mrozikiewicz PM et al. Polymorphisms of drug-metabolizing enzymes CYP2C9, CYP2C19, CYP2D6, CYP1A1, NAT2 and of P-glycoprotein in a Russian population. Eur J Clin Pharmacol 2003; 59(4):303-312.
92. Anitha A, Banerjee M. Arylamine N-acetyltransferase 2 polymorphism in the ethnic populations of South India. Int J Mol Med 2003; 11(1):125-131.
93. Chen B, Zhang WX, Cai WM. The influence of various genotypes on the metabolic activity of NAT2 in a Chinese population. Eur J Clin Pharmacol 2005; 62(5):355-359.
94. Lilla C, Verla-Tebit E, Risch A et al. Effect of NAT1 and NAT2 genetic polymorphisms on colorectal cancer risk associated with exposure to tobacco smoke and meat consumption. Cancer Epidemiol Biomarkers Prev 2006; 15(1):99-107.
95. Lee EJ, Zhao B, Seow-Choen F. Relationship between polymorphism of N-acetyltransferase gene and susceptibility to colorectal carcinoma in a Chinese population. Pharmacogenetics 1998; 8(6):513-517.
96. Cascorbi I, Brockmoller J, Mrozikiewicz PM et al. Homozygous rapid arylamine N-acetyltransferase (NAT2) genotype as a susceptibility factor for lung cancer. Cancer Res 1996; 56(17):3961-3966.
97. Chiou HL, Wu MF, Chien WP et al. NAT2 fast acetylator genotype is associated with an increased risk of lung cancer among never-smoking women in Taiwan. Cancer Lett 2005; 223(1):93-101.
98. Hou SM, Falt S, Yang K et al. Differential interactions between GSTM1 and NAT2 genotypes on aromatic DNA adduct level and HPRT mutant frequency in lung cancer patients and population controls. Cancer Epidemiol Biomarkers Prev 2001; 10(2):133-140.
99. Bouchardy C, Mitrunen K, Wikman H et al. N-acetyltransferase NAT1 and NAT2 genotypes and lung cancer risk. Pharmacogenetics 1998; 8(4):291-298.
100. Seow A, Zhao B, Poh WT et al. NAT2 slow acetylator genotype is associated with increased risk of lung cancer among nonsmoking Chinese women in Singapore. Carcinogenesis 1999; 20(9):1877-1881.
101. Mackenzie PI, Miners JO, McKinnon RA. Polymorphisms in UDP glucuronosyltransferase genes: Functional consequences and clinical relevance. Clin Chem Lab Med 2000; 38(9):889-892.
102. Tukey RH, Strassburg CP. Human UDP-glucuronosyltransferases: Metabolism, expression, and disease. Annu Rev Pharmacol Toxicol 2000; 40:581-616.
103. Owens IS, Ritter JK. Gene structure at the human UGT1 locus creates diversity in isozyme structure, substrate specificity, and regulation. Prog Nucleic Acid Res Mol Biol 1995; 51:305-338.
104. Strassburg CP, Kneip S, Topp J et al. Polymorphic gene regulation and interindividual variation of UDP-glucuronosyltransferase activity in human small intestine. J Biol Chem 2000; 275(46):36164-36171.

105. Bosma PJ, Seppen J, Goldhoorn B et al. Bilirubin UDP-glucuronosyltransferase 1 is the only relevant bilirubin glucuronidating isoform in man. J Biol Chem 1994; 269(27):17960-17964.
106. Beutler E, Gelbart T, Demina A. Racial variability in the UDP-glucuronosyltransferase 1 (UGT1A1) promoter: A balanced polymorphism for regulation of bilirubin metabolism? Proc Natl Acad Sci USA 1998; 95(14):8170-8174.
107. Ando Y, Ueoka H, Sugiyama T et al. Polymorphisms of UDP-glucuronosyltransferase and pharmacokinetics of irinotecan. Ther Drug Monit 2002; 24(1):111-116.
108. Balram C, Sabapathy K, Fei G et al. Genetic polymorphisms of UDP-glucuronosyltransferase in Asians: UGT1A1*28 is a common allele in Indians. Pharmacogenetics 2002; 12(1):81-83.
109. Huang CS, Luo GA, Huang ML et al. Variations of the bilirubin uridine-diphosphoglucuronosyl transferase 1A1 gene in healthy Taiwanese. Pharmacogenetics 2000; 10(6):539-544.
110. Lampe JW, Bigler J, Horner NK et al. UDP-glucuronosyltransferase (UGT1A1*28 and UGT1A6*2) polymorphisms in Caucasians and Asians: Relationships to serum bilirubin concentrations. Pharmacogenetics 1999; 9(3):341-349.
111. Ando Y, Chida M, Nakayama K et al. The UGT1A1*28 allele is relatively rare in a Japanese population. Pharmacogenetics 1998; 8(4):357-360.
112. Koiwai O, Nishizawa M, Hasada K et al. Gilbert's syndrome is caused by a heterozygous missense mutation in the gene for bilirubin UDP-glucuronosyltransferase. Hum Mol Genet 1995; 4(7):1183-1186.
113. Zhou Q, Sparreboom A, Tan EH et al. Pharmacogenetic profiling across the irinotecan pathway in Asian patients with cancer. Br J Clin Pharmacol 2005; 59(4):415-424.
114. Han JY, Lim HS, Shin ES et al. Comprehensive analysis of UGT1A polymorphisms predictive for pharmacokinetics and treatment outcome in patients with nonsmall-cell lung cancer treated with irinotecan and cisplatin. J Clin Oncol 2006; 24(15):2237-2244.
115. Yamamoto K, Sato H, Fujiyama Y et al. Contribution of two missense mutations (G71R and Y486D) of the bilirubin UDP glycosyltransferase (UGT1A1) gene to phenotypes of Gilbert's syndrome and Crigler-Najjar syndrome type II. Biochim Biophys Acta 1998; 1406(3):267-273.
116. McLeod HL, Siva C. The thiopurine S-methyltransferase gene locus — implications for clinical pharmacogenomics. Pharmacogenomics 2002; 3(1):89-98.
117. Krynetski EY, Tai HL, Yates CR et al. Genetic polymorphism of thiopurine S-methyltransferase: Clinical importance and molecular mechanisms. Pharmacogenetics 1996; 6(4):279-290.
118. Honchel R, Aksoy IA, Szumlanski C et al. Human thiopurine methyltransferase: Molecular cloning and expression of T84 colon carcinoma cell cDNA. Mol Pharmacol 1993; 43(6):878-887.
119. McLeod HL, Lin JS, Scott EP et al. Thiopurine methyltransferase activity in American white subjects and black subjects. Clin Pharmacol Ther 1994; 55(1):15-20.
120. Huang MJ, W Lou YL, Cheng MX. Comparison of thiopurine methyltransferase activity between Chinese and Caucasian populations. Chin J Cancer 2000; 19:858-861.
121. Krynetski EY, Schuetz JD, Galpin AJ et al. A single point mutation leading to loss of catalytic activity in human thiopurine S-methyltransferase. Proc Natl Acad Sci USA 1995; 92(4):949-953.
122. Tai HL, Krynetski EY, Yates CR et al. Thiopurine S-methyltransferase deficiency: Two nucleotide transitions define the most prevalent mutant allele associated with loss of catalytic activity in Caucasians. Am J Hum Genet 1996; 58(4):694-702.
123. Loennechen T, Yates CR, Fessing MY et al. Isolation of a human thiopurine S-methyltransferase (TPMT) complementary DNA with a single nucleotide transition A719G (TPMT*3C) and its association with loss of TPMT protein and catalytic activity in humans. Clin Pharmacol Ther 1998; 64(1):46-51.
124. McLeod HL, Pritchard SC, Githang'a J et al. Ethnic differences in thiopurine methyltransferase pharmacogenetics: Evidence for allele specificity in Caucasian and Kenyan individuals. Pharmacogenetics 1999; 9(6):773-776.
125. Collie-Duguid ES, Pritchard SC, Powrie RH et al. The frequency and distribution of thiopurine methyltransferase alleles in Caucasian and Asian populations. Pharmacogenetics 1999; 9(1):37-42.
126. Kham SK, Tan PL, Tay AH et al. Thiopurine methyltransferase polymorphisms in a multiracial Asian population and children with acute lymphoblastic leukemia. J Pediatr Hematol Oncol 2002; 24(5):353-359.
127. Lennard L, Lilleyman JS, Van Loon J et al. Genetic variation in response to 6-mercaptopurine for childhood acute lymphoblastic leukaemia. Lancet 1990; 336(8709):225-229.
128. Ayrton A, Morgan P. Role of transport proteins in drug absorption, distribution and excretion. Xenobiotica 2001; 31(8-9):469-497.
129. Greiner B, Eichelbaum M, Fritz P et al. The role of intestinal P-glycoprotein in the interaction of digoxin and rifampin. J Clin Invest 1999; 104(2):147-153.

130. Bell DR, Trent JM, Willard HF et al. Chromosomal location of human P-glycoprotein gene sequences. Cancer Genet Cytogenet 1987; 25(1):141-148.
131. Kroetz DL, Pauli-Magnus C, Hodges LM et al. Sequence diversity and haplotype structure in the human ABCB1 (MDR1, multidrug resistance transporter) gene. Pharmacogenetics 2003; 13(8):481-494.
132. Ito S, Ieiri I, Tanabe M et al. Polymorphism of the ABC transporter genes, MDR1, MRP1 and MRP2/cMOAT, in healthy Japanese subjects. Pharmacogenetics 2001; 11(2):175-184.
133. Stein CM, Sadeque AJ, Murray JJ et al. Cyclosporine pharmacokinetics and pharmacodynamics in African American and white subjects. Clin Pharmacol Ther 2001; 69(5):317-323.
134. Hoffmeyer S, Burk O, von Richter O et al. Functional polymorphisms of the human multidrug-resistance gene: Multiple sequence variations and correlation of one allele with P-glycoprotein expression and activity in vivo. Proc Natl Acad Sci USA 2000; 97(7):3473-3478.
135. Balram C, Sharma A, Sivathasan C et al. Frequency of C3435T single nucleotide MDR1 genetic polymorphism in an Asian population: Phenotypic-genotypic correlates. Br J Clin Pharmacol 2003; 56(1):78-83.
136. Kim RB. Organic anion-transporting polypeptide (OATP) transporter family and drug disposition. Eur J Clin Invest 2003; 33(Suppl 2):1-5.
137. Kullak-Ublick GA, Ismair MG, Stieger B et al. Organic anion-transporting polypeptide B (OATP-B) and its functional comparison with three other OATPs of human liver. Gastroenterology 2001; 120(2):525-533.
138. Konig J, Cui Y, Nies AT et al. Localization and genomic organization of a new hepatocellular organic anion transporting polypeptide. J Biol Chem 2000; 275(30):23161-23168.
139. Niemi M, Schaeffeler E, Lang T et al. High plasma pravastatin concentrations are associated with single nucleotide polymorphisms and haplotypes of organic anion transporting polypeptide-C (OATP-C, SLCO1B1). Pharmacogenetics 2004; 14(7):429-440.
140. Tirona RG, Leake BF, Merino G et al. Polymorphisms in OATP-C: Identification of multiple allelic variants associated with altered transport activity among European- and African-Americans. J Biol Chem 2001; 276(38):35669-35675.
141. Nozawa T, Minami H, Sugiura S et al. Role of organic anion transporter OATP1B1 (OATP-C) in hepatic uptake of irinotecan and its active metabolite, 7-ethyl-10-hydroxycamptothecin: In vitro evidence and effect of single nucleotide polymorphisms. Drug Metab Dispos 2005; 33(3):434-439.
142. Xiang XQ, Jada SR, Li H et al. Pharmacogenetics of SLCO1B1 gene and the impact of *1b and *15 haplotypes on irinotecan disposition in Asian cancer patients. Pharmacogenet Genomics 2006; 16(9):683-691.
143. Nozawa T, Nakajima M, Tamai I et al. Genetic polymorphisms of human organic anion transporters OATP-C (SLC21A6) and OATP-B (SLC21A9): Allele frequencies in the Japanese population and functional analysis. J Pharmacol Exp Ther 2002; 302(2):804-813.
144. Lee E, Ryan S, Birmingham B et al. Rosuvastatin pharmacokinetics and pharmacogenetics in white and Asian subjects residing in the same environment. Clin Pharmacol Ther 2005; 78(4):330-341.

Chapter 10

Pharmacogenetics in Chinese Population

Hong-Hao Zhou* and Wei Zhang

Abstract

Genetic variations of drug-metabolizing enzymes, receptors and transporters have been recognized as major causes of the interindividual variability in drug responses. As a result of the genotype effect, interindividual variations in drug response may result from interindividual differences in composition of a particular polymorphic allele and mutations that code for enzymes, receptors and transporters with abnormal activity or sensitivity, which occur with altered frequency in the individuals. China is a multinational country with 55 ethnic minorities besides the Han majority. The relatively unique genetic, cultural, dietetic and environmental characteristics of each of the nationalities should affect the function of catalytic activity of drug-metabolizing enzymes. We also confirmed that the environmental factors contribute to the phenotype of the enzymes or transporters in a gene dosage manner. In this review, we summarize studies in our laboratory on the individual variations in drug disposition and response mediated by polymorphic CYP450s as well as drug receptors and transporters, and the role of genotype and environmental factors in the metabolism and disposition of certain currently used drugs.

Pharmacogenetic studies have established the importance of polymorphic drug metabolizing enzymes, receptors and transporters in the differential response of patients to drugs. Among these target genes, cytochrome P450 (CYP) has become the most important subject of extensive studies concerning individual variation of drug metabolism and disposition. The association between decreased drug clearance and activity of several kinds of CYP enzymes, the inherited nature of the deficiency, and their frequencies and clinical importance were evaluated extensively. During the past 20 years, the genetic polymorphisms of these genes have been studied at the protein and gene level. Analysis of allele frequencies in different populations revealed individual and interethnic differences that contribute to the molecular mechanisms responsible for interindividual variations in drug metabolisms and responses. In recent years, we also demonstrated that genetic differences among racial and ethnic groups usually reflect differences in the distribution of polymorphic traits, which occur at different frequencies in different populations.

Genetic Polymorphism of Phase I Drug Metabolizing Enzymes or Receptors in Chinese Population

Genetic Polymorphism of CYP2C19 in Chinese Ethnic Populations

The genetic polymorphism of CYP2C19 was elucidated in several different Chinese ethnic populations. The two defective alleles were found together firstly in PMs of Chinese Dong

*Corresponding Author: Hong-Hao Zhou—Pharmacogenetics Research Institute, Institute of Clinical Pharmacology, Central South University, Changsha, China.
Email: hhzhou2003@163.com

Pharmacogenomics in Admixed Populations, edited by Guilherme Suarez-Kurtz.

Table 1. **CYP2C19** ***allele frequencies in Chinese subjects***

	Han (n=202)	Dong (n=148)	Dai (n=386)	Bai (n=404)
*CYP2C19*1*	0.559(113)	0.541(80)	0.666*(257)	0.688*(278)
*CYP2C19*2*	0.336(74)	0.399(59)	0.303(117)	0.257*(104)
*CYP2C19*3*	0.074(15)	0.060(9)	0.031*(12)	0.052(21)
*CYP2C19*5*	ND	ND	ND	0.0025(1)

*P < 0.05 vs. Han and Dong.

nationality and accounted for 100% of the PM alleles. The *CYP2C19*2* and *CYP2C19 *3* represent 86.8% and 13.2% of the mutant alleles in this ethnic group, respectively. Similar results were then defined in the Chinese Han majority (83.2% and 16.8%, respectively) and Chinese Bai minority (82.5% and 16.7%, respectively). In a group of Han subjects (n = 101), 19.8% of them were classified as PMs phenotypically and 100% of these phenotypes could be explained by *CYP2C19*2* and *CYP2C19*3*. Several other studies[1,2] confirmed that the *CYP2C19*2* accounts for most mutant CYP2C19 alleles in Han, and that *CYP2C19*3* can partially explain these mutant alleles in Chinese Bai subjects, 27 of the 202 (13.4%) Chinese Bai subjects were classified as PMs phenotypically, and only one appeared to be an outlier. The outlier was finally found to be a heterozygote with a *CYP2C19*2* and a new mutant allele consisting of a C→T mutant allele at bp 1297 in exon 9. This mutation, which is designated as *CYP2C19*5*, results in the substitution of Arg433→Trp433 in the heme-binding region and may produce an inactive protein.[3] It was also defined that the *CYP2C19*2* and *CYP2C19*3* account for all mutant alleles in the Chinese Dai minority (90.7% and 9.3%, respectively). It was noticed that other rare mutations of *CYP2C19* found in Caucasian populations including *CYP2C19*6*, *CYP2C19*7* and *CYP2C19*8* were not detected in Chinese populations. The *CYP2C19* allele frequencies in these Chinese ethnic groups are presented in Table 1, in which significant differences in the frequencies were clearly shown.

CYP3A Single Nucleotide Polymorphisms in a Chinese Population

Human cytochrome P450 3A evolved to catalyze the metabolism of numerous common therapy drugs and endogenous molecules. Members of the CYP3A are the majority expressed in human liver and intestine, and there are marked interindividual differences in their protein expression and activity. The activity of CYP3A enzyme in Chinese is highly variable, exceeding 14-fold, and contributes greatly to variation in oral bioavailability and systemic clearance of CYP3A substrates. The genetic factors play an important role in the interindividual variability in CYP3A activity. Detection of CYP3A5 and CYP3A4 variant alleles and knowledge about their allelic frequency in specific ethnic groups are important to lead to individualized drug dosing and improved therapeutics. In a group of 302 unrelated Chinese healthy volunteers, the frequency of the *CYP3A5*3* and *CYP3A4*18* variant allele were 77.8% and 1%, respectively.[4] These *CYP3A4*18* and *CYP3A5*3* allelic frequencies are similar to that reported previously in Chinese resident in Taiwan.[5] The frequency of the *CYP3A5*3* allele in Chinese population is similar to the Japanese but lower than Caucasians. Meanwhile, approximate 62% of the Chinese population carrying *CYP3A5*3/*3* genotype may appear not to express CYP3A5 protein and exhibit a lower total expression of CYP3A.

Polymorphism of CYP2A13 in a Chinese Population

Human cytochrome P450 2A13 (CYP2A13) is involved in the activation of numerous toxicants and carcinogens, especially in the metabolic activation of 4-(methyl-nitrosamino)-1-(3-pyridyl)-1-butanone (NNK), a major tobacco-specific carcinogen. A functionally significant coding SNP (C3375T) in exon 5 of *CYP2A13*, which results in an amino acid substitution of Arg 257 to Cys, has been recently reported to exist in White, Black, Hispanic, and Asian individuals, with the variant 3375T allele frequencies being 1.9%, 14.4%, 5.8% and 7.7%, respectively.[6] In 258 healthy Chinese Han volunteers, 27 (10.5%) heterozygotes and 1 (0.4%) homozygote for the 257Cys allele were detected. The frequency of the variant 257Cys allele in Chinese population was 5.6%.[7] The *CYP2A13* Arg257Cys variant represents a relatively common polymorphism in Chinese, with the 257Cys allele frequency being similar to the Hispanic and Asian groups, but significantly lower than the Black.

The Histamine N-methyltransferase Gene Mutations in Chinese Population

Histamine N-methyltransferase (HNMT) plays an important role in the metabolism of histamine, ambiogenic amine that has many physiologic and pathological roles in human tissues. A total of 11 SNPs were identified in 192 unrelated healthy Chinese adults, among which six SNPs had variant allele frequencies greater than 5%. Of the six common SNPs, three (21637T>C, 2463T>C and 2411 C>T) were located in 5'-FR, one (314C>T) in coding exons, and two (939A>G and 1097A>T) in the 3'-untranslated region (3'-UTR). Most of these common SNPs were in linkage disequilibrium. Genotype-phenotype correlation analyses were performed for those common SNPs in 5'-FR and 3'-UTR. In males, no significant association was found between HNMT activity and these noncoding SNPs. However, in females, the 21637T>C or 2463T>C tended to be associated with decreased HNMT activity, whereas the 939A>G or 1097A>T appeared to be correlated with increased enzymatic activity.[8] *HNMT* polymorphisms differ considerably between Chinese and American. The common SNPs in 5'-FR (21637T>C and 2463T>C) and 3'-UTR (939A>G and 1097A>T) might conditionally regulate the activity of HNMT, or might be genetically linked to unknown mutation(s) underlying the HNMT phenotypic variance.

The C314T missense mutation (Thr105Ile) in the *HNMT* gene has been identified to represent a common functional polymorphism in Caucasians,[9] whereas an A595G (Ile199Val) variant has been reported in one HNMT cDNA from a Japanese subject.[10] The point mutations C314T and A595G within *HNMT* were both investigated in 352 unrelated Chinese Han subjects.[11] None of the 352 subjects contained the A595G mutation, whereas 40 (11.6%) heterozygotes and 1 (0.3%) homozygote for the variant T314 allele were detected. The frequency of the variant T314 allele in this Chinese population was 6%, not different from Japanese (5%) but lower than American Caucasians (9%).

Distribution of NAT2 Genetic Polymorphism in Chinese Population

Frequency of the slow acetylator phenotype in large sample of healthy Han Chinese population (n = 3516) is approximately 19.9 ± 4.0%. In addition, frequencies of the slow acetylator phenotype differ between the different minorities in Chinese populations and the range was between 3.2% and 50.6%, with a mean value of 20.6 ± 12.9% in a total of 1842 individuals from 17 Chinese minorities. In addition, there was no significant heterogeneity in overseas Chinese between the probe drugs isoniazid and sulfadimidine or sulfamethazine, and the mean value of slow acetylator phenotype incidence was 24.5%, consistent with that of the native Chinese. As expected, frequency of the slow acetylator genotypes in Chinese populations was 25.4% (112/441; 95%CI: 21.3%-29.5%), which was in accordance with that of the slow acetylator phenotype in native or overseas Chinese. For all genotypes. **4/*4* (29.9%, 132/441), **4/*6A* (27.4%, 121/441), **4/*7A* (12%, 53/441) and **6A/*6A* (11.3%, 50/441) represent 80.6%, whereas genotypes **5/*7A* (0.2%, 1/441), **5/*5A* (1.1%, 5/441) and **7A/*7A* (1.8%, 8/441) were not frequently found. From this report, the genotype frequencies of homozygous rapid acetylator, heterozygous rapid acetylator, and homozygous slow acetylator were found to be 29.9% (132/441), 44.7% (197/441) and 25.4% (112/441), respectively. Furthermore, both **4* (52.3%; 95%CI: 49-56%) and **6A* (30.5%; 95%CI: 28-34%) were major *NAT2* alleles, while **7A* (11.2%; 95%CI: 9-13%) and **5A*

(6%; 95%CI: 4-8%) were uncommonly present. Frequency of the mutant alleles was observed at 47.7% (421/882 alleles). The *7A constituted 23.5% (99/421) of slow acetylator alleles in Chinese populations, showing that this point mutation also exists frequently in Chinese populations.[12]

102T>C Polymorphism of 5-HT2A Receptor in Chinese Population

The 5-HT2A receptor belongs to the G-protein superfamily. It plays an important role in vascular regulation. Previous reports in the UK have indicated that there is an association of the 102T>C genetic polymorphism of the 5-HT2A receptor with hypertension.[13] The 102T>C polymorphism analysis was conducted on 198 Chinese hypertensive patients and 164 healthy subjects. The C allele frequency of the 5-HT2A receptor genetic polymorphism was 34.3% and 39.3% in Chinese hypertensive patients and in healthy subjects, respectively. The allele frequency was not different between these two populations (x^2 = 1.922; P = 0.166; OR = 0.807).[14] Thus no correlation exists between the 102T>C genetic polymorphism and hypertension in Chinese.

Racial Differences in Drug Response Reflect Differences in Distribution of Polymorphic Traits

Ethnic differences exist in both pharmacodynamics and pharmacokinetics of many drugs that are well documented by the comparison studies of propranolol, atropine and morphine between Chinese and White healthy subjects.[15-17] Such differences among racial and ethnic groups usually reflect differences in the distribution of polymorphic traits, which occur at different frequencies in different races. Therefore, the plausible biological justification for making racial differences in drug response is genetic polymorphism of drug metabolizing enzymes, transporters and receptors. Taking drug metabolizing enzymes as examples, polymorphisms of the enzymes responsible for drug metabolism are distributed differently among different racial and ethnic groups, so the proportion of people with impaired metabolism differs among these groups. Genotype analysis indicated a different frequency for the mutant alleles in different ethnic populations and results in variations in the frequency of subjects who are homozygous for the mutant allele among the extensive metabolizers (EMs) in different ethnic populations.

Frequencies of CYP2C19 Mutations Determine the Racial Difference in S-Mephenytoin Hydroxylation

Humans can be characterized as poor or extensive metabolizers with use of racemic mephenytoin as a phenotyping drug, and CYP2C19 has been identified as the major S-mephenytoin hydroxylase in human. This polymorphism affects the metabolism of other clinically important drugs such as omeprazole, diazepam, imipramine, propranolol, and chloroguanide. Genetic deficiency of *CYP2C19* can be explained by a limited number of SNPs, namely, alleles *CYP2C19*2* (a splice site mutation), *CYP2C19*3* (a premature stop codon), *CYP2C19*4* (an A to G transition in the initiation codon), and *CYP2C19*5* (an amino acid mutation Arg 433 Trp). Two mutations, *CYP2C19*2* and **3*, have been shown to explain almost 100% of Asian and 85% of white PM alleles. Two other defective alleles (*CYP2C19*4* and **5*) were reported to contribute to the PM phenotype in white subjects and Chinese Bai population. Genotype analysis indicated a higher frequency for the *CYP2C19*2* (20.9-39% vs 11.0-16.1%) and *CYP2C19*3* (7.4% vs 0.7%) alleles in Chinese than in Caucasians. The higher frequencies of mutation result in an increase in the frequency of subjects who are homozygous for the *CYP2C19*1* allele among Caucasian EMs compared with Chinese, which is in accordance with the relatively low metabolizing activity in CYP2C19 on average, with 2% to 5% PMs in white populations and 13% to 23% PMs in Asian populations.

Frequencies of CYP2D6 *Polymorphisms Are the Determinant of Average CYP2D6 Activity between Populations*

Chinese have greater sensitivity than Caucasians to the effects of propranolol on heart rate and blood pressure.[15] Decreased protein binding and increased sensitivity to plasma renin activity suppression in Chinese can only partially explain their increased sensitivity to propranolol. CYP2D6 is one of the most important drug metabolizing enzymes in human liver, which is mainly responsible for the propranolol clearance. In Caucasians, 71% of *CYP2D6* alleles code for normal enzyme, while 26% are nonfunctional, are mainly *CYP2D6*4* (15.6%), *CYP2D6*5* (6.9%) and *CYP2D6*3* (1.6%). Except for *CYP2D6*5*, which exhibits a similar frequency in Caucasians, Japanese and Chinese, the other functionally defective alleles of *CYP2D6* are very rare in Chinese.[18] However, Chinese metabolize CYP2D6-mediated drugs more slowly than Caucasians, which may be due predominantly to high frequencies of variants of *CYP2D6*10*, a reduced function allele. The frequency of mutant allele *CYP2D6*10* that encodes a reduced activity of the enzyme is even higher than that of *CYP2D6*1* (50-% vs 26.9%), which allows more Chinese patients to carry this allelic variant. As a result, the average activity value of CYP2D6 is therefore lower in Chinese than other racial populations whose frequencies of *CYP2D6*10* allele are remarkable lower than in Chinese. Therefore, Chinese subjects exhibit a relatively lower metabolizing activity of CYP2D6 on average and may be more sensitive than Caucasians to the effect of CYP2D6 substrates, such as propranolol, metoprolol, timolol and so on.

ORM (Orosomucoid, or α(1)-acid glycoprotein) Genotype Determines Plasma Concentration of Basic Drugs

In pharmacological terms, only unbound drug, rather than protein bound drug, in the plasma could be transported to its site of action or be subject to metabolic alteration or excretion from the body. The degree of protein binding of drugs is a major determinant of the intensity and duration of pharmacological action, especially for the drugs with high affinity to plasma proteins. Orosomucoid (ORM), also called α1-acid glycoprotein, is one of the most important glycoprotein components of blood plasma that binds to basic drugs and some endogenous substances. The *ORM1* locus is highly polymorphic in all populations, and the three codominant alleles are *ORM1*F1*, *ORM1*F2* and *ORM1*S*. Isolated *ORM1*S* displays a higher affinity for various basic drugs, for instance methadone and warfarin, than *ORM1*F* since the number of binding sites of *ORM1*S* is remarkably higher than that of *ORM1*F*.

Very recently, an in vivo study carried out in Chinese population (unpublished data) found that *ORM1*S/*S* and *ORM1*F1/*S* subjects showed significantly lower AUC0-∝ value of free nortritpyline compared with that of *ORM1*F1/*F1* subjects (51.4 ± 23.2 ng·mL-1·h and 42.4 ± 11.6 ng·mL-1·h vs 119.1 ± 74.4 ng·mL-1·h). The allele frequency of *ORM1*S* exhibits significant racial difference, being 27.5% in Chinese population, much less than that in Caucasians (37.3~38.6%). (unpublished data) The higher frequency of *ORM1*S* mutation in Caucasians may result in an increase in the frequency of subjects who are homozygous for the *ORM1*S* allele among Caucasian compared with Chinese. We found that the percentages of unbound diphenhydramine (26.40% ± 6.46% versus 18.30% ± 4.31D) and propranolol (13.81% ± 1.33% vs. 11.68% ± 2.37) were significantly ($P < 0.05$) higher in Chinese subjects compared to Caucasians.[19] A 30% difference was also observed in the nonlinear binding of disopyramide. The lower binding was associated with a lower plasma concentration of the acute-phase reactant in Chinese subjects. Kinetic analysis of the disopyramide binding isotherm was also suggestive of reduced binding capacity with no change in binding affinity. The reason for the racial difference in the α1-acid glycoprotein level might be caused by the relatively lower frequency of *ORM1*S* allele in Asians.

Gene Dosage Determines the Drug Metabolism and Disposition

Gene dosage has been demonstrated as a major factor that determines the drug metabolism or response as shown in a number of drugs mediated via CYP2C19. As a result of the gene dosage effect interindividual variations in drug disposition and response may result from interindividual differences in composition of a particular polymorphic allele and mutations, which code for the special proteins with abnormal activity.

CYP2C19

It has been found that there is a large ethnic difference in the amount of CYP2C19 protein in microsomes. The amount of CYP2C19 in liver microsomes from the Japanese and Caucasians is 0.8% and 1.4% of total CYP450, respectively.[20] Moreover, the amount of CYP2C19 in microsomes is highest in the individuals who are homozygous for the normal gene (*CYP2C19*1/*1*) and lowest in those homozygous for the mutation gene (*CYP2C19*2/ *2* or *CYP2C19*3/*3*) compared with heterozygous (*CYP2C19*1/*2* and *CYP2C19*l/*3*). Studies where we investigated the relation of gene dose and ratio of S/R mephenytoin in Chinese Han, Dong, Dai and Bai subjects showed that heterozygous extensive metabolizers (*CYP2C19*1 /*2* and *CYP2C19*1/*3*) had significantly higher S/R ratios compared with homozygous wild-type extensive metabolizers (*CYP2C19*1/*1*). This clearly indicated that gene dosage has an effect on mephenytoin metabolism.

Diazepam is one of the most commonly prescribed sedative drugs for the treatment of anxiety, convulsions and muscle spasms. N-demethylation is the major metabolic pathway of diazepam in vivo at therapeutic doses. There is evidence that in white and Korean populations the metabolism of both diazepam and its N-demethylated metabolite desmethyldiazepam cosegregates with the S-mephenytoin hydroxylation polymorphism.[21] However the data from a Chinese population conflict with the findings in white and Korean populations. It has been suggested that this discrepancy might be related to the proportion of heterozygotes in Chinese vs. Caucasian EMs. Our studies showed that the presence of mutations of the CYP2C19 gene cosegregates with the impaired metabolism of diazepam among Chinese subjects in a manner suggesting a gene-dosage effect. A significant difference in the half-lives (t1/2) existed between the heterozygous EMs (*CYP2C19*1/*2*) and homozygous EMs (*CYP2C19*1 *1*). The slowest mean clearance of diazepam was seen in the *CYP2C19*2/*2* subjects, and the fastest in the *CYP2C19*1/*1* subjects, with the *CYP2C19*1/ *2* heterozygotes having an intermediate value.

Figure 1 shows the demographic characteristics and pharmacokinetic parameters of oral diazepam in healthy Han Chinese subjects with different *CYP2C19* genotypes. There were significant inter-genotypic differences in the plasma half-lives of diazepam and its active metabolite demethyldiazepam between the different genotypic groups, with the subjects homozygous for the *CYP2C19*2* having the longest t1/2 and the wild-type homozygotes *CYP2C19*1/ *1* having the shortest t1/2 for both compounds. As expected, the slowest mean clearance of diazepam existed in the subjects with *CYP2C19*2/*2*, and the fastest was observed in the wt/ wt subjects. The heterozygotes (*CYP2C19*1/*2*) had intermediate values of systemic clearance and elimination half-life. This study was the first definite evidence that the gene-dosage of CYP2C19 can markedly affect the metabolism and disposition of diazepam and desmethyl-diazepam in humans.

CYP1A2

Either G-2964 or A734 in the human *CYP1A2* gene was confirmed to be associated with high inducible enzyme activity in smokers, but not in nonsmokers.[22] An association between phenotypes and genotypes of CYP1A2 with respect to these two genetic polymorphisms was observed in 163 healthy Chinese volunteers living in Qidong.[23] The ratio of plasma 17X/ 137X at 6 h after oral administration of 300 mg caffeine was employed in CYP1A2 phenotyping analysis, while genotyping analysis was carried out by polymerase chain reaction restriction

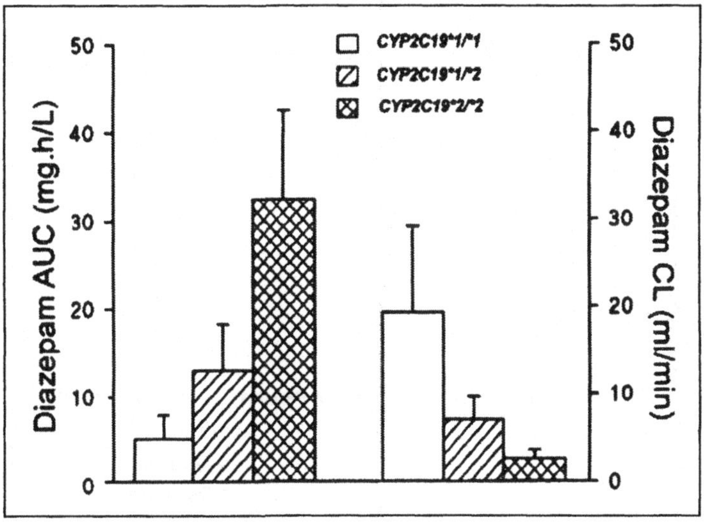

Figure 1. The pharmacokinetic parameters of oral diazepam in healthy Han Chinese subjects with different *CYP2C19* genotypes.

fragment length polymorphism. The allele frequencies of A at -2964 and A at 734 in 139 nonsmoking subjects were 25% and 67%, respectively. The A/A-2964C/C734, G/A-2964C/C734 or A/A-2964C/A734 genotypes that were thought to have lower inducibility/activity of CYP1A2 than the other genotypes did not exist in the tested Chinese subjects. The ratio of 17X/137X was 0.46 ± 0.26 in G/G-2964A/A734 genotypes (n = 22) and 0.36 ± 0.19 in non-G/G-2964A/A734 (n = 117). In addition, there was significant difference between them (P = 0.036). A similar result was also achieved in 24 smokers. Since Qidong is a special region with particularly high incidence of hepatocellular carcinoma in China, the association of phenotypes with genotypes of CYP1A2 in the Qidong population might result from some inducible environmental factors such as cigarettes in smokers.

β1-Adrenergic Receptor

The human β1-adrenergic receptor, an important therapeutic target in cardiovascular diseases, has 2 common functional polymorphisms (Ser49Gly and Gly389Arg). The association between β1-adrenergic receptor polymorphisms and the blood pressure response to metoprolol monotherapy were examined in Chinese population with hypertension. Sixty-one patients with certain β1-adrenergic receptor diplotypes, 18 for 49Ser389Arg/49Ser389Arg, 15 for 49Ser389Arg/ 49Gly389Arg, 19 for 49Ser389Gly/ 49Gly389Arg, and 9 for 49Ser389Gly/ 49Ser389Gly were selected. After administration of 25 mg metoprolol twice daily for consecutive 4 weeks, the descent of systolic blood pressure after metoprolol administration was significantly different among genotype groups (10.4% ± 4.0%, 2.8% ± 4.7%, and 1.1% ± 1.5% for Arg389Arg, Gly389Arg, and Gly389Gly patients, respectively; $P<0.001$) (Fig. 2). There exhibited a similar difference in changes of diastolic blood pressure (6.1% ± 4.3%, 2.2% ± 4.2%, and 0.9% ± 4.0%, respectively; $P < 0.001$) and mean arterial pressure (8.1% ± 3.5%, 2.5% ± 3.0%, and 1.0% ± 2.5%, respectively; $P > 0.001$) for Arg389Arg, Gly389Arg, and Gly389Gly patients. The Gly389Arg variance played an important role in the blood pressure response to metoprolol monotherapy, while the Ser49Gly variance exhibited a smaller contribution to the antihypertensive effect of metoprolol. Systolic blood pressure decreased significantly in Ser49 homozygous patients compared with Ser49Gly patients (8.4% ± 3.2% versus 5.3% ± 5.2%, P =0.047). There was a highly significant relationship between diplotype and blood pressure

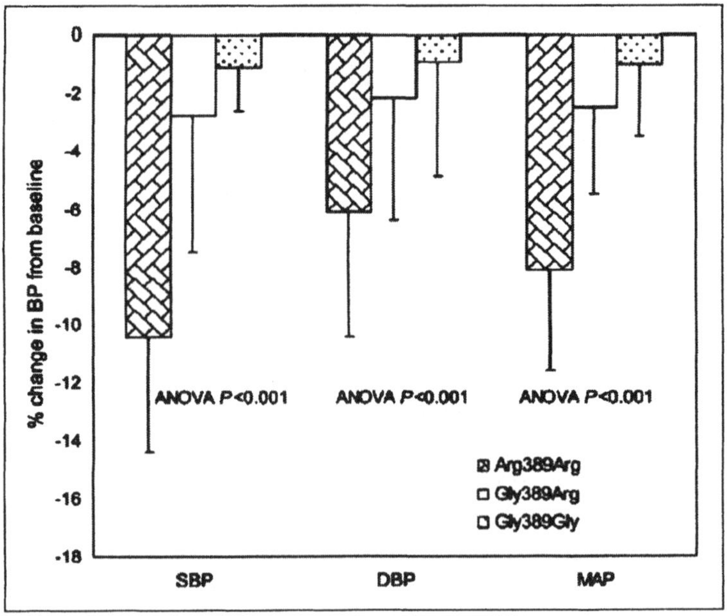

Figure 2. Blood pressure response to metoprolol monotherapy in patients with hypertension stratified according to β1-adrenergic receptor Gly389Arg genotypes (Arg389Arg, n = 33; Gly389Arg, n = 19; Gly389Gly, n = 9). SBP, Systolic blood pressure; DBP, diastolic blood pressure; MAP, mean arterial pressure.

during treatment. Systolic blood pressure significantly decreased in 49Ser389Arg/ 49Ser389Arg (12.0% ± 3.8%, $P < 0.001$) and 49Ser389Arg/ 49Gly389Arg (8.4% ± 5.5%, $P < 0.001$) patients, with the decrease in the former being more pronounced ($P = 0.023$). We also found a significant decrease in diastolic blood pressure (6.5% ± 4.7% versus 5.7% ± 3.2%, respectively; both $P < 0.001$) and mean arterial pressure (8.8% ± 3.2% versus 6.9% ± 3.7%, respectively; both $P < 0.001$) in 49Ser389Arg/49Ser389Arg and 49Ser389Arg/49Gly389Arg patients. Our results suggested that 49Ser389Arg/49Ser389Arg and 49Ser389Arg/ 49Gly389Arg patients were good responders to metoprolol therapy; 49Ser389Arg/49Ser389Arg patients had a larger systolic blood pressure reduction than 49Ser389Arg/49Gly389Arg patients did. 49Ser389Gly/ 49Gly389Arg and 49Ser389Gly/49Ser389Gly patients were nonresponders to metoprolol antihypertensive therapy.[24]

OATP1B1

SLCO1B1 (encoding the human organic anion transporting polypeptide 1B1, OATP1B1) is recently identified as a major determinant of the pharmacokinetics of phenylalanine derivatives, for instance repaglinide. Genetic polymorphisms of *SLCO1B1*, causing reduced transport activity both in vitro and in vivo, have recently gained increasing interest. Several studies in Japanese, German and Finnish subjects[25] proved that *SLCO1B1*5*, *SLCO1B1*15* (a haplotype that consists of *SLCO1B1*1b* and **5* SNPs) and *SLCO1B1*17* (a haplotype that consists of *SLCO1B1*15* and -11187G>A SNPs) were associated with markedly increased pravastatin plasma concentration compared with those homozygous for the wild-type **1a* or **1b* alleles. Because the allelic frequency of the 521T>C (Val174Ala) SNP, which defines the *SLCO1B1*5*, **15* or **17* haplotypes, is relatively high (about 14%) in Chinese subjects, (unpublished data) these findings would suggest that in this population, a large number of substrates of OATP1B1 may have plasma concentrations of the drug higher than expected. Seventeen Chinese healthy volunteers with three different *SLCO1B1* genotypes (11 were 521TT wild type homozygotes, 4 were 521TC heterozygotes and 2 were 521CC mutant homozygotes) were enrolled.[26] After administration of a single oral dose of 90 mg nateglinide, the AUC(0-∞) value of nateglinide was 82% ($P < 0.001$) higher in the *SLCO1B1* 521TC subjects, and 108% ($P < 0.001$) higher in the *SLCO1B1* 521CC subjects than in the *SLCO1B1* 521TT subjects, respectively, which suggest that OATP1B1-mediated hepatic uptake of nateglinide may be the prior step for its metabolism and elimination and that *SLCO1B1* 521T>C SNP might play an important role in the transportation of nateglinide in a gene-dosage manner.

Role of Environmental Factors on the Activity of Phase I Drug Metabolizing Enzyme

The enzyme induction and inhibition could not only contribute to the interindividual variation in drug metabolism, but also involve the drug interaction which is always a major concern in medicine for clinician and patients. Understanding of determinant roles of genetic polymorphism in the enzyme induction and inhibition may provide a tool to predict the possibility and magnitude of drug-drug interaction in different individuals with different genetic backgrounds.

Induction of CYP2C19

Rifampicin is a potent unspecific inducer of many CYP450 isoforms. Treating EMs and PMs of S-MP 4'-hydroxylation with rifampicin, using MP as a probe, the enzyme activity of CYP2C19 was inducible in EMs. In a study after treatment with rifampicin daily for 22 days, the S/R ratios in the PMs with *CYP2C19*2* was decreased by 9.6 ± 5.7% (P <0.05), and the amount of 4'-OH-MP excreted in the urine was increased by 80.1 ± 48.0% (P <0.05). In this study it was also found that the amount of 4'-OH-MP excreted in the urine in homozygous EMs was increased by 203.9 ± 42.5%, while that in heterozygous was only increased by 69.6 ± 4.1%, suggesting the effect of gene dose on the inducibility of CYP2C19. The relation of induction effect of rifampicin on CYP2C19 to *CYP2C19*2* and gene dose represents a good example of the cooperation between genetic and nongenetic factors to determine the activity of drug-metabolizing enzyme.

Inhibition of CYP2C19

Most CYP450 isoforms can not only be induced but also be inhibited by certain foreign compounds including clinically used drugs. Some in vivo and in vitro drug-drug interactions suggest that fluvoxamine, a widely used drug in the treatment of major depression, may have an inhibitory effect on CYP450s activity. Using mephenytoin (MP) and metoprolol as probe drugs, we studied the effect of fluvoxamine on the activities of CYP2C19 and CYP2D6 in Chinese healthy subjects. Administration of a therapeutic dose of fluvoxamine caused a significant increase in the S/R ratio of MP and a reduction in the excretion of 4'-OH-MP in 0-8 h urine. In contrast, fluvoxamine had no effect on either the 0-8 h urinary metoprolol/α-hydroxymetoprolol ratio or the 0-8 h urinary recovery of α-hydroxymetoprolol. These results indicate that fluvoxamine is an inhibitor of CYP2C19 but not CYP2D6 in vivo.

CYP2C19-Mediated Herb-Drug Interaction

St John's wort, an extract of the medicinal plant *Hypericum perforatum*, is widely used as an herbal antidepressant. More recently, its ability to induce both CYP3A4- and CYP2C19-mediated reaction has been well established. Twelve Chinese healthy adult men (6 *CYP2C19*1/*1*, 4 *CYP2C19*2/*2* and 2 *CYP2C19*2/*3*) were enrolled in a 2-phase randomized crossover design. After a 14-day treatment with St John's wort, substantial decreases in plasma concentrations of omeprazole were observed. The peak plasma concentration (Cmax)

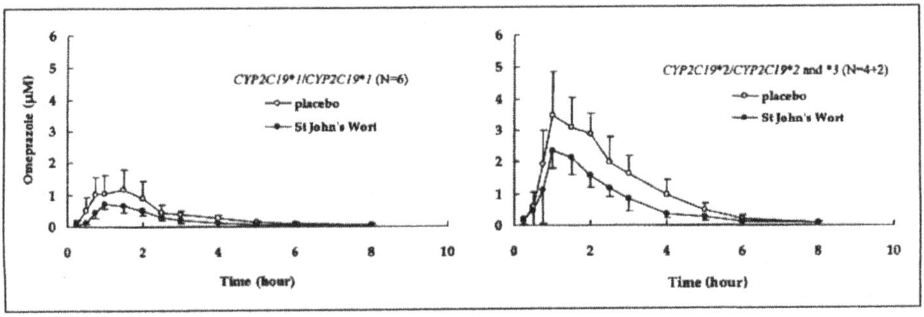

Figure 3. Mean plasma concentrations of omeprazole in *CYP2C19*1/*1* and *CYP2C19*2/*2* or **3* subjects who received a single oral 20-mg dose of omeprazole (capsule) during placebo and *St John's wort* treatments.

significantly decreased by 37.5% ± 13.3% (P< 0.001) in *CYP2C19*2/ *2* or **3* and by 49.6% ± 20.7% (P = 0.017) in *CYP2C19*1/*1*; the area under the concentration-time curve extrapolated to infinity [AUC(0-∞)] decreased by 37.9% ± 21.3% (P = 0.014) and 43.9% ± 23.7% (P = 0.011) in *CYP2C19* mutant and wild genotypes, respectively. Moreover, the Cmax and AUC(0-∞) of omeprazole sulfone increased by 160.3% ± 45.5% ($P < 0.001$) and by 136.6% ± 84.6% (P =.014), 155.5% ± 58.8% ($P < 0.001$), and 158.7% ± 101.4% (P = 0.017) in mutant and wild genotypes, respectively. St John's wort increased the Cmax of 5-hydroxyomeprazole by 38.1% ± 30.5% ($P = 0.028$) and the AUC(0-∞) by 37.2% ± 26% ($P > 0.005$) in *CYP2C19* wild-type subjects, whereas it did not produce any significant alterations to the corresponding pharmacokinetic parameters in subjects with variant genotypes (Fig. 3). The genotype related herb-drug interaction should be taken into account when St John's wort is coadministrated.[27]

Final Considerations

In summary, our knowledge on pharmacogenetic-based tailor-made medication has increased greatly within the last two decades. However, application of pharmacogenetic knowledge to clinical treatment is still limited in current practice. Today, great progress has been made in the field of genetic technologies, which made fast and efficient high-throughput genotyping at lowest cost possible. Our laboratory is now attacking individualized therapy by adopting genotyping microarray. Once such a step has been successfully taken, drug therapy in Chinese population could well become more prevention-directed and patient-tailored than it is possible today.

Acknowledgements

This project was financially supported by research grants from the National Natural Science Foundation of China (C30200346), and by the China Medical Board of New York (grants 01-755).

References

1. Qin XP, Xie HG, Wan W et al. Effect of the gene dosage of CYP2C19 on diazepam metabolism in Chinese subjects. Clin Pharmacol Ther 1999; 66:842-846.
2. Wan J, Xia H, He N et al. The elimination of diazepam in Chinese subjects is dependent on the mephenytoin oxidation phenotype. Br J Clin Pharmacol 1996; 42:471-475.
3. Xiao ZS, Goldstein JA, Xie HG et al. Differences in the incidence of the CYP2C19 polymorphism affecting the S-mephenytoin phenotype in Chinese Han and Bai populations and identification of a new rare CYP2C19 mutant allele. J Pharmacol Exp Ther 1997; 281:604-609.

4. Hu YF, He J, Chen GL et al. CYP3A5*3 and CYP3A4*18 single nucleotide polymorphisms in a Chinese population. Clin Chim Acta 2005; 353(1-2):187-92.
5. Shin PS, Huang JD. Pharmacokinetics of midazolam and 1'-hydroxymidazolam in Chinese with different CYP3A5 genotypes. Drug Metab Dispos 2002; 30:1491-6.
6. Zhang X, Su T, Zhang QY et al. Genetic polymorphisms of the human CYP2A13 gene: Identification of single-nucleotide polymorphisms and functional characterization of an Arg257Cys variant. J Pharmacol Exp Ther 2002; 302:416-23.
7. Cheng XY, Chen GL, Zhang WX et al. Arg257Cys polymorphism of CYP2A13 in a Chinese population. Clin Chim Acta 2004; 343(1-2):213-6.
8. Chen GL, Xu ZH, Wang W et al. Analysis of the C314T and A595G mutations in histamine N-methyltransferase gene in a Chinese population. Clin Chim Acta 2002; 326(1-2):163-7.
9. Castro J, Telleria JJ, Blanco-Quiros A. Susceptibility genes for asthma and allergy: Hits and questions. J Investig Allergol Clin Immunol 2001; 11(2):73-8.
10. Pang YP, Zheng XE, Weinshilboum RM. Theoretical 3D model of histamine N-methyltransferase: Insights into the effects of a genetic polymorphism on enzymatic activity and thermal stability. Biochem Biophys Res Commun 2001; 287:204-8.
11. Chen GL, Xu ZH, Wang W et al. Analysis of the C314T and A595G mutations in histamine N-methyltransferase gene in a Chinese population. Clin Chim Acta 2002; 326(1-2):163-7.
12. Xie HG, Xu ZH, Ou-Yang DS et al. Meta-analysis of phenotype and genotype of NAT2 deficiency in Chinese populations. Pharmacogenetics 1997; 7(6):503-14.
13. Liolitsa D, Powell JF, Prince M et al. Association study of the 5-HT(2A) receptor gene polymorphism, T102C, and essential hypertension. J Hum Hypertens 2001; 15:335-9.
14. Yu BN, Wang A, Zhou G et al. T102C genetic polymorphism of the 5-HT2A receptor in Chinese hypertensive patients and healthy controls. Clin Exp Pharmacol Physiol 2004; 31(12):847-9.
15. Zhou HH, Koshakji RP, Silberstein DJ et al. Altered sensitivity to and clearance of propranolol in men of Chinese descent as compared with American whites. N Engl J Med 1989; 320(9):565-70.
16. Zhou HH, Adedoyin A, Wood AJ. Differing effect of atropine on heart rate in Chinese and white subjects. Clin Pharmacol Ther 1992; 52(2):120-4.
17. Zhou HH, Sheller JR, Nu H et al. Ethnic differences in response to morphine. Clin Pharmacol Ther 1993; 54(5):507-13.
18. Johansson I, Oscarson M, Yue QY et al. Genetic analysis of the Chinese cytochrome P4502D locus: Characterization of variant CYP2D6 genes present in subjects with diminished capacity for debrisoquine hydroxylation. Mol Pharmacol 1994; 46(3):452-9.
19. Zhou HH, Adedoyin A, Wilkinson GR. Differences in plasma binding of drugs between Caucasians and Chinese subjects. Clin Pharmacol Ther 1990; 48(1):10-7.
20. Zhou HH, Liu ZQ. Ethnic differences in drug metabolism. Clinical Chemical Laboratory Medicine 2000; 38:899-903, [in Chinese].
21. Bertillson L, Kalow W. Why are diazepam metabolism and polymorphic S-meophenytoin hydroxylation associated with each other in white and Korean populations but not in Chinese populations. Clin Pharmacol Ther 1993; 53:608-610.
22. Sachse C, Brockmoller J, Bauer S. Roots I: Functional significance of a C→A polymorphism in intron 1 of the cytochrome P450 CYP1A2 gene tested with caffeine. Br J Clin Pharmacol 1999; 47:445-449.
23. Han XM, Ou-Yang DS, Lu PX et al. Plasma caffeine metabolite ratio (17X/137X) in vivo associated with G-2964A and C734A polymorphisms of human CYP1A2. Pharmacogenetics 2001; 11(5):429-35.
24. Liu J, Liu ZQ, Yu BN et al. β1-Adrenergic receptor polymorphisms influence the response to metoprolol monotherapy in patients with essential hypertension. Clin Pharmacol Ther 2006; 80(1):23-32.
25. Kim RB. 3-Hydroxy-3-methylglutaryl-coenzyme A reductase inhibitors (statins) and genetic variability (single nucleotide polymorphisms) in a hepatic drug uptake transporter: What's it all about? Clin Pharmacol Ther 2004; 75(5):381-5.
26. Zhang W, He YJ, Han CT et al. Effect of SLCO1B1 genetic polymorphism on the pharmacokinetics of nateglinide. Br J Clin Pharmacol 2006; 62(5):567-72.
27. Wang LS, Zhou G, Zhu B et al. St John's wort induces both cytochrome P450 3A4-catalyzed sulfoxidation and 2C19-dependent hydroxylation of omeprazole. Clin Pharmacol Ther 2004; 75(3):191-7.

Chapter 11

Pharmacogenetics in Admixed Polynesian Populations

Rod A. Lea* and Geoffrey K. Chambers

Abstract

Collectively, the islands of the remote Pacific Ocean form the last geographic region on earth to be colonised by humans. The region is known as Polynesia and is defined by a triangular boundary joining, Hawaii in the North, Easter Island in the East and New Zealand in the South. Polynesia contains many island populations each with a fascinating genetic history that has been shaped by unique evolutionary forces. Such forces have created substantial population differences in genetic variation which may explain, at least in part, the variable disease and drug response characteristics of Polynesian populations. This chapter summarises some of the genetic research that has been conducted in the largest Polynesian population—The Maori of New Zealand—including measurements of genetic admixture and studies of drug metabolising genes and drug response traits.

A Genetic History of the Polynesian Migrations

The history of the Austronesian-speaking peoples of Remote Oceania (aka Polynesia plus Micronesia) is a history of gene flow currents producing characteristic founder effects and admixture (Fig. 1). Molecular genetics provides a powerful means of confirming accounts of migrations informed by other disciplines such as Archaeology and Linguistics. But it can also provide valuable new information about the specific direction and magnitude of effects of intermarriage between differentiated populations. In short, two major episodes of admixture in Austronesians have been clearly demonstrated by using genetic data. First, the blending of two long distinct gene pools occurred when a voyaging population of Neolithic Austronesian-speaking farmers moved south through island southeast Asia and encountered Palaeolithic Paupuan-speaking hunter-gatherers resident in Near Oceania (effectively today's Melanesia) during the Lapita period. Second, and much later, further admixture took place with European settlers in New Zealand, French Polynesia, Hawaii and elsewhere.

The presence of a physically and culturally (and now also known to be genetically) related set of human populations dispersed across vast distances of the Pacific Ocean has been an intriguing mystery and the subject of academic inquiry, since first European contact. One of the earliest ideas to capture the popular imagination was that these people had come from South America on rafts.[1] This idea is now thoroughly discounted since genes that characterise native South Americans[2] are not widespread in the Pacific and vice versa. The only report of a minor contribution[3] is disputed[4] and, even if real, probably only reflects return trading voyages

*Corresponding Author: Rod A. Lea—Institute of Environmental Science and Research Ltd, 34 Kenepuru Drive, Porirua, New Zealand. Email: Rod.Lea@esr.cri.nz

Pharmacogenomics in Admixed Populations, edited by Guilherme Suarez-Kurtz.

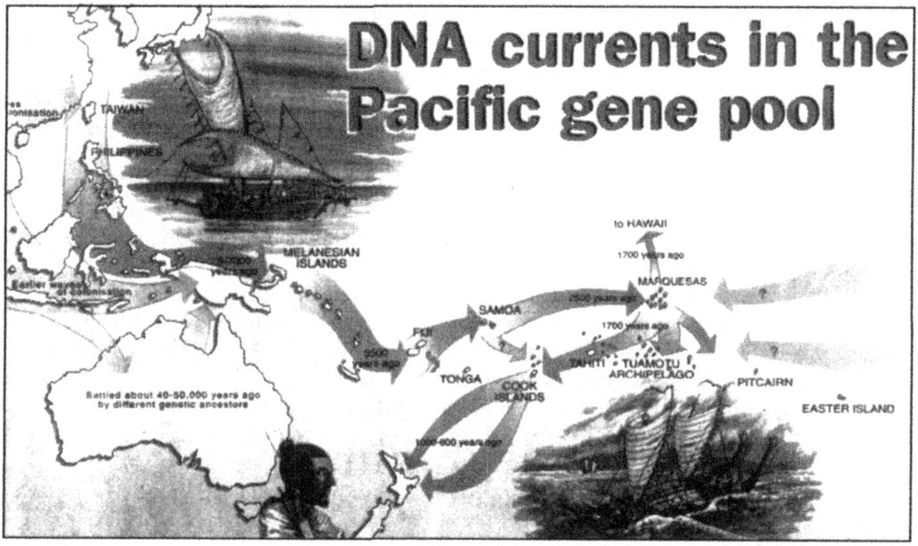

Figure 1. Around 10000 years ago, a small group of people migrated from mainland Asia and settled in Taiwan. They became a great seafaring culture and from their Taiwanese homeland they travelled down past Papua New Guinea. Genetics suggests they took on board local "male" Melanesian guides with whom they married and had children. This mix of Asian and Melanesian seafarers migrated onward to Fiji and surrounding islands. Because of their adventurous nature and/or necessity (e.g., lack of resources) these peoples eventually island hopped out into remote Oceania reaching Hawaii in the North, Easter Island in the East and finally settling in New Zealand between 500 and 1000 years ago (Picture reprinted with permission from Dominion Post, New Zealand).

by Polynesians.[5] Heyerdahl's ingenious suggestion was replaced by a more robust account based on the archaeological chronology of Southeast Asia[6] popularly known as "The Express Train to Polynesia".[7] This new model described the southwards movement of Austronesian farmers from Taiwan starting around 6000 ybp moving rapidly through The Philippines, Indonesia to reach the Bismark Archipelago north of Papua New Guinea around 3500 ybp. The migration paused here before spreading out to fill up the entire "Polynesian Triangle" bounded by Hawaii, New Zealand and Easter Island. This process resulted in effective, but not exclusive, cultural and linguistic replacement along the migration route. It invoked limited, but not zero, genetic admixture between voyagers and residents[8] all contrary claims from other commentators[9] not withstanding.

The position of Taiwan as the "Proto-Austroneisian Homeland in Asia"[10] has been questioned,[11,12] but now seems fairly secure following criticism of the original objections[13] and the discovery of archaic mitochondrial DNA (mtDNA) lineages in aboriginal populations in Taiwan.[14] Indeed, these authors have even identified the Ami tribe as the most likely founders of the Austronesia expansion. The general concept of rapid movement and replacement seemed to be bourne out by early observations that the populations of coastal and highland Papua New Guinea were genetically differentiated.[15] However, it has become increasingly clear that interactions between the two peoples have actually resulted in fairly extensive genetic admixture between them.[16,17] For instance, studies of globin gene deletions have shown that the common $\alpha^{3.7III}$- IIIa haplotype found right across Oceania is the result of a recombination event during the period before voyaging extended from island Papua New Guinea to Vanuatu, Fiji and beyond (see Jobling et al, 2004 for a full account ref. 18).

The most complete evidence comes from studies of the mitochondrial DNA Control Region segments, HVS-I and HVS-II. All of the mtDNA haplotypes so far found in Oceanian populations have been classified into just three haplogroups or clusters Groups I to III.[19] All subsequent work has borne out the utility of this classification as an indicator of genetic origin, with Group I firmly associated with southeast Asia.[3,20] Taken together, the data suggest an 85:15 split of the maternal lineages found in Remote Oceania between Asia and Melanesia.[21] This is in marked contrast with the distribution of Y chromosome haplotypes which show closer to a 50:50 split.[22] The explanation for this discrepancy lies in cultural practice resulting in gender-biased geneflow.[23] This view is well supported by accounts of widespread matrilocal marriage practice in Melanesia.[24,25]

The subsequent spreading out of the proto-Polynesian peoples across Remote Oceania has resulted in a loss of genetic diversity from west to east. This is apparent in mtDNA haplotypes,[26] nuclear markers[27] and diversity at simple tandem repeat loci (STR) associated with SNPs on the Y chromosome.[22,28] The effect has been termed "The Genetic Bottleneck in Polynesia"[29] and has potentially significant consequences for medical science. The reduction in genetic diversity is due to repeated founder effects[30] and has been effective in remodelling and refining the various Polynesian gene pools, despite relatively large-scale migrations.[5]

The final chapter in the Pacific story involves a second round of genetic admixture with the recolonisation of many Pacific Islands by Europeans and subsequent widespread intermarriage with Polynesians.

The Maori of New Zealand are the largest of the Polynesian populations and the group that we will focus on here. The gene pool of contemporary Maori is derived principally from three important sources; ancient Austronesia ancestors from southeast Asia, more recent admixture with Melanesians and intermarriage with Europeans within the last two hundred years. Recognising these facts must have significant consequences for the general future of Maori health, as medical science becomes increasingly a genetic-based science. This observation clearly applies to obvious cases such as tissue transplant technology, but equally to shaping individual responses to chemotherapy and pharmaceuticals in general and also to environmental and public health hazards such as alcohol and tobacco.

Genetic Structure of the New Zealand Maori Population

To date, molecular genetic studies aimed at understanding disease susceptibility in New Zealand Polynesians have largely focussed on rare diseases that have been observed to segregate in large Maori families. Such studies have successfully identified the genetic mutations causing familial gastric cancer[31] and malignant hyperthermia[32] in the affected pedigrees, respectively. Population-based epidemiological studies, which incorporate genetic admixture and/or account for population stratification, can be a powerful strategy for identifying loci that influence common traits in heterogeneous populations.[33] Before such studies can be properly designed, it is important to understand the genetic history of the target population and to use this knowledge to characterise genetic structure of the population. Such population genetic studies will provide important baseline data for designing epidemiological studies aimed at identifying common disease genes in predisposed New Zealanders with Maori (Polynesian) ancestry.

The Maori population of New Zealand represents the final link in a long chain of island-hopping voyages beginning in Taiwan and stretching across the South Pacific—the last of the great human migrations (see Fig. 1). It is generally accepted that the Maori population originated about 800 years ago when several small tribal groups (perhaps totalling 200) departed the Cook Islands in large Waka (canoes) and, through careful planning and skilled navigation, landed on the Northern Island of Aotearoa - the land of the long white cloud (New Zealand). For around 600 years after the arrival of the founding ancestors the Maori population underwent rapid growth in geographic isolation from other populations and reached an estimated size of 100,000. In 1769, James Cook rediscovered New Zealand during the flagship Endeavour's tour of the South Pacific. Thereafter the British began colonising New Zealand. Over the next

50 years the Maori population would be reduced to ~50,000 due to infectious disease and war with Europeans and among tribes. Despite the effects of colonisation this resilient population would recover and has grown to well over 500,000 today. However, this growth was no longer in isolation and over the past 200 years there has been widespread intermarriage between the indigenous Maori and European[a] colonisers resulting in genetic admixture.

Measuring the genetic contribution of Caucasians to the Maori population has received relatively little research attention making accurate estimates of admixture difficult to obtain. Katayama et al, analysed the self-reported ancestry data in the 1976 New Zealand census and reported that the genetic contribution of European Caucasians to the Maori population was 37.3%.[34] The New Zealand census no longer collects ancestral information so up-to-date admixture statistics based on self-report are not available for the Maori population. Genetic analyses have offered some understanding of Caucasian admixture in Maori. Sex-specific genetic markers within a sample of subjects who identify themselves as Maori showed about 15% of mtDNA and around 50% of Y chromosomal were of European descent.[5,28] A more genome-wide measure of genetic admixture for the modern Maori population can be estimated using unlinked autosomal DNA markers. A statistical analysis of autosomal short tandem repeat (STR) polymorphism data representing ~10,000 individuals declaring Maori ancestry allow an estimate of 43% (± 0.2) Caucasian admixture to be derived.[35,36] It is therefore reasonable to expect that between 40% and 50% of the Maori gene pool in the 21 century is comprised of Caucasian genomic material of mainly United Kingdom origin.

It is important to note that despite the extensive admixture in the Maori population this indigenous group still seems to be genetically distinguishable from the Caucasian population of New Zealand with an average Fst genetic distance of 0.02 separating the two groups.[35] Estimates of individual ancestry proportions using unlinked STRs in a small sample of Maori (shown in Fig. 2) also support the idea that New Zealand Maori remain a distinctive group of people.

Disease and Gene Frequencies in the Maori Population

Identifying the genetic determinants of common disease susceptibility and drug response that relate to specific and/or mixed ethnic groups is of substantial public health importance because mode of medical treatment may ultimately depend on it. New Zealand has the largest Polynesian population in the world (including Maori and other Pacific Island Polynesians). According to the 2001 census, a significant proportion of the New Zealand population (~20% or ~800,000 individuals) identified themselves as being Maori or Pacific Islander, or as having mixed "EuroPolynesian" ancestry. Many disease traits are at higher prevalence in Polynesian groups compared with the Caucasian population. These include traits associated with cardiovascular, metabolic and addictive disorders. Conversely, some traits are, or seem to be, at decreased prevalence in Polynesian groups including skin cancer, multiple sclerosis and migraine.[37] The differential risk of these conditions in Polynesians may be partly due to higher (or lower) frequencies of susceptibility genes in these subgroups and/or genetic interaction arising as a result of intermarriage with Caucasians (admixture). Although there have been relatively few genetic epidemiologic studies of disease in the Maori population, several studies have surveyed samples of Maori for genetic variants thought to be of medical importance. The allele frequency estimates for these studies are listed in Table 1 and show that differences, often quite large, exist for many variants when Maori groups are compared to Caucasian. It is plausible that such differences in allele frequencies may partly explain differences in associated disease prevalence between these two ancestral groups.

[a] The vast majority of intermarriage in New Zealand over the past 200 years has occurred between native Maori and British colonisers. However, more recent mixing has occurred with Scottish, German and Dutch immigrants. In this report we will refer to Europeans as "Caucasian" meaning they are non-African and non-Asian/Pacific islander.

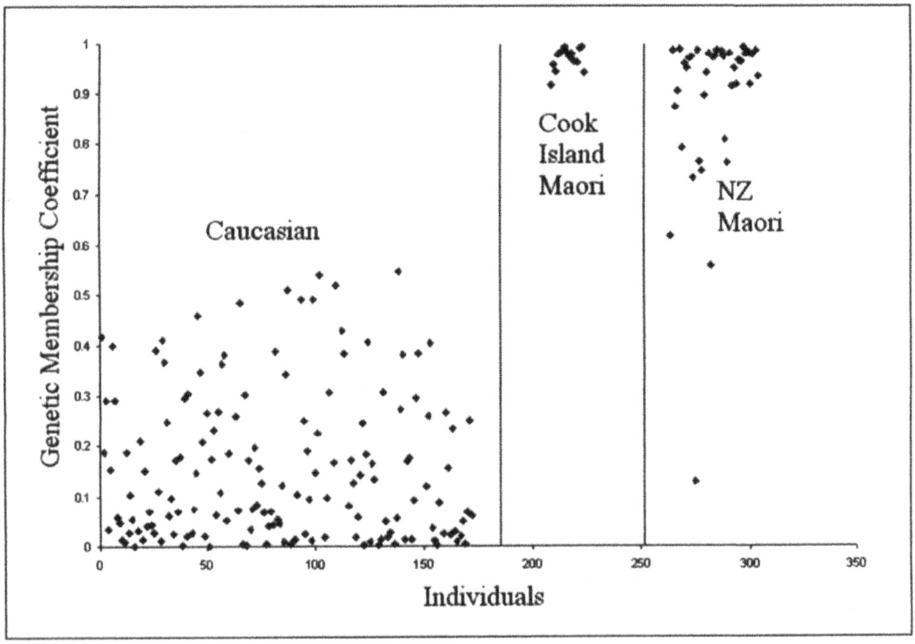

Figure 2. Genetic cluster analysis of individual ancestry proportions based on allele frequencies of 17 unlinked STR polymorphisms shows that individuals with Maori ancestry are clearly distinguishable from those with Caucasian Ancestry ($P < 0.001$). Nb. Cook Island Maori are the putative parental population of the modern New Zealand Maori population.

Alcohol Dehydrogenase Genes and Maori

Alcohol is the most commonly used behaviour-altering recreational drug in New Zealand. Patterns of drinking appear to differ between Maori and the general New Zealand population. A recent survey of Maori drinking patterns found that although 20% of Maori abstain from drinking alcohol, larger quantities of alcohol were consumed annually by Maori drinkers compared to Caucasian.[46] Sarfati (1999) noted Maori drinkers typically drink less often but consume more per session and that Maori have greater potential for hazardous drinking behaviour compared to Caucasian.[47] Similar drinking patterns have been observed in other Austronesian populations including Taiwanese aborigines - the putative origin population of Polynesian peoples. This raises the intriguing hypothesis that the genetic variants, which influence alcohol response and drinking behaviour in New Zealand Maori and contribute to differences between Caucasian, originated in Asia and were shaped by the Polynesian migrations across the Pacific Ocean.

Alcohol is primarily metabolised by alcohol dehydrogenase (ADH) enzymes in the liver. The hepatic alcohol dehydrogenases (EC 1.1.1.1) are a family of catabolic enzymes that oxidise alcohols, including the oxidation of ethanol to form acetaldehyde.[48] The ADH subunits (α, β, γ, π, χ and μ) are encoded by a cassette of linked genes (named *ADH1A, ADH1B, ADH1C, ADH4, ADH5, ADH6,* and *ADH7* respectively) positioned on the long arm of chromosome 4 (4q21-4q23). Variation in the enzyme activity of ADH influences metabolic response to ingested alcohol and susceptibility to abuse behaviour. This variation is due in part to single nucleotide polymorphisms (SNPs) within the genes that encode the enzymes.

One commonly studied exonic SNP allele in *ADH1B* (referred to as Arg47His, *ADH1B*2*, or *ADH2*2*, but called *ADH1B*47His* here) has been associated with protection against alcohol dependence in numerous studies,[43,51,52] and is common in Asian and Pacific populations

Table 1. Allele frequency estimates for Maori and Caucasian groups at genetic loci thought to be of medical importance

Gene Name	Variant	Disease	Allele Frequency Estimates (%)*		Reference
			Maori	Caucasian	
α 1 Antitrypsin	A1-AT(MS)	Respiratory	1	7	38
Phenylalanine hydroxylase	PAH-2	Phenolketonuria	0	1	39
Human Leukocyte Antigen	A1	Immunity	8	19	40
	A3		5	13	
	A9		30	10	
	A11		18	7	
	A19		68	20	
	B7		13	26	
	B8		11	25	
	B12		10	33	
	B16		18	5	
	B22		39	7	
	B40		51	16	
T Cell Receptor (α and β chain loci)	Cα (7kb)	Immunity	32	34	41
	Vα (1.4kb)		12	47	
	Vβ8		19	42	
	Vβ7		0	19	
Angiotensinogen	235T		79	41	42
Angiotensin Converting Enzyme	ACE-D	Cardiovascular disease	18	54	
Alcohol Dehydrogenase	ADH2-2	Alcoholism	42	3	43
CARD15		Crohn's Disease	12	23	44
MAO-A	3 repeat allele	Addiction	56	30	45

*Percentage values are for the rarer of the two alleles. Based on two-by-two contingency table analysis and the chi-squared statistic almost all frequency differences between Maori and Caucasian produce a P-value lower then 0.05. However, because frequencies are estimates from different studies (i.e., are based on different samples of variable size) care should be taken when interpreting these data.

(see Table 2).[43,51] Chambers et al[43] compared ADH genotype frequencies between alcoholics and nonalcoholics in the young male Polynesian population of New Zealand. They found that the *ADH2*2* (*ADH1B*47His*) frequency in known Maori alcoholics (0.15) was much less than that of Maori controls (0.42, $p < 0.01$).[43] *ADH3*1* was also found to have a low frequency in alcoholics, but it has been suggested by Chambers et al[43] that this is probably the result of linkage disequilibrium (LD) between *ADH2*2* and *ADH3*1*.

Our more recent studies of ancestry informative markers (AIMs)[b] spanning the ADH gene region suggest that the Maori population has a different haplotype signature at the ADH gene region compared to Caucasians (see Fig. 3), including the presence of an extended haplotype block of ~5.8Mb. These findings probably reflect the unique gene flow history of this genomic

[b] Ancestry informative markers for Maori are polymorphisms that have markedly different allele frequencies between Maori and European populations and as such can probabilistically distinguish ancestry.

Table 2. Estimates of ADH1B*47His allele frequency in different groups

Population (Sample Size)	ADH1B*47His (%)	Reference
Caucasian		
Mixed European	4	43
Asian		
Japanese (97)	69	49
Chinese - Han (48)	68	50
Chinese - Korean (50)	71	
Chinese - Mongolian (35)	63	
Chinese - Elunchun (37)	38	
Austronesian		
Taiwanese - Atayal (65)	91	51
Taiwanese - Ami (23)	70	52
Taiwanese - Bunun (51)	84	
Taiwanese - Paiwan (34)	86	
Polynesian		
New Zealand Maori (56)	46	43

region in the Maori sample and should be beneficial for designing future genetic association studies, including mapping genes by admixture LD, for alcohol response traits and associated disorders in Polynesians.[53]

The *CYP2A6* Gene and Nicotine Metabolism in Maori Smokers

In New Zealand the prevalence of cigarette smoking is around 22% for the general population. However the smoking rates are markedly higher for Maori (46%) compared to Caucasians (20%) and, for reasons that are unclear, the rates for female Maori are among the highest in the world—52% nationally and up to 60% in some regions.[54] Extensive targeted campaigns for smoking cessation in the 1990s has led to a reduction in tobacco consumption in New Zealand, yet the high prevalence of smoking for Maori has decreased only marginally according to the 2005 census.

High rates of smoking are associated with elevated rates of smoking-related diseases, and it has been estimated that smoking is responsible for around 30% of Maori deaths compared to about 17% nationally.[55] As these statistics emphasize, identifying the determinants of the high smoking prevalence in Maori and using this information to develop new targeted cessation strategies is of major public health importance in New Zealand. Whilst cultural and economic factors contribute to ethnic differences these do not explain all of the prevalence disparity between Maori and Caucasians. Data from the 2001 New Zealand census show that ethnic differences for smoking prevalence exist across all socioeconomic strata.[56] Identifying the underlying metabolic and/or genetic differences between groups with different ancestral backgrounds may be important, since this knowledge could provide new insights into the most effective ways of reducing tobacco-related disease in Maori.

Nicotine is the primary, although not the sole, compound for initiation and maintenance of sustained smoking behaviour.[57] Smokers tend to consume a regular number of cigarettes per day presumably to maintain the desired pharmacological effects of nicotine.[58] The sustained daily levels of nicotine ingested by smokers are partly determined by the rate at which nicotine is metabolised in the liver.[59] Variation in the rate of nicotine metabolism rate is also thought to influence an individual's initial risk of becoming a smoker as well as their degree of dependence on tobacco. Benowitz and colleagues have shown that nicotine metabolic rate varies widely among individuals, and among the major ethnic/racial groups in the United States i.e., Asian,

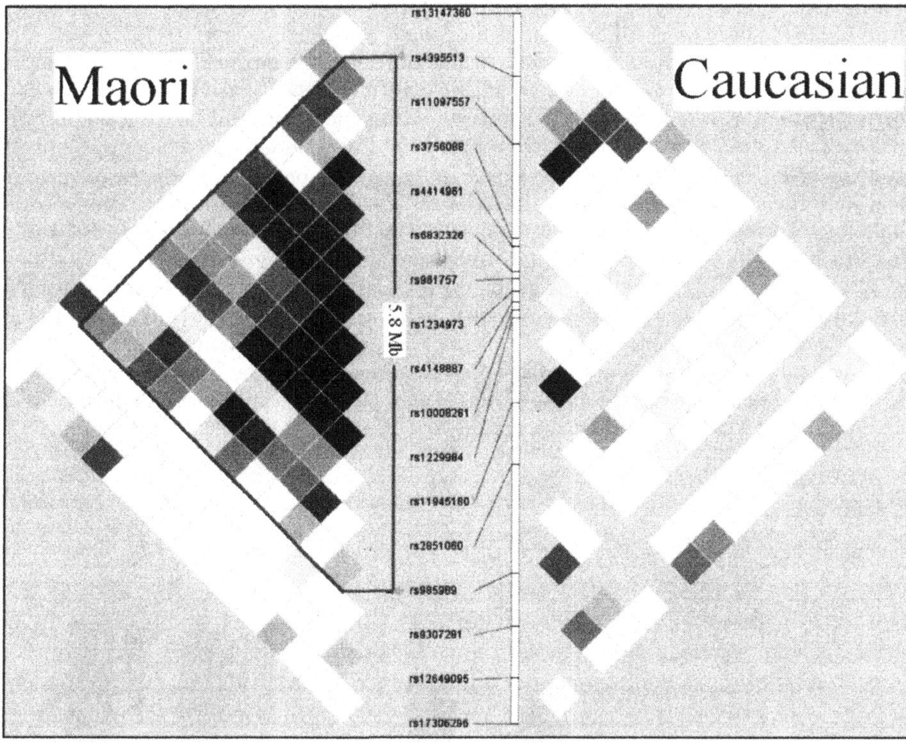

Figure 3. LD patterns observed across an 8.3Mb region on Chr 4 in Maori (n=80) and Caucasian (from Hapmap). Each of the markers (listed in centre) has been correlated with every other marker and the strength of the correlation (D′ statistic) colour coded. Black = strong LD, grey = moderate LD and white = no LD between marker alleles. Moderate admixture LD (D′ > 0.4) extends as far as 5.8Mb in Maori. In contrast there is very little in Caucasian.

African and Caucasians.[60,61] In particular, these researchers showed that Chinese-American smokers exhibit on average a 35% reduction in nicotine metabolic rate and take in less nicotine per cigarette compared to European-American smokers.[60] Information on ethnic differences in nicotine metabolism may have important implications for smoking cessation programs since slower metabolic rate implies that lower optimal dosages of nicotine replacement therapy (NRT) may be required for certain populations of Asian origin.[60]

After nicotine is absorbed through the lungs by cigarette smoking it is primarily (~80%) metabolized to cotinine (COT) by the liver enzyme—Cytochrome P-450 2A6 (CYP2A6).[57] COT is subsequently metabolized by CYP2A6 to trans- 3'-hydroxycotinine (3-HC). The ratio of the 3-HC and COT concentration (3-HC:COT ratio) in saliva is highly correlated with oral clearance of COT in smokers (r > 0.9), which in turn reflects intrinsic metabolic clearance of nicotine by the liver via the CYP2A6 enzyme.[62] Therefore, a single 3-HC:COT ratio derived from a saliva sample taken first thing in the morning can be considered a reliable index of CYP2A6 activity and hence the rate of hepatic metabolism of nicotine.[62] COT concentration on its own is highly correlated with plasma cotinine (r = 0.99) and is a widely used biomarker for the dose of inhaled or ingested nicotine (i.e., nicotine intake).[58,63]

Variation in CYP2A6 enzyme activity (i.e., nicotine metabolic rate) is strongly influenced by genetics with a heritability of ~60% in Caucasians.[64] Several DNA sequence polymorphisms in the *CYP2A6* gene have been associated with nicotine metabolic rate, degree of tobacco dependence and susceptibility to smoking-related disease.[65] Large variation in *CYP2A6* allele frequencies has been observed between ethnic groups worldwide.[66] Thus, *CYP2A6* gene variants are potentially useful biomarkers for ethnic differences in nicotine metabolism and tobacco dependence. Two variants of the *CYP2A6* gene (*CYP2A6*9* and **4* alleles), which have been associated with slow nicotine metabolism, are far more prevalent in Asian populations (Chinese, Japanese, Koreans) compared to Europeans.[66] Specifically, individuals possessing 1 or 2 copies of *CYP2A6 *9* or **4* exhibit significantly reduced, or complete absence of, nicotine metabolism via the CYP2A6 enzymatic pathway.[66,67] Given the putative ancestral (genetic) links between the New Zealand Maori population and South East Asia we suspected similar frequencies might exist for these slow nicotine metabolising gene variants in Maori.

Lea et al conducted a study investigating the prevalence of the *CYP2A6*9* and *4 alleles in the general Maori population and nicotine metabolic rate in a sample of Maori and Caucasian smokers using salivary metabolites as markers of CYP2A6 enzyme activity.[68]

The results of these analyses provide compelling evidence that (a) there is a higher prevalence of the slow nicotine metabolising alleles (**4* and **9*) in the general Maori population (Fig. 4), and (b) that Maori smokers on average are slow nicotine metabolisers (~35%) through the

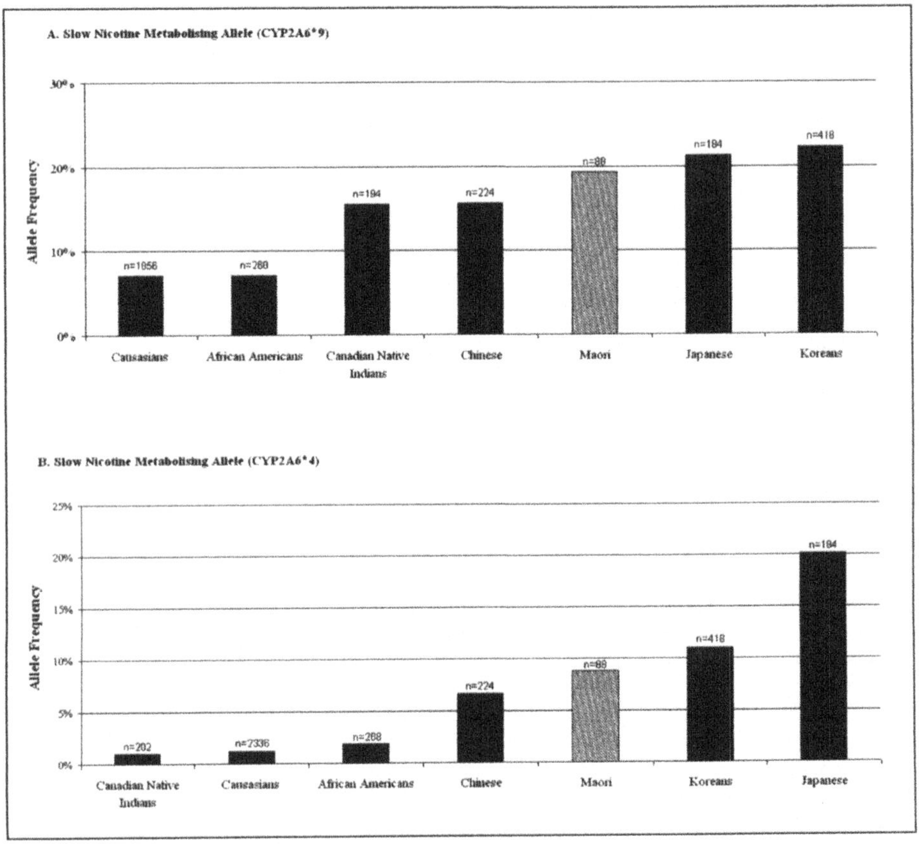

Figure 4. Frequency of slow nicotine metabolising alleles *CYP2A6*9* (A) and *CYP2A6*4* (B) among different worldwide ethnic groups. n = number of alleles tested in each group. Non Maori frequencies were obtained from reference 66. Figure reproduced with permission from the New Zealand Medical Journal.

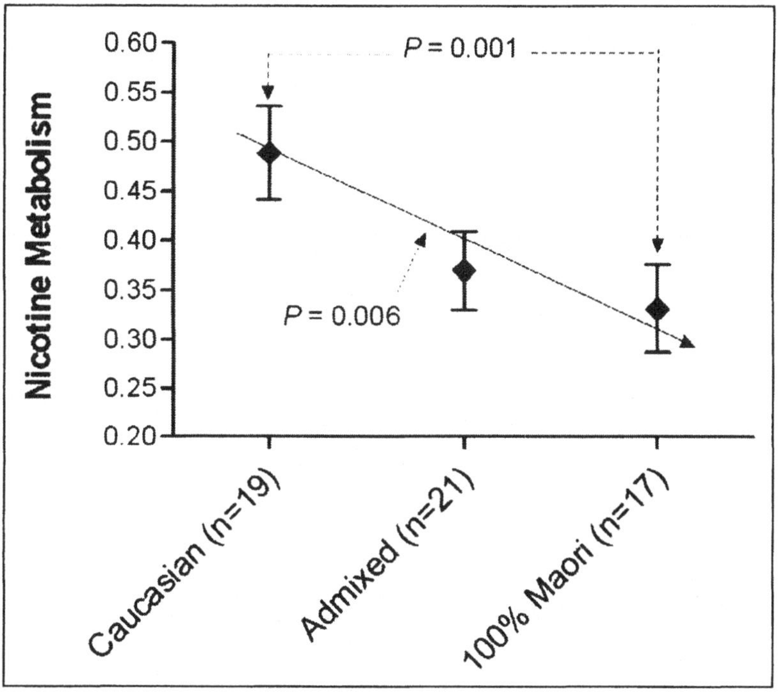

Figure 5. Y-axis shows nicotine metabolism estimated by salivary 3-HC:COT ratio and X-axis shows ancestral groups with sample size in brackets. These data provide evidence that a) Maori smokers have significantly slower nicotine metabolic rates compared to Caucasian smokers and b) that there is a significant linear correlation between nicotine metabolic rate and varying degree of Maori ancestry (results updated from Lea et al, 2005 ref. 68).

CYP2A6 enzyme compared to Caucasians (Fig. 5). In addition, nicotine metabolic rate in smokers seems to correlate with variation in self-reported Maori ancestry suggesting that genetic differences between the two ancestral groups contribute to the trait (Fig. 5).

There is evidence for a higher frequency of *CYP2A6* alleles (**4* and **9*) in Maori compared to Caucasian. These alleles have been previously associated with slower acting CYP2A6 and nicotine clearance in smokers—~80% of nicotine is metabolised via CYP2A6.[67] There is also evidence that slow acting CYP2A6 may modify response to nicotine patch therapy and likelihood of smoking cessation.[69] Therefore, knowledge of increased frequency of the "slow-acting" *CYP2A6* alleles and associated nicotine metabolic rate in Maori might benefit the smoking cessation programmes and clinicians when screening smokers for likely success using standard dose nicotine replacement patches.

Drug Metabolising Genes in New Zealand Maori

The CYP enzymes play a central role in the metabolism of widely used drugs such as antidepressants, beta blockers, and antipsychotics. Many of the CYP genes are known to contain polymorphisms that are associated with differential drug response among individuals. Allele frequencies of some of these polymorphisms are known to vary considerably across different ethnic groups, and therefore may be important for understanding variation in drug response among patients with different ancestral backgrounds.

Little research has been conducted on pharmacokinetics or pharmacogenetics of drugs in the Maori population. In 1995, Wanwimolruk et al, phenotyped CYP2D6 and CYP2C19 in a

Table 3. Frequency estimates of variant CYP alleles in Maori and Caucasian groups

		Allele Frequency (%)		Reference for Caucasian Frequency
Gene	**Variant**	**Maori***	**Caucasian**	
CYP2C9	*2	1.7	11#	72
	*3	0.8	7.4#	
CYP2C19	*2	24	15	73
	*3	1.7	0.4	
CYP2D6	*3	0.9	1	74
	*4	7.9	19.5	
	*5	1.8	4.1	
	*6	0	1.3	
	*9	0	2	
	*10	6.1	2	
	*41	3.5	20#	

* Values for Maori sample are from reference.[71] # Values are the lower of the range published for European from[72] (i.e., British, Italian, Spanish and Swedish).

sample of Maori using debrisoquine and proguanil as substrate drugs, respectively. These researchers found that the prevalence of poor metabolisers (PMs) for debrisoquine was not higher in Maori compared to Caucasians. However the frequency of the PM phenotype for proguanil was increased in Maori.[70] These findings suggest that genetic variants, influencing CYP2C19 activity may play a role in predicting relevant drug response in Maori.

Lea et al[71] have estimated the population prevalence of functionally relevant alleles in the *CYP2C9*, *CYP2C19*, *CYP2D6* and *CYP2A6* genes in the Maori population and compared these frequencies to those reported for Caucasians.

This study involved a sample of DNAs (n = 60) derived from a preexisting bank of DNA from unrelated Maori living in the Wellington region. The Maori population of this urban city represents all tribal groups in New Zealand. Participants were classified as "Maori" by self-report using i) the 2001 census definition for ethnicity and ii) an ancestral definition—i.e., having 8 Maori great grandparents. As such, this sample is considered fairly representative of the ancestral (pre-admixed) Maori population. The DNA samples were tested for the commonly studied *CYP2* variants listed in Table 3.

Table 3 shows the *CYP* allele frequencies observed for the Maori sample compared to previously published estimates in Caucasian samples (see references in Table 3 for detail). Across all variants tested the absolute difference values ranged from less than 1% to 16%. The largest differences were observed for *CYP2C9*2*, *CYP2D6*4* and *CYP2A6*9* (>11%). In Maori, the PM alleles (i.e., *2 and *3) were less prevalent for *CYP2A9* and more prevalent for *CYP2C19* compared to Caucasian. The distributions for *CYP2D6* alleles was different between the groups due to lower frequency of *4 and *41 alleles and higher frequencies of *10 alleles in Maori.

This study estimated allele frequencies for functionally relevant *CYP* gene variants in a sample of Maori considered to be fairly representative of the nonadmixed Maori population. The rationale for the research is based on the premise that the unique genetic history of Maori has significantly modified the genetic structure at these loci, particularly compared to Caucasian, and that this may partially explain differential drug response of this indigenous population.

The results showed that substantial differences exist for alleles of *CYP2C9*, *CYP2C19* and *CYP2D6* polymorphisms between Maori and Caucasian groups compared. The increased prevalence of PM alleles for *CYP2C19* in Maori is consistent with the phenotypic results of Wanwimolruk et al (1995). These findings may ultimately have implications for clinicians prescribing commonly used drugs metabolised via these enzymes such as fluoxetine and warfarin. For example, if a patient with Maori ancestry has a different likelihood of possessing variant CYP450 alleles this might alter their risk of adverse events or otherwise influence successful treatment outcomes.

Disease Gene Mapping and the Admixed Maori Population

Today in New Zealand there are many diseases (e.g., diabetes, heart disease and gout) that are much more prevalent in the Maori ethnic group compared to Caucasians, and visa versa.[37] If these diseases are genetically influenced it is plausible that at least part of the ethnic disparity is due to differences in genetic makeup between Maori and Caucasian populations, although nonbiological factors such as cultural, and economic variables will also play a role.[37]

The striking disease disparities that exist between Maori and Caucasian, coupled with the genetic admixture between these two groups, suggests that the present day New Zealand population is an excellent candidate for identifying unknown disease genes using a concept called admixture gene mapping.[33] The underlying premise of admixture gene mapping is that when two genetically distinct populations interbreed an unusual relationship is created among polymorphisms across the genome. This phenomenon results in the formation of long segments of DNA (haplotypes) that have distinguishable ancestral origins.[75] This natural genomic process provides researchers with a powerful approach to localising (or mapping) the genes underlying diseases that differentially effect human sub-populations (Fig. 6). Whilst the admixture mapping concept is not new it has only recently begun to be explored following the construction of large polymorphism databases as well as the advent of high throughput genotyping and biocomputing technologies. The admixture gene mapping approach has been successfully applied in the admixed African American population to understand the genetics of Multiple Sclerosis (MS) and hypertension.[76,77]

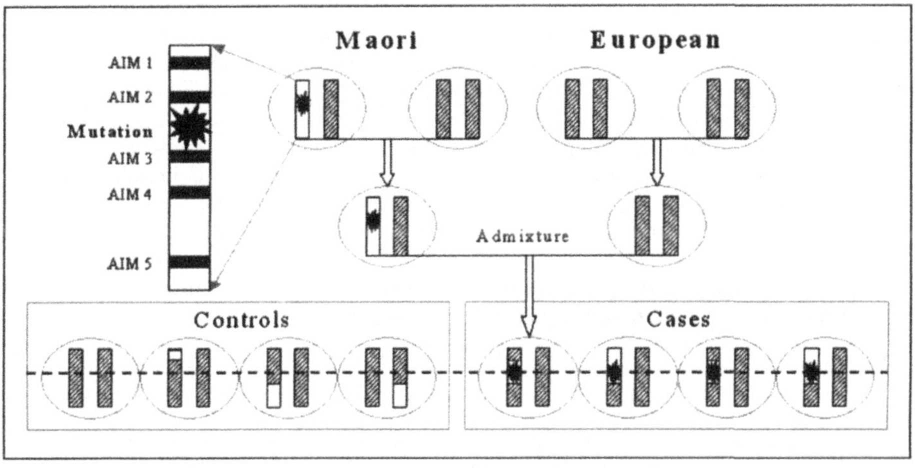

Figure 6. A simplified schematic of a case control admixture mapping design. Both cases and control groups have Maori ancestry. Circles represent individuals. Inheritance of chromosomal regions inherited specifically from Maori ancestors are shown as unshaded boxes. Recombination confines the Maori-specific mutation or disease influencing variant (grey star) to a haplotype which is only present in disease cases (right box). Adjustment to correct for background variation in Maori-ancestry should be made using unlinked genomic control markers.

The first step toward localising genes by admixture mapping is to identify a genome-wide panel of SNPs that have markedly different allele frequencies between the "parental" populations,[78] which for the present day admixed Maori population are precolonisation Maori and UK Caucasians. Such SNPs are referred to as ancestry informative markers (AIMs) because they can distinguish the ancestral origin of the haplotype on which they reside (Fig. 6).[79] Due to ancient gene flow between human groups it is rare to find SNP alleles that are 100% informative of ancestry i.e., SNP alleles that are private to one parental population. However, many studies have shown that SNPs with allele frequency differences of ≥0.3 (30%) can be informative for ancestry.[80] Based on this accepted threshold, our analyses of over 50 unlinked polymorphisms in Maori have shown that >70% are informative for ancestry and as few as 17 can easily distinguish individuals with Maori and/or European ancestry (see Fig. 2).

The number of AIMs required for an admixture genome scan will naturally depend on the size of the admixture-derived haplotypes. Because there have only been 8-10 generations since Maori and Europeans came into contact there will have been relatively little recombination between ancestral chromosomes. This means that ancestry-specific haplotypes should tend to be at least several million base pairs in length in Maori. In support of this hypothesis, our unpublished studies of AIMs on chromosome 4 have indicated that moderate LD in admixed Maori (n = 80 individuals) extends across a region of 5.8 Mb (see Fig. 3). If this pattern is typical of the entire Maori genome we can expect to be able to fully characterise the admixture in Maori and perform a whole genome scan using <3000 SNPs. This is ~500 times fewer than would be required for a whole genome association scan in general European populations[33] and thus highlights the cost-effectiveness of our approach. A case control admixture mapping design is illustrated in Figure 6.

Closing Remarks

The present day Maori population exhibits significant genetic admixture largely as a result of intermarriage with Caucasians of European (UK) origin. Although less well characterised such admixture is also likely to be present in other Polynesian populations of the Pacific Islands e.g., Hawaiians, Samoans, Tahitians. This unique genetic structure presents potential problems and advantages for pharmacogenomics studies involving Polynesian populations. When conducting genetic association studies or clinical trials of drug response care should be taken to control for variation in genetic ancestry within the study cohort to avoid false or misleading results. This is best achieved using genomic control markers (i.e., DNA markers that can distinguish Polynesian and Caucasian ancestry), or if DNA is not available, as may be the case for clinical trials, we suggest self reported ancestry of the patient based on grandparental information could be used. Alternatively, variation in Polynesian ancestry may, in itself, be a clinically useful predictor of drug response. Admixed populations such as Maori also present genetic researchers with an extraordinary opportunity to map genetic variants associated with differential drug response or disease susceptibility using admixture LD. Applying research strategies that address genetic admixture should ultimately lead to more personalised pharmacological treatment and better health outcomes for Polynesian people.

References

1. Heyerdahl T. The Kon-Tiki expedition: By raft across the south seas. London: Allen and Unwin, 1953.
2. Torroni A, Schurr TG, Cabell MF et al. Asian affinities and continental radiation of the four founding Native American mtDNAs. Am J Hum Genet 1993; 53:563-590.
3. Sykes B, Leiboff A, Low-Beer J et al. The origins of the Polynesians: An interpretation from mitochondrial lineage analysis. Am J Hum Genet 1995; 57:1463-1475.
4. Bonatto SL, Redd AJ, Salzano FM et al. Lack of ancient Polynesian-Amerindian contact. Am J Hum Genet 1996; 59:253-256.
5. Whyte ALH, Marshall SJ, Chambers GK. Human evolution in Polynesia. Hum Biol 2005; 77:157-177.

6. Bellwood P. The Austronesian dispersal and the origin of languages. Scientific American 1991; 264:70-75.
7. Diamond J. Express train to Polynesia. Nature 1988; 336:307-308.
8. Bellwood P, Diamond J. On explicit 'replacement' models in Island Southeast Asia: A reply to Stephen Oppenheimer. World Archaeol 2005; 34:503-506.
9. Oppenheimer S. The 'Express train from taiwan to polynesia': On the congruence of proxy lines of evidence. World Archaeol 2004; 36:591-600.
10. Melton T, Clifford T, Martinson J et al. Genetic evidence for the Proto-Austronesian homeland in Asia: mtDNA and nuclear DNA variation in Taiwanese aboriginal tribes. Am J Hum Genet 1998; 63:1807-1823.
11. Richards M, Oppenheimer S, Sykes B. mtDNA suggests Polynesian origins in eastern Indonesia. Am J Hum Genet 1998; 57:1463-1475.
12. Oppenheimer S, Richards M. Fast trains, slow boats, and the ancestry of the Polynesian islanders. Sci Prog 2001; 84:157-181.
13. Cox MP. Indonesian mitochondrial DNA and its opposition to Pleistocene era origin of Proto-Polyneisans in island Southeast Asia. Hum Biol 2005; 77:179-188.
14. Trejaut JA, Kivisild T, Loo JH et al. Traces of archaic mitochondrial lineages persist in Austronesian-speaking Formaosan populations. PLoS Biol 2005; 3:e247.
15. Stoneking M, Jorde LB, Bhatia K et al. Geographic variation in human mitochondrial DNA from Papua New Guinea. Genetics 1990; 124:717-723.
16. Merriwether DA, Friedlaender JS, Mediavilla J et al. Mitochondrial DNA variation is an indicator of Austronesian influence in island Melanesia. Am J Phys Anthropol 1999; 110:243-270.
17. Kayser M, Brauer S, Weiss G et al. Melanesian origin of Polynesian Y chromosomes. Curr Biol 2000; 10:1237-1246.
18. Jobling MA, Hurles ME, Tyler-Smith C. Human evolutionary genetics: Origins, peoples, and disease. New York: Garland, 2002.
19. Lum JK, Rickards O, Ching C et al. Polynesian mitochondrial DNAs reveal three deep maternal lineage clusters. Hum Biol 1994; 66:567-590.
20. Melton T, Peterson R, Redd AJ et al. Polynesian genetic affinities with Southeast Asia as identified by mtDNA analysis. Am J Hum Genet 1995; 57:403-414.
21. Lum JK, Cann RL, Martinson JJ et al. Mitochondrial and nuclear genetic relationships among Pacific Island and Asian populations. Am J Hum Genet 1998; 63:613-624.
22. Kayser M, Bauer S, Weiss G et al. Reduced Y-chromosome, but not mitochondrial DNA, diversity in human populations from West New Guinea. Am J Hum Genet 2003; 72:281-302.
23. Chambers GK. Polynesian genetics and Austronesian prehistory: An invited paper given at the international symposium on the dispersal of austronesians and the ethnogenesis of the peoples in the indonesian archipelago held at solo. Central Java 2005, (In press).
24. Hage P, Marck J. Matrilineality and the Melanesian origin of the Polynesian Y chromosomes. Current Anthropology 2003; 44(suppl.):S121-S127.
25. Bellwood P, Snachez-Mazas A. Human migrations in continental east Asia and Taiwan: Genetic linguistic and archaeological evidence. Curr Anthropol 2005; 46:480-484.
26. Hagelberg E, Kayser M, Nagy M et al. Molecular genetic evidence for the human settlement of the pacific: Analysis of mitochondrial DNA, Y chromosome and HLA markers. Phil Trans R Soc Lond 1999; 354:141-152, (B.).
27. Clark AG, Hamilton JF, Chambers GK. Inference of population subdivision from the VNTR distributions of New Zealanders. Genetica 1995; 96:37-49.
28. Underhill PA, Passarino G, Lin AA et al. Maori origins, Y-chromosome haplotypes, and implications for human history in the Pacific. Hum Mutat 2001; 17:271-280.
29. Flint J, Boyce AJ, Martinson JJ et al. Population bottlenecks in Polynesia revealed by minisatellites. Human Genetics 1989; 83:257-263.
30. Marshall SJ, Whyte ALH, Hamilton JF et al. Austronesian prehistory and Polynesian genetics: A molecular view of migration across the Pacific. An invited contribution to New Zealand Science Review 2005; 63:75-80.
31. Richards FM, McKee SA et al. "Germline E-cadherin gene (CDH1) mutations predispose to familial gastric cancer and colorectal cancer." Hum Mol Genet 1999; 8(4):607-10.
32. Brown RL, Pollock AN et al. "A novel ryanodine receptor mutation and genotype-phenotype correlation in a large malignant hyperthermia New Zealand Maori pedigree. Hum Mol Genet 2002; 9(10):1515-24.
33. Smith MW, O'Brien SJ. Mapping by admixture linkage disequilibrium: Advances, limitations and guidelines. Nat Rev Genet 2005; 6(8):623-32, (Review).

34. Katayama K, Tagaya A, Yamada H et al. Cook Islands populations today, in relation to European mixtures and inter-island cross-breedings. J Anthrop Soc Nippon 1988; 96:47-59.
35. Walsh SJ, Triggs CM, Curran JM et al. Evidence in support of self-declaration as a sampling method for the formation of sub-population DNA databases. J Forensic Sci 2003; 48(5):1091-3.
36. Law B. PhD Thesis: Statistical Interpretation of DNA Evidence. Auckland and Hall D: Department of Statistics, University of Auckland, 2002, (2004 (Honours Thesis): Medical genetic variation in Polynesian Populations, in School of Biological Sciences, Victoria University of Wellington).
37. Abbott W, Scragg R, Marbrook J. Differences in disease frequency between Europeans and Polynesians: Directions for future research into genetic risk factors. Pac Health Dialog 2001; 8(1):129-56, (Review).
38. Janus ED, Joyce PR et al. "Alpha-1-antitrypsin variants in New Zealand. N Z Med J 1975; 82(551):289-91.
39. Hertzberg M, Jahromi K et al. Phenylalanine hydroxylase gene haplotypes in Polynesians: Evolutionary origins and absence of alleles associated with severe phenylketonuria. Am J Hum Genet 1989; 44(3):382-7.
40. Henry S, Perry H et al. HLA class I gene, antigen and haplotype frequencies in New Zealand Maori and Europeans. N Z Med J 1994; 107(975):119-21.
41. Abbott WG, Geursen A et al. "Search for linkage disequilibrium between alleles in the T cell receptor alpha and beta chain loci and susceptibility to rheumatic fever." Immunol Cell Biol 1995; 73(4):369-71.
42. Kyle CV, Abbott W et al. "Angiotensin-1-converting enzyme and angiotensinogen gene polymorphisms in Maori and Pacific Island people in New Zealand." Intern Med J 2001; 31(2):116-8.
43. Chambers GK, Marshall SJ, Robinson GM et al. The genetics of alcoholism in Polynesians: Alcohol and aldehyde dehydrogenase genotypes in young men. Alcohol Clin Exp Res 2002; 26:949-955.
44. Gearry RB, Lea RA, Roberts RL et al. CARD15 allele frequency differences in New Zealand Maori: Ancestry specific susceptibility to Crohn's disease in New Zealand? Gut 2006; 55(4):580.
45. Lea RA, Chambers GK. Monoamine Oxidase, the "Warrior" gene hypothesis and addiction traits in Maori. NZMJ Editorial March 2nd 2007; Vol. 120, No. 1250.
46. Barnes HM, McPherson M, Bhatta K. Te Ao Waipiro 25000: Maori national alcohol survey. Auckland, New Zealand: Whariki Research Group, 2003:460-3.
47. Sarfati D. Alcohol use, in Taking the Pulse: The 1996/97 New Zealand Health Survey. Wellington, New Zealand: Ministry of Health, 1999:69-86.
48. Wagner FW, Burger AR, Vallee BL. Kinetic properties of human liver alcohol dehydrogenase: Oxidation of alcohols by class I isoenzymes. Biochemistry 1983; 22:1857-1863.
49. Nakamura K, Iwahashi K et al. Characteristics of Japanese alcoholics with the atypical aldehyde dehydrogenase 2*2. I. A comparison of the genotypes of ALDH2, ADH2, ADH3, and cytochrome P-4502E1 between alcoholics and nonalcoholics. Alcohol Clin Exp Res 1996; 20(1):52-5.
50. Shen YC, Fan JH et al. Polymorphism of ADH and ALDH genes among four ethnic groups in China and effects upon the risk for alcoholism. Alcohol Clin Exp Res 1997; 21(7):1272-7.
51. Chen WJ, Loh EW, Hsu YPP et al. Alcohol dehydrogenase and aldehyde dehydrogenase genotypes and alcoholism among Taiwanese aborigines. Biol Psychiatry 1997; 41:703-709.
52. Osier MV, Pakstis AJ, Soodyall H et al. A global perspective on genetic variation at the ADH genes reveals unusual pattern of linkage disequilibrium and diversity. Am J Hum Genet 2002; 71:84-99.
53. Hall DA, Chambers GK et al. Haplotype analysis at the alcohol dehydrogenase gene region in New Zealand Maori. J Hum Genet 2006, (In press).
54. Tobacco facts (Ministry of health) and public health association of New Zealand (Policy on tobacco), 2003.
55. Lambe J. Smoking kills 31% of Maori. N Z Med J 1998; 111(1072):327.
56. Borman B, Wilson N, Mailing C. Socio-demographic characteristics of New Zealand smokers: Results from the 1996 census. N Z Med J 1999; 112(1101).
57. Nicotine addiction in britain: A report from the tobacco advisory group of the royal college of physicians, (http://www.rcplondon.ac.uk/pubs/books/nicotine/).
58. Jaakkola MS et al. Determinants of salivary cotinine concentrations in Chinese male smokers. Prev Med 2003; 36(3):282-90.
59. Benowitz NL, Pomerleau OF, Pomerleau CS et al. Nicotine metabolite ratio as a predictor of cigarette consumption. Nicotine Tob Res 2003; 5(5):621-4.
60. Benowitz NL et al. Slower metabolism and reduced intake of nicotine from cigarette smoking in Chinese-Americans. J Natl Cancer Inst 2002; 94(2):108-15.

61. Perez-Stable EJ et al. Nicotine metabolism and intake in black and white smokers. Jama 1998; 280(2):152-6.
62. Dempsey DTP, Jacob P, Allen F et al. Nicotine metabolite ratio as an index of CYP2A6 metabolic activity. Clin Pharmacol Ther 2004; 76(1):64-72.
63. Jarvis MJ, Primatesta P, Erens B et al. Measuring nicotine intake in population surveys: Comparability of saliva cotinine and plasma cotinine estimates. Nicotine Tob Res 2003; 5(3):349-55.
64. Swan GE, Benowitz NL, Lessov CN et al. Nicotine metabolism: The impact of CYP2A6 on estimates of additive genetic influence. Pharmacogenet Genomics 2005; 15(2):115-25.
65. Malaiyandi V, Sellers EM, Tyndale RF. Implications of CYP2A6 genetic variation for smoking behaviors and nicotine dependence. Clin Pharmacol Ther 2005; 77(3):145-58, (Review).
66. Schoedel KA, Hoffmann EB, Rao Y et al. Ethnic variation in CYP2A6 and association of genetically slow nicotine metabolism and smoking in adult Caucasians. Pharmacogenetics 2004; 14(9):615-26.
67. Yoshida R et al. Effects of polymorphism in promoter region of human CYP2A6 gene (CYP2A6*9) on expression level of messenger ribonucleic acid and enzymatic activity in vivo and in vitro. Clin Pharmacol Ther 2003; 74(1):69-76.
68. Lea R, Benowitz N, Green M et al. Ethnic differences in nicotine metabolic rate among New Zealanders. N Z Med J 2006; 119(1228):U1825.
69. Malaiyandi V, Lerman C et al. "Impact of CYP2A6 genotype on pretreatment smoking behaviour and nicotine levels from and usage of nicotine replacement therapy." Mol Psychiatry 2006; 11(4):400-9.
70. Wanwimolruk S, Pratt EL, Denton JR et al. Evidence for the polymorphic oxidation of debrisoquine and proguanil in a New Zealand Maori population. Pharmacogenetics 1995; 5(4):193-8.
71. Lea RA, Roberts RL, Green MR et al. Allele Frequency Differences of Cytochrome P450 Polymorphisms. New Zealand, Maori: Submitted Ethnicity and Disease, 2006.
72. Suarez-Kurtz G. Pharmacogenomics in admixed populations. Trends Pharmacol Sci 2005; 26:196-201.
73. Xie HG, Stein CM, Kim RB et al. Allelic, genotypic and phenotypic distributions of S-mephenytoin 4'hydroxylase(CYP2C19) in healthy Caucasian populations of European descent throughout the world. Pharmacogenetics 1999; 9(5):539-49.
74. Griese EU, Ilett KF, Kitteringham NR et al. Allele and genotype frequencies of polymorphic cytochromes P4502D6, 2C19 and 2E1 in aborigines from western Australia. Pharmacogenetics 2001; 11(1):69-76.
75. McKeigue PM. Prospects for admixture mapping of complex traits. Am J Hum Genet 2005; 76(1):1-7, (Review).
76. Reich D, Patterson N, De Jager PL et al. A whole-genome admixture scan finds a candidate locus for multiple sclerosis susceptibility. Nat Genet 2005; 37(10):1113-8.
77. Zhu X, Luke A, Cooper RS et al. Admixture mapping for hypertension loci with genome-scan markers. Nat Genet 2005; 37(2):177-81.
78. Smith MW, Patterson N, Lautenberger JA et al. A high-density admixture map for disease gene discovery in African Americans. Am J Hum Genet 2004; 74(5):1001-13.
79. Hoggart CJ, Shriver MD, Kittles RA et al. Design and analysis of admixture mapping studies. Am J Hum Genet 2004; 74(5):965-78.
80. Nievergelt CM, Schork NJ. Admixture mapping as a gene discovery approach for complex human traits and diseases. Curr Hypertens Rep 2005; 7(1):31-7.

CHAPTER 12

Pharmacogenetics, Ethnic Differences in Drug Response and Drug Regulation

Rashmi R. Shah*

Abstract

The two key components in the pathway between the administration of a drug and t clinical response it elicits are the dose-concentration (pharmacokinetic) and/ concentration-response (pharmacodynamic) relationships of the drug. Both these coı ponents are subject to genetic influences that account for a substantial fraction of inter-individu variability in drug response. Arising from inter-ethnic differences in the frequency of the varia alleles that exert these genetic influences, it is intuitive to anticipate inter-ethnic differences pharmacokinetics, pharmacodynamics and dose-response relationships of a drug. Th frequently translate into differences in drug response. For a variety of reasons, these differen have not hitherto been investigated systematically. The problem is further compounded difficulty in characterizing ethnicity because of ethnic admixture. This admixing exaggera any dissociation of genes that determine drug response from those that determine skin col and other anthropological features. Nevertheless, there are examples of significant inter-ethı differences in drug response as illustrated by ibufenac, clioquinol and gefitinib. Increasing drug development programs are undertaken at a global level with the aims of reducing cos expediting the development process and addressing issues arising from global prescribing drugs. Drug regulatory authorities have responded to these challenges by promulgating a numb of guidelines that recommend sponsors of new drugs to explore the role of genetic variations differences in drug response between individuals and between ethnic populations. Wh inter-ethnic differences are anticipated, bridging studies may be required before data from o ethnic population can be extrapolated to another. The nature of bridging studies is determin on a case-by-case basis and may include (1) pharmacokinetic studies, (2) pharmacodynan studies, (3) dose-response studies and/or (4) pivotal Phase III studies for either safety and/ efficacy. The recent approval by the US FDA of "BiDil" for the treatment of cardiac failure self- identified black patients is a spectacular verification of the regulatory desire to addr ethnicity-related issues. Furthermore, regulatory determination for promoting safe and effect prescribing of drugs and improving pharmacovigilance at a global level without compromisi efficient drug development requires the sponsors to discuss proactively the target populatic that have not been adequately studied during the preapproval period. These inclu sub-populations carrying known and relevant genetic polymorphism and patients of differe racial and/or ethnic origins. Regulatory authorities may refuse to accept an application wh issues arising from ethnicity are not adequately addressed. With increasing global migrati and the resulting admixing of different ethnic populations, the challenges in the future will even greater.

*Rashmi R. Shah—Former Senior Clinical Assessor, Medicines and Healthcare products Regulatory Agency, London. U.K. Email: clin.safety@lineone.net

Pharmacogenomics in Admixed Populations, edited by Guilherme Suarez-Kurtz.

Introduction

Variation in drug response between individuals arises from differences in the dose-concentration (pharmacokinetic) and/or the concentration-response (pharmacodynamic) relationships of a drug. Either of these two relationships can be modulated by a number of genetic and nongenetic factors. Depending on the drug, genetic factors have been suggested to account for 20 to 95 % of the variability in drug disposition and clinical effects.[1,2] This variation extends not only between individuals but also between ethnic groups.

The notion that genetic differences may alter drug response evolved following demonstration that an inherited deficiency of glucose-6-phosphate dehydrogenase was responsible for severe hemolysis observed in some patients when exposed to primaquine, an antimalarial drug.[3] This deficiency, and hence the abnormal clinical response, occurred with a much higher frequency in the African Americans, and was observed only rarely in Caucasians.[4]

Acetylation Polymorphism

An important foundation to the current pharmacogenetics is the early observations on the pharmacokinetics and response to therapy with isoniazid—a drug used to treat tuberculosis. Isoniazid is eliminated by acetylation by N-acetyltransferase (NAT-2). The rate of isoniazid acetylation was shown to have a bimodal distribution. Later, it was demonstrated that this bimodality was genetically determined, with the division of a given population into rapid and slow acetylating phenotypes.[5,6] To date, more than 20 variants of NAT-2 have been described. The wild-type allele responsible for rapid acetylation is NAT-2*4, present in at least 98% of the rapid acetylators. The major variants responsible for slow acetylation are NAT-2*5B, NAT-2*6A and NAT-2*7B. The pharmacokinetic consequences of NAT-2 polymorphism differ between drugs but in general, there is a 3- to 6-fold difference in the metabolic elimination between rapid and slow acetylators. In terms of drug response phenotype, many studies have documented genotype-phenotype relationships for a number of NAT-2 substrate drugs. For example, slow acetylators are more prone to isoniazid-induced neuropathy or hepatitis, procainamide- or hydralazine-induced systemic lupus and sulphapyridine- or dapsone-induced hematological reactions. In contrast, rapid acetylators have generally inadequate therapeutic responses to the drugs concerned. Phenytoin-induced toxicity is more frequent in slow acetylators when isoniazid is coadministered since isoniazid inhibits the enzyme that metabolizes phenytoin.[7-9] The frequencies of variant alleles responsible for slow acetylation are different in different ethnic populations.[10-12]

Apart from slow acetylation status, another important source of genetic susceptibility to isoniazid-induced hepatotoxicity is CYP2E1 polymorphism. CYP2E1 mediates the final step in the metabolism of isoniazid, which generates a hepatotoxic hydrazine metabolite.[13] Drug metabolizing activity of CYP2E1 is also genetically determined, with considerable inter-ethnic differences in the frequency of variant alleles. This may explain why the risk of isoniazid-induced hepatotoxicity does not correlate more closely with the frequency of slow acetylation status alone. Systematic studies have revealed that isoniazid-induced hepatotoxicity is more frequent in older black women.[14]

A vast number of studies over the later half of the last century, beginning with description of genetic polymorphism in drug acetylation, has collectively led to the conclusion that genetic factors modulate the activity of a number of drug metabolizing enzymes and other drug clearance processes as well as the responsiveness of pharmacological targets. Consequent upon inter-ethnic differences in the frequency of variant NAT-2 alleles and NAT-2 genotype determined inter-individual differences in clinical response to drugs metabolized by NAT-2, it is not surprising that there are expectations of inter-ethnic differences in pharmacokinetics, pharmacodynamics, dose-response relationships and clinical responses to other drugs.

Increasingly, drug development programs now are undertaken at a global level. The objectives are to reduce the costs, expedite the drug development process and address the issues arising from global prescribing of drugs. Clearly, there are expectations of extrapolating data from one population to another. Drug regulatory authorities have responded to the challenges arising from these developments by promulgating a number of guidelines that recommend sponsors of new drugs to explore the role of genetic (and nongenetic) variations in differences in drug response between individuals and between ethnic populations.[15] This chapter will summarize the pharmacogenetic basis that may potentially account for inter-ethnic differences in drug response followed by a review of these guidelines.

Inter-Ethnic Differences in Drug Response

Compared to white Caucasians, certain Oriental populations have long been known to have an increased sensitivity to the effects of ethanol.[16] Similarly, there are reports of inter-ethnic differences in response to a number of drugs used clinically. Most studies have concentrated on cardiovascular and neuropsychiatric agents. Analgesics and proton pump inhibitors are other pharmacological agents that have attracted similar interest.[17] Black patients respond poorly to several classes of anti-hypertensive agents such as angiotensin converting enzyme inhibitors [18] and β-blockers.[19] Observations such as these have significant implications for the treatment of potentially fatal diseases such as cardiac failure.[20]

In July 1996, the US Food and Drug Administration (FDA) received an application for BiDil, a fixed-dose combination of 20 mg of isosorbide dinitrate and 37.5 mg of hydralazine hydrochloride, for use in cardiac failure.[21] Although there was no overall significant difference in mortality between placebo and BiDil, a retrospective analysis of the pivotal study (V-HeFT-1) found a favorable trend in black patients. A subsequent mortality study (V-HeFT II) showed the combination to be inferior to enalapril overall, but retrospective analysis showed that the difference was observed in the white population; there was essentially no difference in mortality in the black population. Therefore, a third study (A-HeFT) evaluated BiDil versus placebo in self-identified black patients. After 18 months, the trial was terminated early because of a statistically significant reduction in all-cause mortality and risk of first hospitalization for heart failure in the BiDil treated group. Therefore, although the decision was controversial, the combination was approved in June 2005 for use in the treatment of heart failure as an adjunct to standard therapy in self-identified black patients.[21] This effect has been attributed to a variant of nitric oxide synthase gene in black patients. BiDil represents the first drug to be approved for use in a specific racial or ethnic group.

Why the Limitations in Evaluating Inter-Ethnic Differences?

Until recently, there had been a general aversion to investigating inter-ethnic differences during drug development. Although Hispanics are now the largest ethnic group after Whites in the US, it is interesting that most studies have focused on African Americans, Asians and Whites.[22] The three clinical trials with BiDil were systematic and driven by pragmatic subgroup analyses. For most drugs, however, an evaluation of the full extent of inter-ethnic differences in clinical responses is hampered by lack of adequate data.

A review of the representation of American blacks in 50 clinical trials of new drugs concluded as long ago as 1989 that the proportion of black subjects enrolled was also less than their proportion in the general population.[23] This situation had not improved over the following decade. A retrospective analysis[22] of racial and ethnic group participation in clinical trials and race-related labeling of 185 new molecular entities approved during a five-year period (1995-1999) by the US FDA found that African Americans participated in trials to the greatest extent. However, their participation steadily declined from 12% in 1995 to 6% in 1999. Labeling of 84 (45%) products contained some statement about race, although only 15 (8%) of these included 18 race-related differences. Of these 18 statements, 9 (50%) related to pharmacokinetics, 7 (39%) to efficacy and 2 (11%) to safety. Only one product label recommended a change in dosage based on racial differences. Another study found that although the total number of trial participants increased during the study period, the representation of ethnic minorities decreased.[24] Non-Caucasian patients have also been reported to receive older, less expensive agents, sometimes at higher doses.[25] Often, studies are not controlled for other variables that may influence drug response, such as observer bias, gender, disease chronicity, dietary and environmental factors and exposure to other comedications that induce or inhibit drug metabolizing enzymes. All these factors make it difficult to evaluate the real impact of ethnicity on drug response.

Inter-Ethnic Differences in Safety and Efficacy

Apart from primaquine-induced hemolysis, another earlier example of inter-ethnic difference in drug tolerance concerned chlorpromazine-induced jaundice. This toxic effect was conspicuously absent in the black population and the few cases that occurred were largely confined to the Europeans who actually received lower doses.[26] Another study showed that the recovery of Kenyan African blacks from anesthesia with propofol and fentanyl was much slower, in comparison with Caucasians. The recovery time of Brazilians was much more variable.[27]

Ibufenac and clioquinol are other more striking examples of inter-ethnic differences in drug toxicity. Ibufenac, a nonsteroidal anti-inflammatory drug, was introduced on the UK market in April 1966 and withdrawn from clinical use in February 1968 because of serious and frequent hepatotoxicity.[28,29] Whereas this complication was relatively frequently observed in the UK, it was almost unknown in Japan. Indeed, the drug continued to be available in Japan for some time after it was withdrawn from the UK market. Clioquinol, an anti-diarrhoeal agent, was first introduced in Japan in 1929. During 1950s, reports of clioquinol-induced neuropathy began to appear. Soon, physicians were reporting the appearance in Japan of a new syndrome called subacute myelo-optic neuropathy (SMON) in association with clioquinol.[30] Soon there were some 10,000 cases of SMON in Japan. However, there were only a few cases of clioquinol-induced neuropathy but none of SMON reported in the UK.[31,32] Similarly, despite wide use of clioquinol, there was only a handful of reports of SMON from other parts of the world.[33,34] An inter-ethnic difference in genetic susceptibility to SMON has been suspected.

With regard to efficacy, some of the earliest observations on racial differences in drug response date back to studies in 1921 when it was demonstrated that atropine had differential effect on heart rates in whites and the blacks.[35] Later, it was reported that the mydriatic effects of locally applied ephedrine and related drugs was the greatest in Caucasians, intermediate in Chinese and the least in African Americans.[36] It is only relatively recently that studies have begun to explore racial factors in drug efficacy. For example, the frequency of treatment success in dyslipidaemia is reportedly significantly lower in American Africans than in nonHispanic white patients, with 53.7% and 69.0% respectively achieving their LDL-C goal.[37] Similarly, compared to Caucasian patients, Asian patients appeared to have a lower dosage requirement of clozapine for comparable clinical efficacy.[38,39]

BiDil represents the most dramatic, and probably the first, example of regulatory acknowledgement of inter-ethnic differences in drug response. Gefitinib (IRESSA) is another recent example of marked inter-ethnic differences in drug response. It is an inhibitor of tyrosine kinases, including those associated with the epidermal growth factor receptor (EGFR-TK). EGFR is expressed on the cell surface of many normal and cancer cells. In May 2003, gefitinib was approved as monotherapy for the treatment of patients with locally advanced or metastatic nonsmall cell lung cancer after failure of both platinum-based and docetaxel chemotherapies.[40] The trial population had consisted of about 90% white Caucasians and the approval was based on a tumor response rate of 10.6% in this population. Gefitinib did not significantly prolong survival even when there was a high level expression of EGFR.[40-42] However, a cluster of somatic mutations in the kinase domain of the EGFR was shown to demonstrate a remarkable beneficial response to EGFR-TK inhibitors.[43,44]

The mutation rate was significantly correlated with gender (women 73.3% vs. men 20%) and ethnicity.[45] Importantly, mutation rates are much higher in Japanese (36.8%) and Chinese (48.6%) patients with adenocarcinoma[45,46] compared to other populations such as 3% rate found in the Middle East.[47]Not unexpectedly, responders to gefitinib are more often females of East Asian descent.[48] It may also be noted that whereas gefitinib-induced pulmonary toxicity was observed in 2.0% of the Japanese population, it was observed in only 0.3% of the US population.[40]

BiDil and gefitinib represent modern medicinal products that have been shown conclusively to have ethnicity-specific efficacy. Not surprisingly, the sponsors of new chemical entities (NCEs) are now anxious to address the issues arising from global development and prescription of their drugs. It is evident that explorations of genetic factors that influence pharmacokinetics, pharmacodynamics and dose-response relationships are highly productive lines of investigations in order to uncover inter-ethnic differences in drug response. Only by thoroughly investigating these properties can one promote global development and safe and effective use of drugs clinically.

Inter-Ethnic Differences in Pharmacokinetics

Genetic polymorphisms have been discovered in virtually all the major classes of drug metabolizing enzymes, including the members of the important cytochrome P450 superfamily.[49] As of May 2006, the number of alleles and total variants respectively were 58 and 96 for cytochrome P450 CYP2D6, 21 and 27 for CYP2C19, 10 and 12 for CYP2C8, 24 and 31 for CYP2C9 and 7 and 13 for CYP2E1. Not surprisingly, there are also significant inter-ethnic differences in the frequency of variant alleles responsible for the expressions and activities of these CYP isozymes. For example, the frequency of CYP2D6 poor metabolizers (PM) is much higher in populations of Western Caucasian origin (5-10%) than in Far East and Asian ethnic groups (0-2%).[50-53] The frequency of PMs of CYP2C19 is lower in Western Caucasians (2-4%) compared to the frequencies observed among Orientals (about 15-25%), reaching as high as 60-70% in Vanuatu and other Pacific islands.[17,51-54] Similar differences are observed with regard to CYP2C9 variants[51-55] and CYP2C8 variants.[56-58]

Inter-ethnic differences have also been reported in the frequency of variant alleles of the genes encoding for other nonP450 drug metabolizing enzymes,[12,59-62] P-glycoprotein-mediated efflux transporter (encoded by MDR1, also known as ABCB1)[12,63,64] and members of the organic anion-transporting polypeptides and organic anion transporters (OATP).[65] The functional or clinical significance of most of the alleles of MDR1 is controversial and not yet adequately characterized.[66,67]

Intuitively, this global heterogeneity in allelic frequencies can predictably be expected to result in inter-ethnic differences in pharmacokinetics of drugs. This is an important regulatory issue if data on the dose of a drug from one ethnic group are to be extrapolated to another in different geographical region. However, inter-ethnic differences have also been reported within the same geographical region such as Singapore that includes ethnically heterogeneous (with possibly much admixing) populations such as the Chinese, Malays and Indians.[12]

A number of reviews have emphasized that there are relatively few drugs for which information is available on inter-ethnic differences in their pharmacokinetics.[68-72] Based on the available literature, the drugs most likely to exhibit ethnic differences in their pharmacokinetics are those that undergo significant gut metabolism/transport and/or hepatic first pass metabolism, are highly bound to plasma proteins or have hepatic metabolism as a major route of elimination. Although the appropriate guidelines have been in place for long (see later sections), regulatory authorities have only recently begun to insist on being provided with information on inter-ethnic differences in pharmacokinetics and their potential clinical consequences.

The problem is that virtually all the studies investigating inter-ethnic differences in pharmacokinetics so far have been relatively small, and the ethnicity of the population being investigated is not always clearly defined. The definition of a subgroup as belonging to a distinct race or ethnicity is important not only because of an ever-increasing ethnic admixing but also because it has serious implications for fulfilling the promise of personalized medicine.[73] This admixing invariably results in dissociation of genes for drug response from those that determine skin color and other anthropological features. For example, many geographically Asian populations are lumped together in one ethnic group termed as "Asians" but it is evident that ethnicity so defined may obscure the effects of many ethnicity-related genetic traits.[12,74] Furthermore, whenever inter-ethnic differences in pharmacokinetics are evaluated, there is frequently no evaluation of their corresponding clinical relevance. The following three case studies illustrate the pressing need to investigate the clinical relevance of any observed differences in pharmacokinetics.

Omeprazole concentrations are significantly higher and its oral clearances lower in Chinese subjects than in white subjects.[75] Correspondingly, gastric pH-related increases in exposure to gastrin are also significantly greater in the Chinese than in the white subjects. CYP2C19 metabolic capacity correlates inversely with exposure to omeprazole and the extent of omeprazole-induced hypergastrinaemia. Since proton pump inhibitors such as omeprazole and lansoprazole are primarily metabolized by CYP2C19, it is not surprising that poor metabolizers of CYP2C19 experience greater exposure to these drugs and hence, populations rich in this genotype enjoy higher efficacy rates.[75,76]

A comparison of the pharmacokinetics and pharmacodynamic responses to repaglinide, an oral hypoglycemic drug metabolized by CYP2C9, revealed that the Japanese had significantly higher peak plasma concentrations and systemic exposure compared to the Caucasian subjects.[77] Decrease in blood sugar was more marked in the Japanese. Not surprisingly, hypoglycemic reactions were more common at the highest dose (2.0 mg), where they were observed more frequently in Japanese (7 of the 12 volunteers) than in Caucasian subjects (4 of the 15 volunteers).

Pharmacokinetic studies in healthy volunteers have shown an increase of approximately 2-fold in systemic exposure to rosuvastatin, a very potent HMG-CoA reductase inhibitor, in the Japanese subjects compared with the white subjects. Differences are evident even among the ethnic populations grouped as Asians.[74] Since metabolism plays only a very minor role in rosuvastatin clearance, other mechanisms probably account for these inter-ethnic differences. Multiple OATP family members have been shown to be capable of rosuvastatin transport.[78] Naturally occurring polymorphisms in OATP1B1, including *5, *9, *15, and *18, are associated with profound loss of activity toward rosuvastatin. The limited information that is available suggests that transporter expression and polymorphisms may be key determinants of inter-ethnic variability in response to rosuvastatin and probably other statins.[79] Given that this class of drugs is associated with potentially fatal complication of rhabdomyolysis, it is not surprising that the European regulatory authorities now recommend that Asian patients, as well nonAsians with risk factors, should be started on an ultra low dose of 5mg daily; clearly, Asian ethnicity is a risk factor.[80]

Drug-drug interactions also show dramatic differences between various genotypes of a drug metabolizing enzyme or a transporter. These may then translate into inter-ethnic differences in clinical outcomes from drug-drug interactions. For example, phenytoin-induced toxicity following coadministration of phenytoin with isoniazid is most frequently observed in slow acetylators of isoniazid and is rare in rapid acetylators. High concentrations of isoniazid in slow acetylators readily inhibit phenytoin metabolism.[7] Similarly, CYP2D6 PMs (with alleles expressing no functional enzyme) do not show the drug-drug interactions predicted from in vitro studies.[81] Likewise, ultra-rapid metabolizers (UMs) may also fail to exhibit the expected drug-drug interaction because they have a sufficiently large functional reserve of CYP2D6 activity that they would most probably need much higher (and potentially toxic) doses of the inhibitor to elicit an interaction.[82,83] There are wide inter-ethnic differences in the prevalence of UMs.[84] Under normal conditions of use, the individuals most likely to display a drug interaction are those who have an intermediate or otherwise compromised drug metabolizing capacity (IMs) or those who have inherited CYP2D6 alleles (such as CYP2D6*9, CYP2D6*10 or CYP2D6*17) with reduced or altered affinity for CYP2D6 substrates. The frequencies of these alleles with increased potential for drug interactions vary widely between different ethnic groups.[50,84] CYP2D6 PMs may of course exhibit interactions at alternative competing metabolic pathways.[85] Such genotype-dependent drug interactions have been reported for a number of CYP2D6 and CYP2C19 substrates.[15] Therefore, it is evident that drug-drug interactions may also depend on ethnicity.[86,87] Geographical differences in pattern of drug usage aggravate the resulting inter-ethnic differences in drug interactions. For example, drugs that inhibit CYP2C8 have the effect of converting an individual of extensive metabolizing genotype into a poor metabolizing phenotype. An acyl-glucuronide metabolite of gemfibrozil is a potent inhibitor of CYP2C8 as well as OATP1B1—both of which are polymorphically expressed and responsible for cerivastatin disposition. Whereas cerivastatin was frequently coprescribed with gemfibrozil in the US, this did not occur in Japan since gemfibrozil was not approved there at the time when cerivastatin was withdrawn from the market in August 2001.

Inter-Ethnic Differences in Pharmacodynamics

There is now ample evidence that pharmacological targets of drugs are also polymorphically expressed within a population. Better characterized among these are the pharmacological targets involved in cardiac arrhythmias (mutations of sodium and potassium channels), asthma (mutations of β2-adrenoceptors and of the core promoter of 5-lipoxygenase), cardiac failure (mutations of β2-adrenoceptors) and depression (mutations in the promoter region of serotonin transporter (5-HTT) gene that gives rise to the long and the short alleles). These pharmacodynamic polymorphisms have been shown to be highly relevant in terms of clinical responses to nonantiarrhythmic drugs with QT-prolonging potential,[88] bronchodilatory drugs active at β2-adrenoceptors and ALOX5,[89-91] cardiac failure therapy with carvedilol, a β2-adrenoceptor antagonist,[19,92] and antidepressant efficacy (associated with long allele of 5-HTT) and side effects (associated with short allele) to selective serotonin reuptake inhibitors (SSRI).[93-97]

As with drug metabolizing enzymes, there also exist marked inter-ethnic differences in the frequency of variant pharmacological targets such as β2-adrenoceptors,[98,99] 5-HTT,[93,100-102] cardiac potassium[103] and sodium channels.[104] These differences have implications for potential inter-ethnic differences in concentration-response relationships of drugs acting at these targets.

Of considerable current academic and regulatory interest from drug safety point of view are the mutations of cardiac potassium and sodium channels. These mutations give rise to congenital long QT syndromes that are associated with potentially fatal ventricular arrhythmias such as torsade de pointes (TdP). Drug-induced prolongation of the QT interval is the second leading cause of drug withdrawals.[105] In view of the low penetration of many of these mutations, the size of the population with dysfunctional potassium channels is substantially larger than that diagnosed by ECG recording alone. Because of considerable overlap, measurement of the QT interval alone may not distinguish between individuals who are the carriers and the noncarriers of these mutations.[106-108] Although the affected individuals have a normal ECG phenotype, they have a diminished repolarization reserve nonetheless and are highly susceptible to drug-induced QT interval prolongation and/or TdP, even at the recommended doses that are normally safe. Available evidence suggests that a substantial proportion (10-20%) of the cases of the drug-induced long QT syndrome might represent cases of "forme fruste" of the congenital long QT syndrome. The pharmacogenetic aspects of drug-induced QT interval prolongation and TdP are beginning to attract considerable interest.[88,109-111]

The possibility of an inter-ethnic variation in response to a standard challenge with QT-prolonging drug cannot be ruled out. Following a single 200mg oral dose of quinidine, prolongation of heart rate corrected QT interval from baseline was significantly greater in the healthy white subjects compared to their Nigerian counterparts, although the Nigerians had higher intraerythrocytic quinidine levels.[112] The limitation of this interesting observation is that the subjects were studied in different centers - Nigerians in Lagos and the whites in Liverpool.

Unfortunately, inter-individual and inter-ethnic differences in pharmacodynamic actions of drugs have until recently been a poorly investigated area. Following studies on anticoagulants and antidepressants provide an illustration of studies in this very flourishing and promising area of ethnopharmacogenetics.

Although both VKORC1 and CYP2C9 are polymorphic and present in several haplotypes, VKORC1 is the principal genetic modulator of ethnic differences in warfarin response. In two cohorts of European patients with either increased sensitivity or partial resistance to coumarin therapy, the VKORC1*2 frequency varied highly significantly between the two groups when compared to 200 blood donor controls (coumarin sensitive 96%, coumarin resistant 7%, controls 42%).[113] Analysis of database-derived VKORC1 genotypes reveals marked inter-ethnic differences. For example, VKORC1*2 frequencies are 42% in Europeans, 95% in Chinese and 14% in African Americans.[113] Chinese and Malay subjects have been reported to require lower maintenance doses of warfarin than do the Indians. CYP2C9*3 associated with lower warfarin requirements was less common in Chinese and Malays (7% and 9%, respectively) than in Indians (18%) and could not account for their lower warfarin requirements.[114] However, VKORC1 H1 haplotype (requiring low warfarin doses) was more frequent in Chinese (87%) and Malays (65%) compared to the Indians (12%). In contrast, VKORC1 H7, H8, and H9 haplotypes (requiring high warfarin doses) were rare in Chinese (9%), intermediate in Malays (30%), and very common in Indians (82%).[114] In one multivariate analysis, VKORC1 and CYP2C9 explained 31% and 7.9% respectively of the variability in warfarin dose.[115]

The significance of inter-ethnic differences in the frequencies of the long and short alleles of 5-HTT gene is unclear. As stated earlier, the antidepressant efficacy and safety of SSRIs are associated with long and short alleles respectively, of 5-HTT gene in Caucasians.[93-97]

Response to SSRI drugs also correlates strongly with the presence of long allele in Koreans.[93] In one study, Chinese depressed patients appeared to require lower dosages with consequently lower plasma concentrations of sertraline compared to Caucasian patients to achieve clinical efficacy. Although there were no pharmacokinetic differences between the groups,[116] no correlation has been found between 5-HTT genotype and response to sertraline in these two ethnic groups.[101] Similarly, fluvoxamine was as effective in the Japanese patients as it was in the Caucasians, regardless of the presence or absence of the long allele.[102] A meta-analysis suggested that the effect of the genotype might be different in different ethnic groups.[117] To complicate the matter, recent evidence suggests that polymorphism of the norepinephrine transporter (NET) may be more predictive of response to at least some antidepressants. In a study in Japanese patients, T182C variant of NET was associated with a superior antidepressant response to milnacipran (a serotonin-norepinephrine reuptake inhibitor) whereas the A/A genotype of the NET G1287A polymorphism was associated with a slower onset of therapeutic response. In contrast, no influence of 5-HTT polymorphism was detected on the antidepressant response to this drug.[118]

Similarly, there are significant ethnic differences in response to acute antipsychotic treatment—32% in mixed descent, 24% in black and 9% in white patients[119] but the reasons for this difference have not been fully investigated.

Inter-Ethnic Differences in Dose-Response Relationships

A typical dose-finding program involves approximately 400-600 patients at the most, frequently investigating a fairly narrow range of doses. To compound the problems, these studies unfortunately do not include adequate representation of ethnicity- or age-related variability in pharmacology. Dose-response studies will likely uncover significant inter-ethnic differences if they are designed and adequately powered to identify these. The difference for a given drug will most likely correlate with the difference in frequency of variant alleles that modulate pharmacokinetics and/or pharmacodynamics of the drug concerned. This is intuitive since these properties of a drug are the two key components of its dose-response curve (Fig. 1).

In Japan, the clinical trials advice division of the Organization for Pharmaceutical Safety and Research has been dealing with consultations about studies to bridge data from one population to another since February 1998 (see later section). The most common type of consultations involved the desire by industry to bridge or extrapolate the results of foreign, Phase III, clinical studies by conducting the dose-response studies domestically in the form of bridging studies.[120] Since variations in concentration of a drug can be managed by adjustments in its dose, differences arising from inter-ethnic differences in pharmacokinetically driven dose-response relationships are reflected in regional differences in doses approved. These

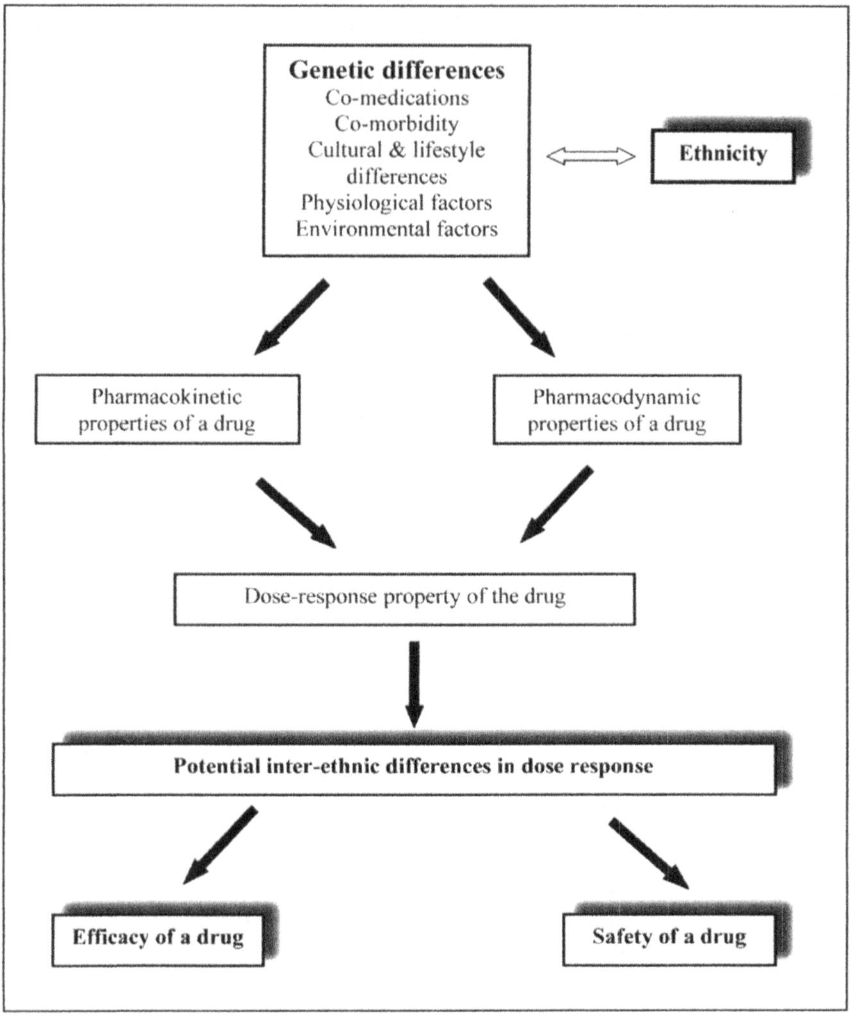

Figure 1. Key determinants of inter-ethnic differences in drug response.

differences in approved dosages of deliberately selected drugs used in Japan, the US and the European Union (EU) have already been highlighted previously.[121] There were differences between the Japanese dosages and the Western dosages as well as between the US and the EU dosages. These limited data suggested that in general, daily dosages tend to be lower in Japanese than in Caucasians but for most medicines, there was no evidence that the Japanese metabolize drugs more slowly than the Caucasians as judged from the exposure to the drug. The data also suggested an important point that inter-ethnic differences were not larger than the intra-ethnic (that is, inter-individual) differences. This is hardly surprising since it is not ethnicity per se but the genotype of the trial population that imposes the hurdle. Individuals of one genotype, for example the PMs of CYP2D6, constitute one distinct subgroup of regulatory interest regardless of their ethnicity. Ethnicity becomes an important issue only when the trial population is not characterized for its genetic profile or inter-genotype differences in pharmacokinetics are not evaluated and the frequency of the variant alleles (e.g., CYP2C19) is substantially different between the trial population (from one ethnic group) and the target population (of another ethnic group). The additional challenges that ethnic admixture poses to the use of ethnicity in promoting personalized medicine are obvious.[73]

When inter-ethnic differences in dose-response relationship of a drug are driven by pharmacodynamic polymorphisms (usually resulting in sub-therapeutic response), the most likely outcome is differential efficacy rates that may impact adversely on risk/benefit assessment of the drug concerned.

Global Drug Development and Regulatory Guidelines

The potential for ethnic differences in drug response and its consequences have been of concern for well over two decades.[122] From the foregoing, it is evident that evaluation of dose, safety and efficacy data from one ethnic population can only be extrapolated to another population if the application for a marketing authorization includes extensive documentation on inter-individual differences in pharmacokinetics, pharmacodynamics, dose-response relationship and drug interactions in the clinical trial population and how these might differ in the target population. Regulatory authorities have long recognized the significance of genetic factors in drug response, especially now that the drug development process extends across the globe. When evaluating new drugs, therefore, they are now increasingly directing their attention to addressing issues that may arise from genetic heterogeneity of the trial and the target patient populations.

In 1993, the FDA published New Drug Evaluation Guidance Document: Refusal to File, on the Agency's use of the refusal-to-file (RTF) option if certain analyses were not performed. The guidance cautions the sponsors that the Agency may exercise its RTF authority if there is "inadequate evaluation for safety and/or effectiveness of the population intended to use the drug, including pertinent subsets, such as gender, age, and racial subsets." In 1998, the FDA published the Demographic Rule, requiring sponsors to (1) tabulate the numbers of participants in clinical trials by age group, gender, and race in investigational new drug applications annual reports and (2) characterize the data in new drug applications according to the same subgroups. Most recently in September 2005, the FDA issued "Guidance for Industry: Collection of Race and Ethnicity Data in Clinical Trials".[123] This document provides guidance on using a standardized approach and race/ethnicity categories for collecting and reporting these data from clinical trials with the aim of improving consistency and comparison across databases.

The International Conference on Harmonization (ICH) is a tripartite body, composed of the regulatory authorities and the industry associations of the EU, Japan and the US with observers from WHO, Health Canada and European Free Trade Area countries, charged with developing guidelines to harmonize global drug development and thus facilitate availability of new medicines efficiently. A key consideration for global drug development and registration therefore is the acceptability of foreign clinical data in the three ICH regions. Two ICH guidelines (E5 and E4) are important in this respect. A number of other regulatory guidelines also make references to the need for thoroughly exploring genetic factors when developing an NCE. These guidelines also impact on acceptability of foreign clinical data.

ICH E5 Guideline on Acceptability of Foreign Clinical Data

In 1998, ICH adopted ICH E5 document entitled "Note for Guidance on Ethnic Factors in the Acceptability of Foreign Clinical Data" for implementation in the ICH regions.[124] In terms of drug development, this guideline is highly significant because (1) there is an increasing globalization of drug development with clinical trials often conducted in a geographical population which may not be the ultimate target of the drug, (2) an increasing number of NCEs is found to be substrates of polymorphic drug metabolizing enzymes and/or active at polymorphic drug targets and (3) that modern drugs are more potent with narrow therapeutic indices and therefore, relatively small differences in either the pharmacokinetics or pharmacodynamics may become highly

Table 1. Features of a medicinal product and likely susceptibility to ethnic differences

Greater Ethnic Sensitivity	Reduced Ethnic Sensitivity
Nonlinear pharmacokinetics	Linear pharmacokinetics
Low bioavailability	High bioavailability
High metabolic clearance	Minimal metabolic clearance
Single metabolic pathway	Multiple metabolic pathways
Polymorphic metabolism	Low protein binding
Steep dose-response curve	Wide dose range
Narrow therapeutic index	Non–systemic mode of action
Wide use of comedication	

relevant clinically. ICH E5 requires sponsors of drugs to submit information on ethnic demography of clinical trial populations and potential ethnic influences on drug response.

ICH E5 recommends the sponsors and the regional regulatory authority in the new region to assess an application for registration "for the ability to extrapolate to new region those parts of the application based on studies from the foreign region".

To this end, it is recommended that the submission should include:

1. adequate characterization of pharmacokinetics, pharmacodynamics, dose-response, efficacy and safety in the population of the foreign region, and
2. characterization of pharmacokinetics, pharmacodynamics and dose-response in the new region.

Important features of a medicinal product that may be indicative of greater or reduced susceptibility to ethnic factors are summarized in Table 1.

When inter-ethnic differences are anticipated, bridging studies may be required. The nature of bridging studies is determined on a case-by-case basis and may include (1) pharmacokinetic studies, (2) pharmacodynamic studies, (3) dose-response studies and/or (4) pivotal Phase III studies for either safety and/or efficacy. The need for bridging studies is determined by the new region to which the foreign data are to be extrapolated and may also arise from differences in choice of dose, comparator(s) used in pivotal trials, unique regional medical practices and novelty of the drug class. Frequently, an expanded safety database may be required if there is a known index case of serious concern in the submission or there is limited database applicable to the new region. In 2003, the ICH also adopted an ICH E5 Questions and Answers document to provide greater clarity to the sponsors of the new drug as well as to ensure that there is consistency in the way this guidance is interpreted and applied by the three regions.[125]

ICH E4 Guideline on Dose-Response Information

The ICH E4 document entitled "Dose-response information to support drug registration" describes how helpful the knowledge of the shape of individual dose-response curves is and it distinguishes individual curves from the population curve.[126] Indirectly, the guideline addresses inter-ethnic differences when it cautions, "Choice of a starting dose might also be affected by potential inter-subject variability in pharmacodynamic response to a given blood concentration level, or by anticipated inter-subject pharmacokinetic differences, such as could arise from nonlinear kinetics, metabolic polymorphisms or a high potential for pharmacokinetic drug-drug interactions". It recommends that in utilizing dose-response information, the influences of various demographic features (including race), individual characteristics (including metabolic differences) and concurrent drugs and diseases should be identified as far as possible.

CPMP Guideline on Pharmacokinetic Studies in Man

The Committee for Proprietary Medicinal Products (CPMP) of the EU has adopted a document entitled "Pharmacokinetic studies in man."[127] Published as early as February 1987, this is probably the first regulatory guideline that includes direct references to the role of genetic factors in determining drug response. This guideline recommends that metabolic studies should indicate whether the metabolism of a drug may be substantially modified in a case of genetic enzyme deficiency and whether saturation of metabolism may occur, thereby resulting in nonlinear kinetics, within the dose levels normally used. The guidelines also cautions that every effort must be made to elucidate the pharmacokinetic mechanism(s) if there is any reason to suspect that the adverse reaction is caused by the altered pharmacokinetics of the substance.

CPMP, FDA and Japanese Guidelines on Drug Interactions

As stated earlier, local patterns of drug usage, and therefore the potential for drug interactions, also impact extrapolation of data from one region to another (and *inter alia*, ethnic populations). Since drug interactions are also genotype-dependent, the CPMP guideline on "Investigation of drug interactions" recommends that when performing mechanism-based in vivo studies, consideration should be given to pharmacogenetic factors.[128] Subjects participating in metabolic in vivo interaction studies should be appropriately genotyped and/or phenotyped (with respect to their drug metabolizing capacity) at the beginning of the study if any of the enzymes mediating the metabolism of the interacting drugs are polymorphically distributed in the population. This CPMP guideline also recommends investigation of drug interactions at sites other than metabolic route such as renal excretion and transport by efflux pumps and P-glycoprotein.

The FDA issued in April 1997 their guidance note "Drug metabolism/drug interaction studies in the drug development process: studies in vitro."[129] This states "Identifying metabolic differences in patient groups based on genetic polymorphisms, or on other readily identifiable factors such as age, race, and gender, could help guide the design of dosimetry studies for such populations groups. This kind of information also will provide improved dosing recommendations in product labeling, facilitating the safe and effective use of a drug by allowing prescribers to anticipate necessary dose adjustments. Indeed, in some cases, understanding how to adjust dose to avoid toxicity may allow the marketing of a drug that would have an unacceptable level of toxicity were its toxicity unpredictable and unpreventable".

The Japanese Ministry of Health, Labor and Welfare's drug regulatory authority (Koseisho, now known as the Pharmaceuticals and Medical Devices Agency, PMDA) has also issued guidelines in June 2001 that recommend genotyping in all drug development programs for drugs that are metabolized by cytochrome P450s.[130,131]

More recently, regulatory authorities have emphasized the significance of drug interactions at transporters and following inhibition of multiple pathways that are responsible for the elimination of a drug.

CPMP Guideline on Bioavailability

The CPMP guidance note on "Investigation of bioavailability and bioequivalence" also recommends that phenotyping and/or genotyping of subjects may be considered for safety or pharmacokinetic reasons.[132]

ICH E14 Guideline on Drug-Induced QT Interval Prolongation

As stated earlier, drug-induced prolongation of QT interval is the second leading cause of drug withdrawal from the market. Regulatory concerns on the ability of an ever-increasing number of nonantiarrhythmic drugs to delay ventricular repolarization and prolong the QT interval have culminated in the adoption of two additional guidelines of relevance to ethnicity. In May 2005, the ICH reached an important milestone when it adopted guidelines on nonclinical

(ICH topic S7B) and clinical (ICH topic E14) strategies by which drugs should be investigated during development for their potential to induce these effects.[133,134] ICH E14 is the first clinical guideline developed by ICH that deals with a specific drug-induced toxicity.

ICH E14 calls for a clinical "thorough QT/QTc study" that is intended to determine whether a drug has a threshold pharmacological effect on QT/QTc interval duration.[134] It speculates that although data are limited, it is not expected that the results of the "thorough QT/QTc study" would be affected by ethnic factors. Although there appear to be no ethnic differences in QTc interval duration at baseline,[135] the possibility of an inter-ethnic difference in response to a standard challenge with QT-prolonging drug cannot be ruled out. As stated earlier, inter-ethnic variations have been demonstrated in the activities of drug metabolizing enzymes as well as in the frequency of pathogenic potassium and sodium channel variants in apparently healthy individuals. Available data suggest that white Caucasians may be more susceptible than the Asians to QT interval prolongation by hERG (human ether-a-go-go) channel blockers. Not surprisingly, the Japanese authority has received only a small fraction of the reports of drug-induced TdP received by the Western authorities from their corresponding regions. ICH E14 recommends that genotyping patients who experience marked prolongation of QTc interval or TdP while on drug therapy should be considered.

ICH E2E Guideline on Pharmacovigilance Specification and Planning

Regulatory authorities have recognized the need for better and earlier planning of pharmacovigilance activities before a product is approved. ICH E2E proposes a structure for a (1) Safety Specification and (2) Pharmacovigilance Plan.[136]

The Safety Specification should summarize important risks identified as well as important potential risks, and important information that is missing. Limitations of the safety database should be considered, and the implications of such limitations with respect to predicting the safety of the product in the marketplace should be explicitly discussed. Particular reference is required regarding populations likely to be exposed during the intended or expected use of the product in medical practice. Regulatory determination for promoting safe and effective prescribing of drugs and improving pharmacovigilance at a global level without compromising efficient drug development requires the sponsors to discuss proactively the target populations that have not been adequately studied during the preapproval period. Therefore, the Specification should discuss the populations that have not been studied or have only been studied to a limited degree in the preapproval phase, especially with respect to the implications regarding predicting the safety of the product in the market. Among the populations that should be considered in this context are:

- Sub-populations carrying known and relevant genetic polymorphism
- Patients of different racial and/or ethnic origins.

Conclusions

In order to expedite and reduce the costs of drug development, new drugs are now frequently developed and investigated for their properties, safety and efficacy in geographical and ethnic populations that may not be the ultimate target of the drugs concerned. There is now a greater appreciation of the risks associated with this strategy and there is little doubt that pharmacoanthropology, also referred to as "ethnopharmacogenetics", will continue to receive much greater regulatory attention than it has hitherto. The sponsors will be required to address issues arising from the global diversity of ethnic populations in terms of how drugs are developed, evaluated, approved, promoted and ultimately prescribed. One of the major issues will be to determine in which cases a complete clinical data package will be sufficient to support regulatory submission, and in which cases additional clinical studies or bridging studies, and what kind of bridging studies will be required.[137] With increasing global migration and the resulting admixing of different ethnic populations, the challenges in the future will be even greater.

References

1. Evans WE, McLeod HL. Pharmacogenomics - Drug disposition, drug targets and side effects. N Eng J Med 2003; 348:538-549.
2. Weinshilboum R. Inheritance and drug response. N Engl J Med 2003; 348:529-537.
3. Alving AS, Carson PE, Flanagan CL et al. Enzymatic deficiency in primaquine-sensitive erythrocytes. Science 1956; 124:484-485.
4. Beutler E. Drug induced haemolytic anaemia. Pharmac Rev 1969; 21:73-103.
5. Evans, DA, Manley KA, McKusick VA. Genetic control of isoniazid metabolism in man. BMJ 1960; 2:485-491.
6. Levy GN, Weber WW. Interindividual variability in arylamine N-acetyltransferase. In: Pacifici GM, Pelkonen O, eds. Interindividual Variability in Human Drug Metabolism. London: Taylor and Francis, 2001:333-357.
7. Kutt H, Winters W, McDowell FH. Depression of prahydroxylation of diphenylhydantoin by antituberculosis chemotherapy. Neurology 1966; 16:594-602.
8. Kutt H, Verebely K, McDowell F. Inhibition of diphenylhydantoin metabolism in rats and in rat liver microsomes by antitubercular drugs. Neurology 1968; 18:706-710.
9. Miller RR, Porter J, Greenblatt DJ. Clinical importance of the interaction of phenytoin and isoniazid: A report from the Boston Collaborative Drug Surveillance Program. Chest 1979; 75:356-358.
10. Bertilsson L, Kalow W. Interethnic differences in drug disposition and effects. In: Pacifici GM, Pelkonen O, eds. Interindividual variability in human drug metabolism. London: Taylor and Francis, 2001:15-74.
11. Lin HJ, Han CY, Lin BK et al. Ethnic distribution of slow acetylator mutations in the polymorphic N-acetyltransferase (NAT2) gene. Pharmacogenetics 1994; 4:125-134.
12. Chowbay B, Zhou S, Lee EJ. An interethnic comparison of polymorphisms of the genes encoding drug-metabolizing enzymes and drug transporters: Experience in Singapore. Drug Metab Rev 2005; 37:327-378.
13. Huang YS, Chern HD, Su WJ et al. Cytochrome P450 2E1 genotype and the susceptibility to antituberculosis drug-induced hepatitis. Hepatology 2003; 37:924-930.
14. Jordan TJ, Lewit EM, Reichman LB. Isoniazid preventive therapy for tuberculosis. Decision analysis considering ethnicity and gender. Am Rev Respir Dis 1991; 144:1357-1360.
15. Shah RR. Pharmacogenetics in drug regulation: Promise, potential and pitfalls. Phil Trans R Soc B 2005; 360:1617-1638.
16. Goedde HW, Agarwal DP. Aldehyde oxidation: Ethnic variation in metabolism and response. In: Kalow W, Goedde HW, Agarwal DP, eds. Ethnic Differences in Reactions to Drugs and Xenobiotics. New York: Alan R Liss, 1986:113-138.
17. Xie HG, Kim RB, Wood AJ et al. Molecular basis of ethnic differences in drug disposition and response. Ann Rev Pharmacol Toxicol 2001; 41:815-850.
18. Exner DV, Dries DL, Domanski MJ et al. Lesser response to angiotensin-converting-enzyme inhibitor therapy in black as compared with white patients with left ventricular dysfunction. N Engl J Med 2001; 344:1351-1357.
19. Yancy CW, Fowler MB, Colucci WS et al. for US Carvedilol Heart Failure Study Group. Race and the response to adrenergic blockade with carvedilol in patients with chronic heart failure. N Engl J Med 2001; 344:1358-1365.
20. Taylor JSW, Ellis GR. Racial differences in responses to drug treatment: Implications for pharmacotherapy of heart failure. Am J Cardiovasc Drugs 2002; 2:389-399.
21. Anonymous: Labelling of 'BiDil' - NDA Number. 020727. Rockville, Maryland, USA: Food and Drug Administration, (http://www.fda.gov/cder/foi/appletter/2005/020727ltr.pdf, http://www.fda.gov/cder/foi/label/2005/020727lbl.pdf, [Accessed on 30 May 2006]).
22. Evelyn B, Toigo T, Banks D et al. Participation of racial/ethnic groups in clinical trials and race-related labeling: A review of new molecular entities approved 1995-1999. J Natl Med Assoc 2001; 93(12 Suppl):18S-24S.
23. Svensson CK. Representation of American blacks in clinical trials of new drugs. JAMA 1989; 261:263-265.
24. Murthy VH, Krumholz HM, Gross CP. Participation in cancer clinical trials: Race-, sex-, and age-based disparities. JAMA 2004; 291:2720-2726.
25. Nasrallah HA. Avoiding side effects in ethnically diverse patients. CNS Spectr 2005; 10(3 Suppl 2):21-29.
26. Lambo TA. Chlorpromazine jaundice. BMJ 1957; iii:1048.
27. Ortolani O, Conti A, Ngumi ZW et al. Ethnic differences in propofol and fentanyl response: A comparison among Caucasians, Kenyan Africans and Brazilians. Eur J Anaesthesiol 2004; 21:314-319.
28. Cahal DA. Jaundice and ibufenac. Lancet 1967; 1:280.
29. Anonymous: Ibufenac (Dytransin) withdrawn. Drug Ther Bull 1968; 6(12):48.

30. Nakae K, Yamamoto S, Shigematsu I. Relation between subacute myelo-optic neuropathy (SMON) and clioquinol: A nationwide survey. Lancet 1973; 1:171.
31. Terry SI. Transient dysaesthesiae and persistent leucocytosis after clioqionol therapy. BMJ 1971; iii:745.
32. Shah RR. Thalidomide, drug safety and early drug regulation in the UK. Adverse Drug React Toxicol Rev 2001; 20:199-255.
33. Ross FC. Clioquinol. In: Kalow W, Goedde HW, Agarwal DP, eds. Ethnic Differences in Reactions to Drugs and Xenobiotics. New York: Alan R Liss, 1986:323-330.
34. Wadia NH. SMON as seen from Bombay. Acta Neurol Scand Suppl 1984; 100:159-164.
35. Paskind HA. Some differences in response to atropine in white and coloured races. JAMA 1921; 76:104-108.
36. Chen KK, Poth EJ. Racial differences as illustrated by the mydriatic action of cocaine, euphthalmine and ephedrine. J Pharmacol Exp Ther 1929; 36:429-445.
37. Clark LT, Maki KC, Galant R et al. Ethnic differences in achievement of cholesterol treatment goals. Results from the National Cholesterol Education Program Evaluation Project utilizing novel e-technology II. J Gen Intern Med 2006; 21:320-326.
38. Matsuda KT, Cho MC, Lin KM et al. Clozapine dosage, serum levels, efficacy, and side-effect profiles: A comparison of Korean-American and Caucasian patients. Psychopharmacol Bull 1996; 32:253-257.
39. Ng CH, Chong SA, Lambert T et al. An inter-ethnic comparison study of clozapine dosage, clinical response and plasma levels. Int Clin Psychopharmacol 2005; 20:163-186.
40. Anonymous: Labelling of 'Iressa' - NDA Number. 021399. Rockville, Maryland, USA: Food and Drug Administration, (http://www.fda.gov/cder/foi/label/2003/021399lbl.pdf, [Accessed on 30 May 2006]).
41. Bell DW, Lynch TJ, Haserlat SM et al. Epidermal growth factor receptor mutations and gene amplification in nonsmall-cell lung cancer: Molecular analysis of the IDEAL/INTACT gefitinib trials. J Clin Oncol 2005; 23:8081-8092.
42. Thatcher N, Chang A, Parikh P et al. Gefitinib plus best supportive care in previously treated patients with refractory advanced nonsmall-cell lung cancer: Results from a randomised, placebo-controlled, multicentre study (Iressa Survival Evaluation in Lung Cancer). Lancet 2005; 366:1527-1537.
43. Lynch TJ, Bell DW, Sordella R et al. Activating mutations in the epidermal growth factor receptor underlying responsiveness of nonsmall-cell lung cancer to gefitinib. N Engl J Med 2004; 350:2129-2139.
44. Paez JG, Janne PA, Lee JC et al. EGFR mutations in lung cancer: Correlation with clinical response to gefitinib therapy. Science 2004; 304:1497-1500.
45. Sasaki H, Shimizu S, Endo K et al. EGFR and erbB2 mutation status in Japanese lung cancer patients. Int J Cancer 2006; 118:180-184.
46. Mu XL, Li LY, Zhang XT et al. Gefitinib-sensitive mutations of the epidermal growth factor receptor tyrosine kinase domain in Chinese patients with nonsmall cell lung cancer. Clin Cancer Res 2005; 11:4289-4294.
47. Al-Kuraya K, Siraj AK, Bavi P et al. High epidermal growth factor receptor amplification rate but low mutation frequency in Middle East lung cancer population. Hum Pathol 2006; 37:453-457.
48. Thomas SK, Fossella FV, Liu D et al. Asian ethnicity as a predictor of response in patients with nonsmall-cell lung cancer treated with gefitinib on an expanded access program. Clin Lung Cancer 2006; 7:326-331.
49. Daly AK. Pharmacogenetics of the major polymorphic metabolizing enzymes. Fundam Clin Pharmacol 2003; 17:27-41.
50. Bradford LD. CYP2D6 allele frequency in European Caucasians, Asians, Africans and their descendants. Pharmacogenomics 2002; 3:229-243.
51. Mizutani T. PM frequencies of major CYPs in Asians and Caucasians. Drug Metab Rev 2003; 35:99-106.
52. Shimizu T, Ochiai H, Asell F et al. Bioinformatics research on inter-racial difference in drug metabolism I. Analysis on frequencies of mutant alleles and poor metabolizers on CYP2D6 and CYP2C19. Drug Metab Pharmacokinet 2003; 18:48-70.
53. Ozawa S, Soyama A, Saeki M et al. Ethnic differences in genetic polymorphisms of CYP2D6, CYP2C19, CYP3As and MDR1/ABCB1. Drug Metab Pharmacokinet 2004; 19:83-95.
54. Kaneko A, Lum JK, Yaviong L et al. High and variable frequencies of CYP2C19 mutations: Medical consequences of poor drug metabolism in Vanuatu and other Pacific islands. Pharmacogenetics 1999; 9:581-590.
55. Xie HG, Prasad HC, Kim RB et al. CYP2C9 allelic variants: Ethnic distribution and functional significance. Adv Drug Deliv Rev 2002; 54:1257-1270.
56. Dai D, Zeldin DC, Blaisdell JA et al. Polymorphisms in human CYP2C8 decrease metabolism of the anticancer drug paclitaxel and arachidonic acid. Pharmacogenetics 2001; 11:597-607.
57. Totah RA, Rettie AE. Cytochrome P450 2C8: Substrates, inhibitors, pharmacogenetics, and clinical relevance. Clin Pharmacol Ther 2005; 77:341-352.

58. Nakajima M, Fujiki Y, Noda K et al. Genetic polymorphisms of CYP2C8 in Japanese population. Drug Metab Dispos 2003; 31:687-690.
59. Guillemette C. Pharmacogenomics of human UDP-glucuronosyltransferase enzymes. Pharmacogenomics J 2003; 3:136-158.
60. Collie-Duguid ES, Pritchard SC, Powrie RH et al. The frequency and distribution of thiopurine methyltransferase alleles in Caucasian and Asian populations. Pharmacogenetics 1999; 9:37-42.
61. Hon YY, Fessing MY, Pui CH et al. Polymorphism of the thiopurine S-methyltransferase gene in African-Americans. Hum Mol Genet 1999; 8:371-376.
62. Carlini EJ, Raftogianis RB, Wood TC et al. Sulfation pharmacogenetics: SULT1A1 and SULT1A2 allele frequencies in Caucasian, Chinese and African-American subjects. Pharmacogenetics 2001; 11:57-68.
63. Ieiri I, Takane H, Otsubo K. The MDR1 (ABCB1) gene polymorphism and its clinical implications. Clin Pharmacokinet 2004; 43:553-576.
64. Kim RB, Leake BF, Choo EF et al. Identification of functionally variant MDR1 alleles among European Americans and African Americans. Clin Pharmacol Ther 2001; 70:189-199.
65. Marzolini C, Tirona RG, Kim, RB. Pharmacogenomics of the OATP and OAT families. Pharmacogenomics 2004; 5:273-282.
66. Eichelbaum M, Fromm MF, Schwab M. Clinical aspects of the MDR1 (ABCB1) gene polymorphism. Ther Drug Monit 2004; 26:180-185.
67. Soranzo N, Cavalleri GL, Weale ME et al. Identifying candidate causal variants responsible for altered activity of the ABCB1 multidrug resistance gene. Genome Res 2004; 14:1333-1344.
68. Wood AJ, Zhou HH. Ethnic differences in drug disposition and responsiveness. Clin Pharmacokinet 1991; 20:350-373.
69. Johnson JA. Predictability of the effects of race or ethnicity on pharmacokinetics of drugs. Int J Clin Pharmacol Ther 2000; 38:53-60.
70. Burroughs VJ, Maxey RW, Levy RA. Racial and ethnic differences in response to medicines: Towards individualized pharmaceutical treatment. J Natl Med Assoc 2002; 94(10 Suppl):1-26.
71. Bjornsson TD, Wagner JA, Donahue SR et al. A review and assessment of potential sources of ethnic differences in drug responsiveness. J Clin Pharmacol 2003; 43:943-967.
72. Kim K, Johnson JA, Derendorf H. Differences in drug pharmacokinetics between East Asians and Caucasians and the role of genetic polymorphisms. J Clin Pharmacol 2004; 44:1083-1105.
73. Suarez-Kurtz G. Pharmacogenomics in admixed populations. Trends Pharmacol Sci 2005; 26:196-201.
74. Lee E, Ryan S, Birmingham B et al. Rosuvastatin pharmacokinetics and pharmacogenetics in white and Asian subjects residing in the same environment. Clin Pharmacol Ther 2005; 78:330-341.
75. Caraco Y, Lagerstrom PO, Wood AJ. Ethnic and genetic determinants of omeprazole disposition and effect. Clin Pharmacol Ther 1996; 60:157-167.
76. Furuta T, Shirai N, Sugimoto M et al. Influence of CYP2C19 pharmacogenetic polymorphism on proton pump inhibitor-based therapies. Drug Metab Pharmacokinet 2005; 20:153-167.
77. Thomsen MS, Chassard D, Evene E et al. Pharmacokinetics of repaglinide in healthy Caucasian and Japanese subjects. J Clin Pharmacol 2003; 43:23-28.
78. Ho RH, Tirona RG, Leake BF et al. Drug and bile Acid transporters in rosuvastatin hepatic uptake: Function, expression, and pharmacogenetics. Gastroenterology 2006; 130:1793-1806.
79. Mwinyi J, Johne A, Bauer S et al. Evidence for inverse effects of OATP-C (SLC21A6) *5 and *1b haplotypes on pravastatin kinetics. Clin Pharmacol Ther 2004; 75:415-421.
80. Anonymous: Rosuvastatin (Crestor): Introduction of 5mg starting dose. Current Problems in Pharmacovigilance. London: Medicines and Healthcare products Regulatory Agency/Commission on Human Medicines, 2006:4.
81. Shah RR. Clinical Pharmacokinetics: Current requirements and future perspectives from a regulatory point of view. Xenobiotica 1993; 23:1159-1193.
82. Dalen P, Dahl ML, Ruiz ML et al. 10-Hydroxylation of nortriptyline in white persons with 0, 1, 2, 3, and 13 functional CYP2D6 genes. Clin Pharmacol Ther 1998; 63:444-452.
83. Dalen P, Dahl M, Andersson K et al. Inhibition of debrisoquine hydroxylation with quinidine in subjects with three or more functional CYP2D6 genes. Br J Clin Pharmacol 2000; 49:180-184.
84. Brockmoller J, Kirchheiner J, Meisel C et al. Pharmacogenetic diagnostics of cytochrome P450 polymorphisms in clinical drug development and in drug treatment. Pharmacogenomics 2000; 1:125-151.
85. Dilger K, Greiner B, Fromm MF et al. Consequences of rifampicin treatment on propafenone disposition in extensive and poor metabolisers of CYP2D6. Pharmacogenetics 1999; 9:551-559.
86. Caraco Y, Tateishi T, Wood AJ. Interethnic difference in omeprazole's inhibition of diazepam metabolism. Clin Pharmacol Ther 1995; 58:62-72.
87. Caraco Y, Sheller J, Wood AJ. Impact of ethnic origin and quinidine coadministration on codeine's disposition and pharmacodynamic effects. J Pharmacol Exp Ther 1999; 290:413-422.
88. Shah RR. Pharmacogenetic aspects of drug-induced torsade de pointes: Potential tool for improving clinical drug development and prescribing. Drug Saf 2004; 27:145-172.

89. Israel E, Drazen JM, Liggett SB et al. The effect of polymorphisms of the β2-adrenergic receptor on the response to regular use of albuterol in asthma. Am J Respir Crit Care Med 2000; 162:75-80.
90. Drazen JM, Yandava CN, Dube L et al. Pharmacogenetic association between ALOX5 promoter genotype and the response to anti-asthma treatment. Nat Genet 1999; 22:168-170.
91. Evans DAP, McLeod HL, Pritchard S et al. Interethnic variability in human drug responses. Drug Metab Dispos 2001; 29:606-610.
92. Kaye DM, Smirk B, Williams C et al. β2-adrenoceptor genotype influences the response to carvedilol in patients with congestive heart failure. Pharmacogenetics 2003; 13:379-382.
93. Kim DK, Lim SW, Lee S et al. Serotonin transporter gene polymorphism and antidepressant response. Neuroreport 2000; 11:215-219.
94. Pollock BG, Ferrell RE, Mulsant BH et al. Allelic variation in the serotonin transporter promoter affects onset of paroxetine treatment response in late-life depression. Neuropsychopharmacology 2000; 23:587-590.
95. Arias B, Catalan R, Gasto C et al. 5-HTTLPR polymorphism of the serotonin transporter gene predicts nonremission in major depression patients treated with citalopram in a 12-weeks follow up study. J Clin Psychopharmacol 2003; 23:563-567.
96. Perlis RH, Mischoulon D, Smoller JW et al. Serotonin transporter polymorphisms and adverse effects with fluoxetine treatment. Biol Psychiatry 2003; 54:879-883.
97. Durham LK, Webb SM, Milos PM et al. The serotonin transporter polymorphism, 5HTTLPR, is associated with a faster response time to sertraline in an elderly population with major depressive disorder. Psychopharmacology (Berl) 2004; 174:525-529.
98. Weir TD, Mallek N, Sandford AJ et al. Beta2-Adrenergic receptor haplotypes in mild, moderate and fatal/near fatal asthma. Am J Respir Crit Care Med 1998; 158:787-791.
99. Drysdale CM, McGraw DW, Stack CB et al. Complex promoter and coding region beta 2-adrenergic receptor haplotypes alter receptor expression and predict in vivo responsiveness. Proc Natl Acad Sci USA 2000; 97:10483-10488.
100. Lotrich FE, Pollock BG, Ferrell RE. Serotonin transporter promoter polymorphism in African Americans: Allele frequencies and implications for treatment. Am J Pharmacogenomics 2003; 3:145-147.
101. Ng CH, Easteal S, Tan S et al. Serotonin transporter polymorphisms and clinical response to sertraline across ethnicities. Prog Neuropsychopharmacol Biol Psychiatry 2006; 30:953-957.
102. Yoshida K, Ito K, Sato K et al. Influence of the serotonin transporter gene-linked polymorphic region on the antidepressant response to fluvoxamine in Japanese depressed patients. Prog Neuropsychopharmacol Biol Psychiatry 2002; 26:383-386.
103. Ackerman MJ, Tester DJ, Jones GS et al. Ethnic differences in cardiac potassium channel variants: Implications for genetic susceptibility to sudden cardiac death and genetic testing for congenital long QT syndrome. Mayo Clin Proc 2003; 78:1479-1487.
104. Ackerman MJ, Splawski I, Makielski JC et al. Spectrum and prevalence of cardiac sodium channel variants among black, white, Asian, and Hispanic individuals: Implications for arrhythmogenic susceptibility and Brugada/long QT syndrome genetic testing. Heart Rhythm 2004; 1:600-607.
105. Shah RR. Can pharmacogenetics help rescue drugs withdrawn from the market? Pharmacogenomics 2006; 7:889-908.
106. Vincent GM, Timothy KW, Leppert M et al. The spectrum of symptoms and QT intervals in carriers of the gene for the long-QT syndrome. N Engl J Med 1992; 327:846-852.
107. Saarinen K, Swan H, Kainulainen K et al. Molecular genetics of the long QT syndrome: Two novel mutations of the KVLQT1 gene and phenotypic expression of the mutant gene in a large kindred. Hum Mutat 1998; 11:158-165.
108. Priori SG, Napolitano C, Schwartz PJ. Low penetrance in the long-QT syndrome: Clinical impact. Circulation 1999; 99:529-533.
109. Kannankeril PJ, Roden DM, Norris KJ et al. Genetic susceptibility to acquired long QT syndrome: Pharmacologic challenge in first-degree relatives. Heart Rhythm 2005; 2:134-140.
110. Fitzgerald PT, Ackerman MJ. Drug-induced torsades de pointes: The evolving role of pharmacogenetics. Heart Rhythm 2005; 2:S30-S37.
111. Aerssens J, Paulussen ADC. Pharmacogenomics and acquired long QT syndrome. Pharmacogenomics 2005; 6:259-270.
112. Olatunde A, Evans DAP. Blood quinidine levels and cardiac effects in white British and Nigerian subjects. Br J Clin Pharmacol 1982; 14:513-518.
113. Geisen C, Watzka M, Sittinger K et al. VKORC1 haplotypes and their impact on the inter-individual and inter-ethnical variability of oral anticoagulation. Thromb Haemost 2005; 94:773-779.
114. Lee SC, Ng SS, Oldenburg J et al. Interethnic variability of warfarin maintenance requirement is explained by VKORC1 genotype in an Asian population. Clin Pharmacol Ther 2006; 79:197-205.

115. Veenstra DL, You JH, Rieder MJ et al. Association of Vitamin K epoxide reductase complex 1 (VKORC1) variants with warfarin dose in a Hong Kong Chinese patient population. Pharmacogenet Genomics 2005; 15:687-691.
116. Ng CH, Norman TR, Naing KO et al. A comparative study of sertraline dosages, plasma concentrations, efficacy and adverse reactions in Chinese versus Caucasian patients. Int Clin Psychopharmacol 2006; 21:87-92.
117. Smits KM, Smits LJ, Schouten JS et al. Influence of SERTPR and STin2 in the serotonin transporter gene on the effect of selective serotonin reuptake inhibitors in depression: A systematic review. Mol Psychiatry 2004; 9:433-441.
118. Yoshida K, Takahashi H, Higuchi H et al. Prediction of antidepressant response to milnacipran by norepinephrine transporter gene polymorphisms. Am J Psychiatry 2004; 161:1575-1580.
119. Emsley RA, Roberts MC, Rataemane S et al. Ethnicity and treatment response in schizophrenia: A comparison of 3 ethnic groups. J Clin Psychiatry 2002; 63:9-14.
120. Naito C. Necessity and requirements of bridging studies and their present status in Japan. Int J Clin Pharmacol Ther 2000; 38:80-86.
121. Naito C. Ethnic effects on pharmacokinetic parameters. In: Walker S, Lumley C, McAuslane, eds. The Relevance of Ethnic Factors in the Clinical Evaluation of Medicines. Lancaster: Kluwer Academic Publishers, 1994:161-177.
122. Lewis P, Rack PH, Vaddadi KS et al. Ethnic differences in drug response. Postgrad Med J 1980; 56(Suppl 1):46-49.
123. Anonymous: Guidance for Industry: Collection of Race and Ethnicity Data in Clinical Trials. Rockville, Maryland, USA: Food and Drug Administration, (http://www.fda.gov/cder/guidance/5656fnl.pdf, [Accessed on 30 May 2006]).
124. Anonymous: Guidance on Ethnic Factors in the Acceptability of Foreign Clinical Data (CPMP/ICH/289/95). London: European Medicines Agency, (http://www.emea.eu.int/pdfs/human/ich/028995en.pdf, [Accessed on 30 May 2006]).
125. Anonymous: Questions and Answers E5 Guidance on Ethnic Factors in the Acceptability of Foreign Clinical Data (CPMP/ICH/5746/03). London: European Medicines Agency, (http://www.emea.eu.int/pdfs/human/ich/574603en.pdf, [Accessed on 30 May 2006]).
126. Anonymous: Guidance on Dose Response Information to Support Drug Registration (CPMP/ICH/378/95). London: European Medicines Agency, (http://www.emea.eu.int/pdfs/human/ich/037895en.pdf, [Accessed on 30 May 2006]).
127. Anonymous: Guidance on Pharmacokinetic Studies in Man (Eudra/C/87/013). The Rules Governing Medicinal Products in the European Union EudraLex Vol 3C 'Guidelines on Efficacy'. Luxembourg: Office for Official Publications of the European Communities, 99-106. (http://pharmacos.eudra.org/F2/eudralex/vol-3/pdfs-en/3cc3aen.pdf, [Accessed on 30 May 2006]).
128. Anonymous: Guidance on the Investigation of Drug Interactions (CPMP/EWP/560/95). London: European Medicines Agency, (http://www.emea.eu.int/pdfs/human/ewp/056095en.pdf, [Accessed on 30 May 2006]).
129. Anonymous: Guidance note on Drug Metabolism/Drug Interaction Studies in the Drug Development Process: Studies In Vitro. Rockville, Maryland, USA: Food and Drug Administration, (http://www.fda.gov/cder/guidance/clin3.pdf, [Accessed on 30 May 2006]).
130. Anonymous: Guidance on Clinical Pharmacokinetic Studies of Pharmaceuticals. ELD Notification No 796 (1 June 2001). Tokyo, Japan: Ministry of Health, Labour and Welfare, [Not available in English].
131. Anonymous: Guidance on Methods of Drug Interaction Studies. ELD Notification No 813 (4 June 2001). Tokyo, Japan: Ministry of Health, Labour and Welfare, [Not available in English].
132. Anonymous: Guidance on the Investigation of Bioavailability and Bioequivalence (CPMP/EWP/QWP/1401/98). London: European Medicines Agency, (http://www.emea.eu.int/pdfs/human/ewp/140198en.pdf, [Accessed on 30 May 2006]).
133. Anonymous: ICH S7B: Nonclinical Evaluation of the Potential for Delayed Ventricular Repolarization (QT Interval Prolongation) by Human Pharmaceuticals (CPMP/ICH/423/02). London: European Medicines Agency, (http://www.emea.eu.int/pdfs/human/ich/042302en.pdf, [Accessed on 30 May 2006]).
134. Anonymous: ICH E14: Clinical Evaluation of QT/QTc Interval Prolongation and Proarrhythmic Potential for NonAntiarrhythmic Drugs (CPMP/ICH/2/04). London: European Medicines Agency, (http://www.emea.eu.int/pdfs/human/ich/000204en.pdf, [Accessed on 30 May 2006]).
135. Mansi IA, Nash IS. Ethnic differences in electrocardiographic intervals and axes. J Electrocardiol 2001; 34:303-307.
136. Anonymous: ICH E2E Note for Guidance on Planning Pharmacovigilance Activities CPMP/ICH/5716/03). London: European Medicines Agency, (http://www.emea.eu.int/pdfs/human/ich/571603en.pdf, [Accessed on 30 May 2006]).
137. Nagata R, Fukase H, Rafizadeh-Kabe JD. East-West development: Understanding the usability and acceptance of foreign data in Japan. Int J Clin Pharmacol Ther 2000; 38:87-92.

CHAPTER 13

Human Genomic Variation Studies and Pharmacogenomics Are Critical for Global Health

Béatrice Séguin, Samina Essajee, Gerardo Jimenez-Sanchez, Peter A. Singer and Abdallah S. Daar*

Introduction

Many people in the world still lack access to essential medicines. The World Health Organization (WHO) has attempted to address this inequity by creating the essential medicines list (See http://www.who.int/medicines/publications/essentialmedicines/en/). This list is intended to help countries with limited resources focus health expenditures on medicines likely to produce the most health benefits. However, as pointed out in a recent report by Marsh et al,[1] if we consider that most of these drugs have been developed and tested in predominantly male Caucasians in industrialized countries, then how can we be sure we are indeed maximizing health benefits globally? We know that there is individual variability in drug response and that human genetic variation is partly responsible for how humans respond to therapeutics. There is also a growing body of evidence demonstrating ethnic variability in drug response. Recent trends in genetic research and the quest for improving drug therapy indicate that, at least in industrialized countries, the adoption of pharmacogenomics is inevitable due to regulatory incentives, increases in biomarker validation, pharmacoeconomic evidence and patient demand.[2]

When it comes to the developing world however, some believe that pharmacogenomics applications will be too expensive to adopt, and argue that the traditional focus on improving environmental conditions is the most practical approach to improve population health.[3-5] While such approaches are important for solving global health inequities, scientific developments can complement this effort and should not be suppressed when they have much potential; it has recently been argued that pharmacogenomics can also be practically applied to improve the health of people in the developing world.[1,2,6] As advances in pharmacogenomics progress, it is likely that new drugs will be developed by—and for—industrialized countries. Therefore, to ensure that developing countries share in the potential social and economic benefits of the genomics revolution and to prevent the emergence of a "pharmacogenomics divide," it seems reasonable to assume that understanding and harnessing genomic variation (SNPs, gene duplications and deletions, mutations in regulatory genes and large-scale copy number variations) in developing world populations will help them improve efficiency of medication use in resource-limited settings, and potentially develop therapeutics that meet their local health needs. To understand how this can become a reality, we need to explore the

*Corresponding Author: Abdallah S. Daar—McLaughlin-Rotman Centre UHN|MCMM at the University of Toronto, Toronto, Ontario, Canada. Email: a.daar@utoronto.ca

Pharmacogenomics in Admixed Populations, edited by Guilherme Suarez-Kurtz.

connection between pharmacogenetics, genotyping projects in developing countries, and the evolution of the pharmaceutical industry in both the developed and developing worlds. Indeed, the stimulus for the large-scale adoption of pharmacogenomics is already apparent in some developing countries, such as Mexico (Table 2), that are initiating their own large-scale genotyping studies.

Although pharmacogenomics may provide benefits to industrialized and developing nations alike, special considerations apply in the case of developing countries, which face increasing challenges in their healthcare systems. Considering ethical, legal and social issues is necessary to facilitate the successful implementation of pharmacogenomics. Table 1 provides a brief overview of these issues, some of which have been extensively reviewed in the literature[7] and will not be revisited in this chapter. Instead, we classify these issues into drivers for—and barriers to—the implementation of pharmacogenomics in the developing world. Using this framework, we will focus the remainder of this chapter on exploring, in more depth, what we perceive to be key benefits to—and challenges for—the adoption of pharmacogenomics in developing countries. Some of these key benefits and challenges relate to three core themes: (1) controversy over growing scientific evidence that there is genetic variation between ethnic groups; (2) cost of implementation of human genomic variation studies may be perceived as a short term barrier; (3) proper scientific, human resource and intellectual property infrastructure needs to be developed in order to make adoption of a genomic medicine platform possible. We thus begin this chapter by discussing the use of "race" in pharmacogenomic studies because, at least in the short term, many genetic studies in both developed and developing countries (including admixture studies) will be making use of "race" or "ethnicity" as a proxy marker and the controversy this has generated merits attention. Secondly, we address the issues of cost and feasibility pertaining to the adoption and implementation of pharmacogenomics in the developing world and finally, we briefly mention some of the regulatory issues that may affect the adoption of pharmacogenomics in the developing world.

The Use of "Race" in Genetic Studies Is Controversial

Due to the ambiguity inherent in "race", we will avoid using the term. We use *geographical ancestry* to refer to population clusters based on genetic differences that some experts attribute to evolutionary pressures. These differences refer to varying frequencies and patterns of certain polymorphisms. Differences based on geographical ancestry would encompass phenotypes such as skin colour, but also include other variations that may not be visible. Outward characteristics do not necessarily correlate with geographical ancestry. These phenotypic differences are often conflated with social constructions so that individuals may identify with a certain group based on culture, geographical location, parentage etc. We use *ethnicity* to refer to these social constructions inherent in self-identification and distinguish them from objective genetic designations. These genetic differences, although they are dwarfed by common genetic heritage, are significant in the field of pharmacogenomics. We will review the relevance of geographical ancestry to genomic medicine, before exploring the difficulties attendant to clustering, based on current knowledge.

Genetic differences among population groups do exist, and it is possible with some degree of accuracy to determine an individual's ancestry by studying genetic variation at a small number of loci.[17,18] Many scientists believe that exploring differences among ancestral groups will provide valuable insights into complex diseases and, ultimately, improve global health.[19] Although categories of social organization do correlate with health outcomes, we need to separate those components that may be attributable to beliefs, practices and healthcare disparities from those that are determined by genetic variation. We should "look through race to ascertain genetic factors" that affect drug response and disease susceptibility.[20] It seems likely that geographical ancestry will continue to be useful as long as such categorization "explains" variation unexplained by other factors and it is likely that more specific knowledge of genetic variation will actually resolve the social issues that arise from the use of ethnicity.

Table 1. Drivers for and barriers to the adoption of pharmacogenomics in developing countries

	Driver	**Barrier**
Ethical	-Need for innovative approaches to reduce healthcare disparities -Responsibility of each country to meet domestic health needs -Need to build local scientific capacity so local health needs can be addressed -Ensuring that developing countries benefit from the "genomics revolution" so as to prevent further increases in health inequities -Importation of products in markets where they have not been tested is unethical	-Public conflates ELSI issues specific to pharmacogenomics with general genetics ELSI issues: Informed Consent, Privacy, Stigmatization, Discrimination, Determinism, and Benefit Sharing[8-11] -New approaches may divert scarce resources from traditional strategies such as environmental or social interventions -Use of socially constructed definitions of race in pharmacogenomics studies
Legal	-TRIPS* implementation will force developing countries to respect patent laws—encouraging them to develop innovative products rather than relying on generics (for example India has become TRIPS compliant as of 2005) -Laws, regulations, and policies which encourage research and clinical developments -Litigation might play a role in driving the adoption of pharmacogenetics not only in industrialized countries but also in emerging markets (see Box 1).	-Weak infrastructure for developing proper IP protection in developing countries
Social	-National pride and identity building -Increasing patient demand for pharmacogenetic testing as positive scientific and economic benefits begin to surface	-Lack of optimal infrastructure to develop genomic medicine -Potential for exploitation of vulnerable groups in research -Social-identity and self-identity may be questioned if results of large-scale human genotyping reveal the existence of ancestral lineages that were previously unknown in certain groups
Economic	-Contracts for research (outsourcing) and collaborative projects with other countries boost local economies and lower implementation costs for developing country partners -Non-communicable diseases are of common interest with industrialized countries -Market segmentation may result in niche market opportunities for small domestic biopharmaceutical companies[12] -Need to generate a knowledge-based economy in the developing world so as to be competitive in the world economy	-Cost of implementation -Market segmentation may lead to expensive therapies

continued on next page

Table 1. Continued

	Driver	Barrier
Scientific	-Genotyping projects are already established and generating data that shows association between certain genotypes and disease predisposition or drug response -Opportunities for research to improve healthcare, design better drugs, and decrease risk in drug development -Improving biomarker validation -Drugs designed by developing countries for their own populations may also be beneficial for minority groups in developed countries, fostering north-to-south collaboration	-Shortage of trained professionals who can work in R&D in pharmacogenomics and who can interpret the results in clinical settings

* The Agreement on Trade-Related Aspects of Intellectual Property Rights (TRIPS) is an international treaty that sets minimum standards for intellectual property rights and their enforcement which apply to all member countries of the World Trade Organization (see http://www.wto.org/english/docs_e/legal_e/legal_e.htm#TRIPs)

The challenge in using geographical ancestry as a useful variable in genomic variation research lies in determining how to define group categories. Grouping subjects according to their outward physical differences, such as skin colour, is problematic.[21,22] As rates of migration and admixture increase, ethnic categories are becoming less accurate as proxies for genetic variants.[23] Although many still doubt the validity, if not the accuracy, of self-reported ethnicity in biomedical research, a recent study demonstrated that self-reported ethnicity can provide a reasonable measure of

Box 1. Cassidy versus SmithKline Beecham. Filed in December 1999 in Pennsylvania, this class action suit was the first of many in which claimants alleged SmithKline Beecham (now GlaxoSmithKline) failed to warn both doctors and the public of the risks associated with their LYMErix vaccine.[13,14] The vaccine was based upon the Osp A surface protein of *Borrelia burgdorferi*, the causative agent of lyme disease.[15] As of the early 1990s, there was speculation that, in individuals positive for the Histocompatibility Antigen DR4 biomarker (roughly one third of the general population), the vaccine may trigger autoimmune arthritis.[13,15] Thus, claimants argued that the manufacturer had a duty to warn genetically susceptible patients and as such, should have provided the genetic screening tests required to determine genetic susceptibility.[13-15]

In response to these adverse event reports, the Food & Drug Administration and the Centre for Disease Control conducted a follow-up study analyzing both adverse reactions reported during the clinical trial and those reported post approval. However, they found no statistical evidence of a causal link between LYMErix and autoimmune disease.[16] Nevertheless, in February 2002, amidst increasing pressure, SmithKline Beecham withdrew LYMErix from the market in response to poor sales.[14,15]

Currently, Baxter Vaccines, in Vienna, is developing a new vaccine for lyme disease based on the same Osp A surface protein. However, the protein region believed to be responsible for inducing autoimmunity has been omitted.[15]

Box 2. BiDil® (hydralazine and isosorbide dinitrate; NitroMed Inc.), a drug to treat heart failure in African-Americans, is the first drug approved by the FDA for a specific sub-population. A previous attempt to patent this fixed combination of drugs was rejected by the FDA, as the initial study in a mixed population showed no demonstrable benefit. A retrospective analysis of the data from this study showed increased efficacy in African-American cohorts resulting in a second clinical trial testing its use in a population of self-identified African-Americans. This subsequent study was halted early for ethical reasons when there was a significant reduction in mortality rate (43%) in those treated with the combination therapy as opposed to the placebo group.[35] Currently, NitroMed is struggling to increase sales in the face of controversy over BiDil. However, BiDil illustrates the potential of pharmacogenomics for different ethnic groups. Although the genetic basis for this difference is still unknown, studies addressing this question are underway. While controversial, BiDil® could be of interest to Africans who share their geographical ancestry with African-Americans.*

Box 3. Losartan® is an anti-hypertensive drug that is metabolized by the enzyme CYP2C9. A defective variant of this enzyme (CYP2C9*5) has been associated with poor response to the drug and is restricted to sub-Saharan Africans and their descendants. The health implications for sub-Saharan Africans who are treated with Losartan® and have this variant would have remained unknown if studies had only been carried out on populations with European geographic ancestry.[23] *

genetic classification.[24] In addition, there is striking evidence of an association between ethnicity and variable drug response. At least 29 different drugs have been reported in the literature as eliciting different drug responses between ethnic groups.[25] Notable examples of these drugs include warfarin, BiDil® (Box 2) and Losartan® (Box 3). One future challenge will be validating and standardizing frequencies of causal genetic variants in these different populations, since values vary.[26-28]

Risks of Using Ethnicity as a Proxy Need to Be Mitigated

The practical benefits of using ethnic categories in research must be weighed against the potential risks to society. The purpose of pharmacogenomics is to stratify a population into smaller genetically homogeneous sub-populations characterized by their different responses to particular drugs. The consequences of this segmentation are unclear for ethnic groups.[29] Some groups might receive unequal access to treatment, since drugs would not be targeted towards them because companies may not view them as large and/or profitable markets.[25] Moreover, as Foster states, "pharmacogenomics might increase the social significance of differential drug response between ethnic groups, potentially leading to reification of health disparities".[30] On the other hand, without pharmacogenomics, drug companies may abandon drugs that actually benefit the majority of the population but have serious adverse effects in a small group. If frequencies of alleles vary across ethnic groups, then the groups included in clinical research will be most likely to benefit from pharmacogenomics.[31] As mentioned in the introduction, inattention to ethnicity in clinical research means most clinical trials were conducted on white males, and the weight-adjusted results have been applied to minorities, women, and children. However, this extrapolation is inaccurate and groups excluded from

*From Daar, Dabu and Seguin's contribution to the technical section of the WHO report, "The Ethical, Legal, and Social Considerations for Pharmacogenomics in the Developing World" (upcoming).

trials are disadvantaged because they receive less effective treatment.[32] Furthermore, there is an ethical imperative to broaden the scope of clinical research to include a variety of ethnic groups. Developing countries are repositories of human genetic diversity that can help to define patient subgroups for pharmacogenomics, and this diversity needs to be explored. Specifically, developing countries could benefit from such exploration in the following ways:[2]

- Pharmaceutical companies in the developing world could license drugs previously withdrawn from the market and develop them both for local populations and for others in the developing world who are either not genetically predisposed to the adverse effects or for whom efficacy can be demonstrated to a greater extent. For example, Iressa™, a drug meant for lung cancer treatment, was found to have low efficacy in the US but appears to significantly improve overall survival in Japanese patients[33] and is now available in several countries around the world, including the US, Canada, Japan and Australia.
- Pharmaceutical companies in developing countries could capitalize on emerging trends in genotyping and their application to understanding variable drug responses and disease susceptibility. Theoretically they could do this by developing drugs specifically for sub-populations that are more likely to benefit without side effects.
- Compounds discovered in the laboratories of developing countries could be of interest to pharmaceutical companies in the developed world if they are relevant to selected minority sub-populations living in developed countries, leading to potentially beneficial partnerships.

In the short term, the use of pharmacogenomics will be most valuable in guiding the use of existing medicines. Pharmacogenomics may help bring back to market previously withdrawn drugs by identifying populations that can benefit from these drugs and are not susceptible to the adverse effects responsible for their original withdrawal (referred to as "drug resuscitation"). While the social implications of the use of ethnicity in pharmacogenomics are debatable, further research in the field should inform the way sub-populations are characterized. This will ultimately contribute to a mature understanding of human diversity and render race, as a useful concept, obsolete.[34]

Implementing Pharmacogenomics Is Feasible for the Developing World

Infrastructure for a Pharmacogenomic Platform Needs to Be Developed

In order to realize the benefits of pharmacogenomics, developing countries will need to surmount certain obstacles.[4,5] Although developing countries vary with regard to their resources and scientific capacity, in general, there is hesitance on the part of developing countries to adopt pharmacogenomics at the expense of more popular public health initiatives that address social and environmental causes of disease. If we deconstruct this hesitance we find that the challenges facing implementation of pharmacogenomics in the developing world relate to a shortage of human resources and to cost. The former may be due to an absence of training facilities, brain drain,[36,37] or high morbidity and mortality as a result of disease epidemics, as is the case in Africa, where a significant number professionals such as nurses, doctors and pharmacists have died from HIV/AIDS. The additional costs associated with the uptake of pharmacogenomics, from training personnel to buying equipment, represent another barrier. Nevertheless, some developing countries are embarking on SNP projects, drug development initiatives, and other research closely related to pharmacogenomics (Table 2). The impact of the Human Genome Project will vary, since countries differ with regard to burden of disease, financial resources, education, scientific and technological capacity, and health systems.[38] Least developed countries, such as those in sub-Saharan Africa, need to be distinguished from developing countries that have a high capacity for science and technology (S&T), such as India, China, Mexico and Brazil.[39] As Table 2 illustrates, countries in the latter group are already conducting genomics research on their own. Although the least developed countries face many more challenges, they may be the ones that need pharmacogenomics the most.

Box 4. The government of India supports human genomic variation research. India is one country which has already started conducting genotyping studies. The Indian Genome Variation Consortium consists of six laboratories of the Council of Scientific and Industrial Research (CSIR) that will conduct research to provide data on validated SNPs or repeats and gene duplications in over one thousand genes. Genes relevant to pharmacogenomics and disease etiology are the focus. The project, which started in 2003, has received tenure for five years from the Government of India, which supports pharmacogenomics and encourages research that can improve the health of the Indian population.[43] The Department of Biotechnology has issued guidelines on pharmacogenomics research on its web-site in response to increasing numbers of research proposals from scientists. These guidelines stress that studies in India should have national relevance. The Central Drugs Standard Control Organization (CDSCO), in its *Guidelines for Bioavailability and Bioequivalence Studies*, suggests that genotyping should be considered for exploratory bioavailability studies, crossover studies, and all studies using parallel group design. It also encourages studies performed on subjects known to have a major genetic polymorphism relevant to drug response. Again, the guidelines state that pharmacogenomic issues should be considered in the context of the Indian population.

Box 5. Mexico is characterizing its genomic variation and developing a platform for genomic medicine. The National Institute of Genomic Medicine (INMEGEN) of Mexico, created by the Mexican Congress, is currently sampling regions in Mexico in order to understand the genetic structure of the population. The Mexican government has made genomic medicine a priority, because new strategies are needed to meet long-term healthcare costs in Mexico. INMEGEN aims to uncover disease-related genetic variation and generate products that will benefit the Mexican population, so that Mexico does not have to import applications developed for other populations. The project is expected to contribute to economic growth and ease the financial burden of healthcare.[41]

While governments of some developing nations (Boxes 4 and 5) are creating their own institutions for genomics research, collaborating with industrialized countries is a strategy developing countries can take to reduce start-up costs. Countries from the developing and developed world are forming partnerships based on common interests. While infectious disease epidemics continue to ravage developing nations, the rates of noncommunicable disease are also spiraling upwards. Currently, one third of all deaths in the world are caused by cardiovascular diseases. Of these deaths, nearly 80% occur in developing nations. By 2020, the number of new cancer cases will increase a staggering 73% in developing countries as opposed to only 29% in developed countries.[44] One example of collaboration between industrialized and developing nations, in order to address this global problem, is a partnership between the Organization for Nucleotide Sequencing and Analysis (ONSA) in Brazil and the Ludwig Institute in Switzerland, which is contributing half of the cost of sequencing human cancer-related genes.[40]

Industrialized countries are also interested in communicable diseases that are devastating populations in the developing world. Public-private partnerships, such as the Medicines for Malaria Venture, have formed to discover other treatments for diseases prevalent in the developing world.[2] In addition to public-private partnerships, there is an increasing trend for pharmaceutical companies to contract research and clinical trials to developing countries, which

Table 2. Initiatives in the developing world aimed at characterizing population-based genetic differences*

Initiative	Description
National Institute of Genomic Medicine (INMEGEN), Mexico	INMEGEN is one of twelve National Institutes of Health in Mexico. Created in 2004, it aims to develop a national platform in genomic medicine focused on national health problems and based on the genomic structure of Mexican populations. Most of Mexico's modern population is considered Mestizo resulting from a dynamic admixture of over 65 ethnic groups, Spaniards and, to a lesser extent, Africans, within the last 500 years. INMEGEN is genotyping 600 Mestizos from different regions of Mexico, analyzing 500,000 to 600,000 Single Nucleotide Polymorphisms (SNPs). This information is triggering a series of disease-related genomic studies in Mexico that will be used to improve healthcare for the Mexican population, and is likely to be useful to other countries in Latin America where the Mestizo admixture is the origin of most of their populations. Investigators expect to complete the study in 2007.[40,41] http://www.inmegen.gob.mx
HUGO Pan-Asian SNP Initiative	The HUGO Pacific Pan-Asian SNP Initiative is composed of a coalition of scientists from China, India, Indonesia, Japan, Korea, Malaysia, Nepal, Philippines, Singapore, Thailand and Taiwan. The researchers will look at 50,000 SNPs in each study participant. The goal of this initiative is to uncover both the breadth of genetic diversity and the extent of genetic similarity in Asia.[42]
The Indian Genome Variation Consortium	This network program was initiated by the Council of Scientific and Industrial Research (CSIR) and is funded by the Government of India. The Indian Genome Variation Consortium aims to create a database of genetic variants inherent to the people of India and make it available to researchers for understanding disease predisposition, adverse drug reaction, population migration etc. Fifteen thousand unrelated individuals of different sub-populations will be sampled.[43]
Thailand SNP Discovery Project	This project is the result of a collaborative effort between the National Center for Genetic Engineering and Biotechnology (BIOTEC, Thailand) and Centre National de Genotypage (CNG, France). The aim of the project is to identify intragenic SNPs frequent in Thai populations. A SNP database will be completed of all the genes identified in the entire human genome and their regulatory regions with allele frequency and linkage disequilibrium (LD) block patterns in Thai and other (French, Japanese and African) populations. http://thaisnp.biotec.or.th:8080/thaisnp
University of Cape Town's Division of Human Genetics/ The Africa Genome Education Institute (South Africa)	The Africa Genome Education Institute is devoted to educating the public about the structure and function of genomes as well as studying the genetic basis of diseases relevant to the South African population. http://web.uct.ac.za/depts/genetics

continued on next page

Table 2. Continued

Initiative	Description
REFARGEN	Rede Nacional de Farmacogenetica/farmacogenomica (REFARGEN) or the Brazilian National Pharmacogenetics/Pharmacogenomics network is a network of Brazilian researchers from various institutions situated in five regions of the country. It aims to create a repository of biological samples for pharmacogenomic studies and archive pharmacogenomic data for the Brazilian population. A significant focus is on its large, admixed population.[22] http://www.refargen.org.br
The Pharmacogenetics for Every Nation Initiative (PGENI)	Based out of the University of North Carolina, PGENI is an interdisciplinary group of collaborators who aim to promote the integration of genetic information into drug formulary decision-making. PGENI will work with developing country partners to sample the population, publicize global genotype profiles, and provide the Ministry of Health department with recommendations. Initially, PGENI is partnering with 108 countries that have moderate to good health system infrastructure, representing 78% of the global population.[1] http://pgeni.unc.edu

*From Daar, Dabu and Seguin's contribution to the technical section of the WHO report, "The Ethical, Legal, and Social Considerations for Pharmacogenomics in the Developing World" (upcoming).

may also help build scientific capacity and reduce implementation costs.[2] Collaborations such as these indicate that pharmacogenomics is also relevant for lesser developed nations that may not be able to conduct genomics research independently on a national scale. As developing countries undertake their own projects and build up infrastructure, they will build up their own capacity in genomic medicine.

Market Segmentation May Lead to Niche Markets

One concern is that pharmacogenomics will lead to market segmentation and result in unaffordable therapeutics with limited accessibility in the developing world. However, this may not necessarily be true; pharmacogenomics research may also create new niche markets and boost the private sector in developing countries. Pharmacogenomics research is expected to uncover between 5000 to 10 000 new drug targets, compared to about 500 molecular targets used currently.[45] These new drug targets could lead to the development of new therapies. Nunnally et al conclude that pharmacogenomics will bifurcate the pharmaceutical industry so that large companies continue to produce blockbuster drugs indicated for high percentages of people, but only for people with common genetic variants underlying their disease or drug response, and this narrower focus will result in fewer failed drugs.[46] These smaller markets would create opportunities for companies in developing countries that are just starting to make significant gains on the global market and if they retain control over intellectual property they will also be able to control pricing in their local markets.[2] If such ventures were applied in developing countries, then business areas that would grow as a result of the use of pharmacogenomics would include bioinformatics, diagnostics, testing services, and medical management. This industrial growth would, in turn, create jobs that require trained individuals thereby strengthening local economies.[12]

*Developing Countries Can Least Afford to Waste Limited Health Resources on Ineffective Therapy**

Governments in developing countries are interested in the potential of the pharmacogenomics approach to reduce overall healthcare costs. According to the Director of Mexico's National Institute of Genomic Medicine (INMEGEN), "New strategies for prevention, early diagnosis, and more effective treatment are essential to meet the mid- and long-term healthcare costs in Mexico".[41] Preliminary studies of the cost-effectiveness of genotyping to guide dosing of some drugs are crucial to understanding how best to use pharmacogenomics. Marra et al found that pretesting for thiopurine-S methyltransferase (TPMT) status to guide azathioprine dosing resulted in some direct reduction in costs.[47] Another study found that prospective genotyping for the HLAB*5701 allele associated with abacavir hypersensitivity is cost-effective in HIV patients as compared to the costs associated with hospitalization due to occurrence of this adverse event.[48] Finally, You et al. estimated that the cost of a major bleeding event associated with warfarin sensitivity (~$6000 US) could be averted with prospective CYP2C9 genotyping based on a simulated decision tree over a twelve month period.[49] These types of cost-saving measures are an important incentive to developing countries, especially those with very small health budgets; some sub-Saharan African countries have annual per capita healthcare expenditures that can be as low as $10-15 US. However drug development is influenced by over-arching regulatory frameworks. Although niche-market companies of all sizes should have more manageable regulatory functions compared to companies following the blockbuster approach, the context in which they operate is crucial for their success and the impact of regulation and intellectual property protection should be considered.[50]

Regulatory Frameworks and Intellectual Property Protection Play a Role

Although intellectual property (IP) protection and regulatory frameworks will impact the implementation of pharmacogenomics in developing countries, these issues are complex and a comprehensive discussion of them is beyond the scope of this chapter. While one concern is that the price of pharmacogenomics drugs might increase as a consequence of market segmentation, the implementation of pharmacogenomics could also widen the gap between rich and poor by encouraging researchers in developing countries who are conducting human genomic variation studies to patent their findings and adversely affect access to patented products. As discussed in the 2005 WHO report on "Genetics, genomics and the patenting of DNA" patents can influence access to genomics products by "improving incentives to develop useful tests; increasing the cost of available services; imposing transaction costs and inconvenience on research and development; impeding the transfer of existing tools and technologies".[40] However, patenting itself may not lead to expensive therapeutics or increases in health inequities, because the licensing practice adopted, rather than the patenting itself, will influence its effect. For example, following the SARS outbreak in 2002, a series of laboratories around the world sequenced the genome of this virus. As a result, many patents on the genomic sequence of SARS were filed. In an attempt to ensure that social benefits accrue from them, these patent rights may be placed into a pool to be licensed on a non-exclusive basis.[51]

Alternatively, earning fees or royalties on patents on an exclusive basis can be seen as an incentive for investing in research and development (R&D) in developing countries. For example, patents can provide opportunities for out-licensing to firms that exist in more profitable markets, thus leading to financial gain via fees or royalties which can be reinvested by the

*From Daar, Dabu and Seguin's contribution to the technical section of the WHO report, "The Ethical, Legal, and Social Considerations for Pharmacogenomics in the Developing World" (upcoming).

patent holder. In addition, patents can be traded in exchange for other technologies or services.[40] Patenting may also facilitate collaboration between the developed and the developing world, since industrialized countries are more willing to transfer technology when they know it will be protected. In fact, technology transfer has often been accompanied by forced reforms designed to strengthen domestic patent protection.[52] Also, patents for drugs withdrawn from the market may have expired, or be near expiry, thus presenting opportunities for biotechnology and pharmaceutical companies in developing countries to license these compounds and target local populations or other markets in the developing world.[2] Ultimately, developing countries need to strengthen intellectual property protection to take their place in a knowledge-based economy. If they build S & T capacity and retain IP rights for their own research, they will have more control over prices downstream for the products they develop, and can ensure that their citizens have access to these products. For example, INMEGEN has partnered with the Mexican Institute of Industrial Property (IMPI) to ensure control over their innovations and ensure that their products will be accessible to Mexicans.

Strengthening IP protection at a national level may not be easy, but the international community is driving a trend towards increasing intellectual property rights (IPR) capacity by requiring developing nations to comply with the Agreement on Trade-Related Aspects of Intellectual Property Rights. TRIPS offers mechanisms to use patents that would be generated from pharmacogenomics research strategically to reduce costs and control drug prices. For example, TRIPS allows researchers in developing countries to use patented products freely for research purposes without infringing the patent. TRIPS also provides an option for compulsory licensing so that governments in developing countries may authorize the use of a patented product without the patent holders' consent. Further research into how developing countries can benefit from these mechanisms, in the field of pharmacogenomics specifically, is necessary to determine how intellectual property protection and regulation can drive or hinder the implementation of pharmacogenomics in the developing world.

Conclusion

As human genomic variation projects in the developing world emerge and grow, careful examination of ethical, social and cultural issues is necessary to guide policy and maximize success of future initiatives in genomic medicine that these countries undertake. We have reviewed the difficulties attendant to the use of race and ethnicity, and explored how economic barriers such as implementation costs and market segmentation may be overcome or transmuted into opportunities. Developing countries have compelling reasons to harness pharmacogenomics to improve the health of their populations. In the global movement to increase access to essential medicines in the developing world, genomics is a useful tool to help determine which medicines are essential for a given country. In particular, a model which focuses on using geographical ancestry as a tool to understand drug response and disease susceptibility offers an opportunity for developing countries to meet their local health needs without depending on knowledge derived in industrialized countries. Pharmacogenomics can generate substantial cost-savings in drug development, reduce healthcare costs, stimulate the private sector, and uncover genetic diversity relevant to drug response or disease predisposition. How they choose to integrate pharmacogenomics into their healthcare systems will depend on their domestic health needs, and different strategies may arise from these efforts. To our knowledge, there are no empirical studies that have looked at exactly how the emerging knowledge of human genomic variation can be practically applied to improving the health of people in the developing world. We are currently conducting a case study of the Mexican National Institute of Genomic Medicine (INMEGEN) and are planning studies of the other four large-scale genomic variation research projects in India, Thailand, South Africa, and Asia. The purpose of our research is to understand the factors driving the adoption of genomic medicine in developing countries and to help create conceptual frameworks for the integration of pharmacogenomics into the health sector in the developing world. Comparing experiences and facilitating cross communication among these developing world programs would be useful in this regard. We hope that our research will produce specific policy recommendations that will assist these countries as they spearhead the adoption of pharmacogenomics to improve global health.

Acknowledgements

We would like to thank Billie-Jo Hardy for her useful comments. We would like to thank Anna Dabu for her contributions to our previous work on the technical section of the WHO report "The Ethical, Legal, and Social Considerations for Pharmacogenomics in the Developing World". Grant support was provided primarily by Genome Canada through the Ontario Genomics Institute (Toronto, Canada). Matching partners can be found at www.geneticsethics.net. ASD and PAS are both supported by the McLaughlin-Rotman Centre, UHN|MCMM at the University of Toronto, Toronto, Ontario Canada.

References

1. Marsh S, Van Booven DJ, McLeod HL. Global pharmacogenetics: Giving the genome to the masses. Pharmacogenomics 2006; 7(4):625-631.
2. Daar AS, Singer PA. Pharmacognetics and geographical ancestry: Implications for drug development and global health. Nat Rev Genet 2005; 6:241-246.
3. Collins FS, McKusick VA. Implications of the Human Genome Project for medical science. JAMA 2001; 285(5):540-544.
4. Royal Society. Personalised medicines: Hopes and realities. London: The Royal Society, 2005.
5. Morris K, Nundy S. A global role for the human genome. Lancet 2001; 357(9255):537.
6. Pang T. Impact of pharmacogenomics on neglected diseases of the developing world. Am J Pharmacogenomics 2003; 3(6):393-398.
7. The Ethical, Legal, and Social Considerations for Pharmacogenomics in the Developing World. WHO Report (upcoming).
8. Paul NW, Roses AD. Pharmacogenetics and pharmacogenomics: Recent developments, their clinical relevance and some ethical, social, and legal implications. J Mol Med 2003; 81:135-140.
9. Silva FG. Ethics of the new biology and genetic medicine (molecular ethics): Brief (re)view from the USA. Pathol Int 2002; 52:555-562.
10. Austin MA, Harding SE, McElroy CE. Monitoring ethical, legal, and social issues in developing population genetic databases. Genet Med 2003; 5(6):451-457.
11. Rothstein MA. Epilogue: Policy prescriptions. In: Rothstein MA, ed. Pharmacogenomics: Social, Ethical, and Clinical Dimensions. Hoboken: John Wiley and Sons, 2003:319-335.
12. Reeder CE, Dickson MW. Economic implications of pharmacogenomics. In: Rothstein MA, ed. Pharmacogenomics: Social, Ethical, and Clinical Dimensions. Hoboken: John Wiley and Sons, 2003:229-250.
13. Marchant GE. Genetic susceptibility and biomarkers in toxic injury litigation. Jurimetrics 2000; 41:67-109.
14. Marchant GE. Genetic data in toxic tort litigation. J Law Policy 2006; 14:7-37.
15. Abbott A. Uphill struggle. Nature 2006; 439:524-525.
16. Lathrop SL, Ball R, Haber P et al. Adverse event reports following vaccination for Lyme disease: December 1998-July 2000. Vaccine 2002; 20:1603-1608.
17. Shriver MD, Smith MW, Jin L et al. Ethnic-affiliation estimation by use of population-specific DNA markers. Am J Hum Genet 1997; 60(4):957-964.
18. Shriver MD, Mei R, Parra EJ et al. Large-scale SNP analysis reveals clustered and continuous patterns of human genetic variation. Hum Genomics 2005; 2(2):81-89.
19. Holden C. Race and medicine. Science 2003; 302(5645):594.
20. Clayton EW. The complex relationship of genetics, groups, and health: What it means for public health. Journal of Law and Medical Ethics 2002; 30(2):290-297.
21. Parra EJ, Kittles RA, Shriver MD. Implications of correlations between skin color and genetic ancestry for biomedical research. Nat Genet Suppl 2004; 36(11):S54-S60.
22. Suarez-Kurtz G. Pharmacogenomics in admixed populations: The Brazilian pharmacogenetics/pharmacogenomics network—REFARGEN. Pharmacogenomics J 2004; 4:347-348.
23. Suarez-Kurtz G. Pharmacogenomics in admixed populations. Trends Pharmacol Sci 2005; 26(4):196-201.
24. Tang H, Quertermous T, Rodriguez B et al. Genetic structure, self-identified race/ethnicity, and confounding in case-control association studies. Am J Hum Genet 2005; 76(268):268-275.

25. Tate S, Goldstein D. Will tomorrow's medicines work for everyone? Nat Genet Suppl 2004; 36(11):S34-S42.
26. Meyer UA. Pharmacogenetics and adverse drug reactions. Lancet 2000; 356:1667-1671.
27. Terra SG, Johnson JA. Pharmacogenetics, pharmacogenomics, and cardiovascular therapeutics: The way forward. Am J Cardiovasc Drugs 2002; 2(5):287-296.
28. Oscarson M. Pharmacogenetics of drug metabolizing enzymes: Importance for personalized medicine. Clin Chem Lab Med 2003; 41(4):573-580.
29. Greely HT. Genome research and minorities. In: Rothstein MA, ed. Pharmacogenomics: Social, Ethical, and Clinical Dimensions. Hoboken: John Wiley and Sons, 2003:67-82.
30. Foster MW. Pharmacogenomics and the social construction of identity. In: Rothstein MA, ed. Pharmacogenomics: Social, Ethical and Clinical Dimensions. Hoboken: John Wiley and Sons, 2003:251-265.
31. Licinio J. Pharmacogenomics and ethnic minorities. Pharmacogenomics J 2001; 1:85.
32. Robertson JA. Constitutional issues in the use of pharmacogenomic variations associated with race. In: Rothstein MA, ed. Pharmacogenomics: Social, Ethical, and Clinical Dimensions. Hoboken: John Wiley and Sons, 2003:291-316.
33. Paez JG, Janne PA, Lee JC et al. EGFR mutations in lung cancer: Correlation with clinical response to gefitinib therapy. Science 2004; 304(5676):1497-1500.
34. Rothstein MA, Epps PG. Pharmacogenomics and the (ir)relevance of race. Pharmacogenomics J 2001; 1(2):104-108.
35. Taylor AL, Ziesche S, Yancy C et al. Combination of isosorbide dinitrate and hydralazine in blacks with heart failure. N Engl J Med 2004; 351(20):2049.
36. Seguin B, State L, Singer PA et al. Scientific diasporas as an option for brain drain: Recirculating knowledge for development. International Journal of Biotechnology 2006; 8(1/2):78-90.
37. Seguin B, Singer PA, Daar AS. Scientific diasporas. Science 2006; 312:1602-1603.
38. Bloom BR, Trach DD. Genetics and developing countries. BMJ 2001; 322(7293):1006-1007.
39. Mashelkar RA. 10th Zuckerman Lecture: Nation Building through Science and Technology: A Developing World Perspective. 10th Zuckerman Lecture ed. London: Royal Society, 2003:1-39.
40. Genetics, genomics and the patenting of DNA. Geneva: World Health Organization: Human Genetics Programme. Chronic Diseases and Health Promotion, 2005.
41. Jimenez-Sanchez G. Developing a platform for genomic medicine in Mexico. Science 2003; 300:295-296.
42. Normile D. Consortium hopes to map human history in Asia. Science 2004; 306.
43. IGV. The Indian Genome Variation database (IGVdb): A project overview. Hum Genet 2005; 118:1-11.
44. Boutayeb A. The double burden of communicable and noncommunicable diseases in developing countries. Trans R Soc Trop Med Hyg 2006; 100(3):191-199.
45. Drews J. Drug discovery: A historical perspective. Science 2000; 287(5460):1960-1964.
46. Nunnally AC, Brown SA, Cohen GA. Intellectual property and commercial aspects of pharmacogenomics. In: Rothstein MA, ed. Pharmacogenomics: Social, Ethical, and Clinical Dimensions. Hoboken: John Wiley and Sons, 2003:109-133.
47. Marra CA, Esdaile JM, Anis AH. Practical pharmacogenetics: The cost effectiveness of screening for thiopurine s-methyltransferase polymorphisms in patients with rheumatological conditions treated with azathioprine. J Rheumatol 2002; 29(12):2507-2512.
48. Hughes D, Vilar F, Ward CC. Cost-effectiveness analysis of HLA B*5701 genotyping in preventing abacavir hypersensitivity. Pharmacogenetics 2004; 14(6):335-342.
49. You JH, Chan FW, Wong RSM et al. The potential clinical and economic outcomes of pharmacogenetics-oriented management of warfarin therapy - A decision analysis. Thromb Haemost 2004; 92(3):590-597.
50. PricewaterhouseCoopers. Personalized medicine: The emerging pharmacogenomics revolution: PricewaterhouseCoopers, 2005.
51. Simon JHM, Claassen E, Correa CE et al. Managing severe acute respiratory syndrome (SARS) intellectual property rights: The possible role of patent pooling. Bull World Health Org 2005; 83:707-710.
52. Carroll AE. A review of recent decisions of the United States Court of Appeals for the Federal Circuit: Comment: Not always the best medicine: Biotechnology and the global impact of U.S. Patent Law. Am Univ Law Rev 1995; 44:2433-2493.

CHAPTER 14

Synopsis and Perspectives

Guilherme Suarez-Kurtz*

A mi edad y con tantas sangres cruzadas, ya no sé a ciencia cierta de dónde soy...
Nadie lo sabe por estos reinos y creo que necesitarán siglos para saberlo.[a]
—Gabriel García Márquez, 1994, Del Amor y otros Demonios

Introduction

"Pharmacogenetics deals with pharmacological responses and their modification by hereditary influences". This definition, offered by Werner Kalow in the first book dedicated to pharmacogenetics,[1] highlights the three pillars of this discipline: pharmacology, genetics and human diversity. Pharmacogenetics has evolved greatly over the 50 years elapsed since Kalow's book was published, was rechristened as pharmacogenomics in the fashion of the "omics" revolution, but its conceptual development and praxis remain contingent upon a better understanding of human genomic diversity and its impact on drug pharmacokinetics and pharmacodynamics. The evolution and structure of human genetic diversity has been reviewed in this book by Sergio Pena (Chapter 2),[2] who presented three models: The first, essentially typological, is based on the partition of humanity into races, visualized as being very different from each other, but internally homogeneous; the second, proposes a division into (continental) populations rather than races; the third, labeled the variable mosaic genome-VMG-paradigm, stresses individuality rather than membership in population categories based on race, ethnicity or geographical origin.[2] I would suggest that most pharmacogenetics/-genomics (PGx) studies continue to be performed, and their data analyzed, reported and fed into databases, in consonance with either the first or the second models described by Pena.[2] A caveat with this approach is that either model is a poor descriptor of admixed populations—which are the focus of this book—since genetic admixture is best modeled as a continuous variable, consistent with the VGM paradigm.[2,3] The individual uniqueness that is central to this paradigm implies that "each person must be treated as an individual. rather than as an exemplar of a race"[4] or in the words of Howard McLeod (Chapter 4),[5] "data from ethnic groups will not be as useful as analysis of individuals patients. While knowledge of ethnic differences may be relevant to much of the world's population, its usefulness is significantly limited in situations of extensive genetic mixing". Rashmi Shah (Chapter 12)[6] predicts that "with increasing global migration and the resulting admixing of different ethnic populations, the challenge in the future will be even greater".

[a] At my age and with so much mixed blood, I no longer know for sure where I belong... Nobody knows it in these lands... and I believe that it would take centuries to know it.

*Guilherme Suarez-Kurtz—Coordenação de Pesquisa, Instituto Nacional de Câncer, Rio de Janeiro, RJ, Brazil. Email: kurtz@inca.gov.br

Pharmacogenomics in Admixed Populations, edited by Guilherme Suarez-Kurtz.

Population Stratification and Structure: Impact on PGx

There is ample documentation in various chapters of this book that allele, genotype or haplotype frequency of polymorphisms in "pharmacogenes" differ among populations categorized by "race", ethnicity, or continental origin. Such differences may be large (>50%) as is the case, for example, of the *CYP3A5*3* and *GSTM3*B* allele frequency between Europeans and sub-Saharan Africans.[7,8] However, only rarely, if ever, common (>10%) polymorphisms present in PGx databases segregate in one population. Although large scale resequencing studies of PGx candidate genes are more likely to identify low frequency polymorphisms that are "private" to a given population group,[5] the implications of these findings in the praxis of PGx are still debatable.

Asia

Most populations covered in this book are structured, i.e., they comprise sub-populations or strata defined according to a variety of criteria. Kumar and Adhitan (Chapter 8)[9] refer to the remarkable number of 4635 recognized ethnic communities in India, as a result of segregation, isolation and the strict practice of endogamy across social ranks. Nevertheless, most PGx studies undertaken to date in India have chosen their subjects based on geographical location, predominantly from the north Indian (Indo-European speakers) and south Indian (Dravidian speakers) populations. North and south Indian sub-groups differ substantially in the prevalence of some common polymorphisms (e.g., *NAT2*5* allele, 50% and 22%, respectively) but not others (e.g., *GSTM1-null, CYP2C9*3* and *CYP2C19**3). A remarkable difference (>10-fold) in frequency of the duplicated *CYP2D6* emerges from comparison of data for Indians from Malaysia (2%)[10] and Singapore (25%).[11] The stratified population of Singapore (77% Chinese, 14% Malay and 8% Indian) provides excellent opportunity for comparative inter-ethnic PGx studies, as reported by Su Pin Choo, Suman Lal and Balram Chowbay (Chapter 7).[11] Their data reveal variable patterns in the distribution of polymorphisms in pharmacogenes among the three population strata: for example, the *CYP3A5*3* allele is nearly twice as frequent in Indians and Malays, compared to Chinese, whereas the *GSTM1-null* genotype occurs in ~62% of the Chinese and Malays but in only 33% of the Indians. Interethnic differences in the prevalence of polymorphims among the peoples of Singapore may have therapeutic implications, as discussed by Choo et al[11] in relation to polymorphisms in the *SLCO1B1* gene that encodes the liver specific OATP1B1 drug transporter. The functional haplotype *SLCO1B1*15* is rare in Indians (~2%) but occurs in ca. 10% of both Chinese and Malays from Singapore. Therefore, the latter two groups are likely to be more susceptible to the impact of the *SLCO1B1*15* polymorphism on the pharmacokinetics of SLCO1B1 substrates, such as the antioneoplastic agent irinotecan. Accordingly, Hong-Hao Zhou and Wei Zhang (Chapter 9)[12] observed in native Chinese individuals that the 521T>C SNP, which defines the *SLCO1B1*5, *15* and **17* haplotypes, modulates the hepatic uptake of repaglinide, in a gene-dose dependent manner. China is a multinational country with 55 ethnic minorities besides the Han majority[12] and evidence for PGx diversity among these minorities is presented, e.g., the 16-fold variation in the frequency (3.2–50.6%) of the NAT2 slow acetylator phenotype among 17 Chinese minorities. Significant, but not quite as dramatic, is the difference in frequency of the defective *CYP2C19*2* allele in Chinese Bai subjects (25.7%) vs. the Dong nationality (39.9%) and the Han majority (33.6%). The high frequency of *CYP2C19*2* (plus the occurrence in 3–7% of Chinese of another defective allele, namely *CYP2C19**3) has been used advantageously by H-H Zhou and colleagues to explore the PGx of CYP2C19, a major pathway in the biotransformation of several therapeutic classes of drugs. Their data show that these alleles have a gene-dosage effect on CYP2C19-mediated metabolism (e.g., mephenytoin and diazepam) and on the inducibility of CYP2C19 by either rifampicin or the medicinal plant *Hypericum perforatum* (St John's wort). Also reviewed by Zhou and Zhang[12] are PGx data on drug transporters, drug receptors, and the drug binding plasma protein, alfa1-glycoprotein or orosomucoid (ORM1) in Chinese. The *ORM1* locus is highly polymorphic, displaying three codominant alleles, *ORM1*F1*, *ORM1*F2* and *ORM1*S* that differ in their affinity for basic drugs. *ORM1*S*, the allele with the largest number of drug binding sites, occurs at lower frequency among Chinese (27.5%) than in Caucasians (37.3 - 38.6%). This provides an explanation for the significantly higher plasma concentrations of unbound diphenydramine and propranolol in Chinese compared to Caucasians.[12]

Collectively, the above examples selected from the data reported in the chapters dedicated to the peoples of India,[9] Singapore[11] and China,[12] point to the remarkable impact that population heterogeneity and stratification in Asia—at the continental, national, regional and community levels—have on PGx targets. Nevertheless, many geographically Asian populations are often lumped together in one ethnic group termed as "Asians", thus obscuring the diversity of the peoples living in the largest continent of the world.

Africa

The second largest continent and the origin of the first human species which gave rise to the modern man, Africa hosts 10% of the world population and is home to more than 1500 languages that are though to represent as much genetic variability.[8] PGx studies often distinguish sub-Saharan populations (Negroids) from those inhabiting the north of Africa down to the border of the Sahara desert (Caucasoids).[8] However, the distribution of polymorphisms in pharmacogenes among the peoples of each of these two sub-groups is quite variable, as shown by Eleni Aklillu and coauthors (Chapter 2).[8] In some cases, a relatively homogenous pattern prevails, e.g., alleles *CYP2C19*3* and *CYP3A4*1B*, which are, respectively, rare (0 - 0.6%) or predominant (72 - 87%) in sub-Saharan countries. Not surprisingly, however, large differences in frequency of other *CYP* alleles—e.g., *CYP2D6*2* (11 - 44%), *CYP2D6*17* (17 - 34%) and *CYP3A5*3* (6 - 31%)—occur among sub-Saharan populations.[8] The duplicated *CYP2D6* gene polymorphism reveals interesting patterns of heterogeneity in Africans: first, its overall frequency is much higher in Ethiopians than in sub-Saharan peoples; second, in Ethiopians, gene duplication involves the functional allele variant *CYP2D6*2*, whereas in Zimbabweans it is the nonfunctional *CYP2D6*4* which is duplicated, and in Tanzanians both alleles **2* and **4* are duplicated, albeit at different frequencies. Aklillu et al[8] revisit their interesting observation that CYP2D6-mediated drug metabolism differs significantly between Ethiopians with the same genotype (whether duplicated *CYP2D6* genes or homozygous for *CYP2D6*1)* living in Ethiopia or in Sweden. This finding highlights the role of environmental factors modulating the clinical consequences of PGx polymorphisms in different populations.

Americas

Sub-Saharan Africans represent one of the three ancestral roots of the people living in the American continent; the other two are the native Amerindians and the European colonizers/immigrants. Five centuries of interethnic crosses between these groups resulted in the heterogeneous, admixed populations of the Americas. The history of admixture in the American continent is reviewed by Esteban Parra (Chapter 3)[13] and by myself and Sergio Pena (Chapter 6). The variable extent and dynamics of the admixture process and the prevailing social environment where this process developed translate into substantial differences in genetic makeup across the American continent, within individual countries and, importantly, within sub-groups categorized by phenotypic, "racial" or ethnic criteria. The availability of ancestry informative markers (AIMs) made it possible to estimate the relative proportion of the Amerindian, European and African ancestry to the individual genetic constitution of the population of the Americas. Parra[13] presents data showing that the percentage of European contribution to several African American communities within the continental US varies 10-fold, from 3.5% in the isolated Gullah-speaking Sea Islanders from South Carolina to 35% in Seattle. McLeod[5] refers to studies on CYP2D6 polymorphisms to illustrate the PGx counterpart of this diversity: the frequency of *CYP2D6* poor metabolizing alleles in the African American population ranges from 0–8%, depending on the region of the US in which the study was performed. The same heterogeneity is true for the frequency of the *CYP2D6* copy number variation, and "likely would be found for most any pharmacogenetic example".[5] By contrast, there is little diversity in genetic background among white US individuals. Indeed, the different population structures in white Americans versus African Americans has been recently described as a "dichotomy", such that 93% of self-reported whites have no African ancestry and 94% of self-reported blacks display a broad range of African ancestry, indicative of mixed race and ancestral origin.[15] However, such a dichotomy is not verified in other populations of the Americas, which nevertheless share the same ancestral roots of the US population, such as the present-day 188 million Brazilians. Data shown in Chapter 6 (Suarez-Kurtz and Pena)[7] for self-reported urban white and black Brazilians indicate that regardless of their skin color, most individuals have a significant degree of both African and European ancestry, and many have also substantial Amerindian ancestry. We have recently extended these observations to another group of 353 Brazilians from the city of Rio de Janeiro, self-reported as white, black or intermediate ("pardo" in Portuguese),[16] and used the data to construct the histograms of the frequency distribution of the individual African component of ancestry (ACA), shown in Figure 1. It is apparent that the ACA is well described by a normal distribution in each self-reported category in Brazilians. Clearly, the dichotomy observed between white and African Americans (see above)[15] does not apply to Brazilians, in whom the African and the European ancestries are distributed as a continuous variable, irrespective of self-reported "color" categories. This pattern might be a more a realistic model for Latin America and other regions of the world where there occurred extensive admixture from different ancestral roots.

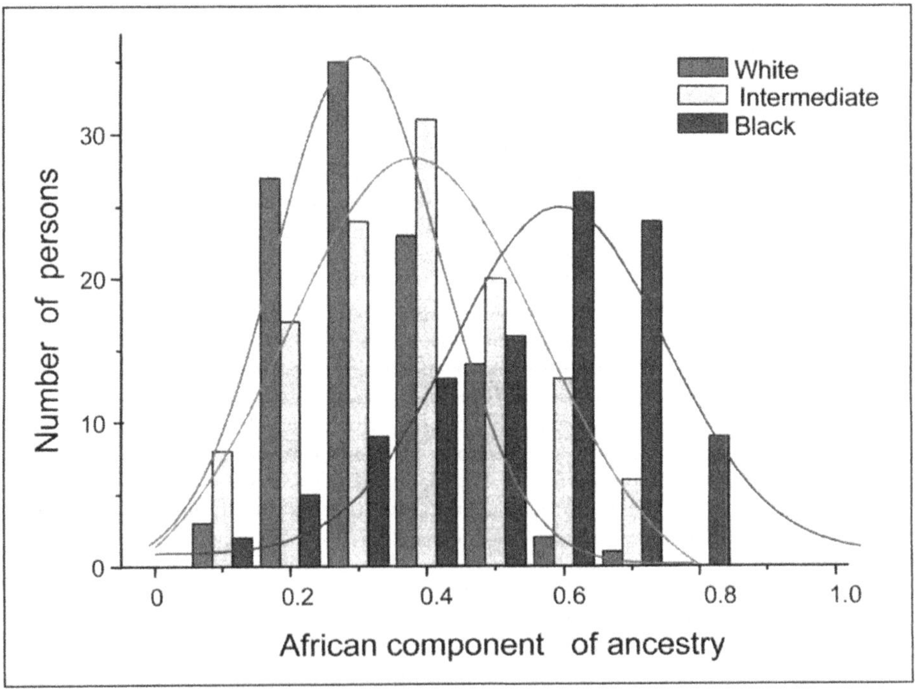

Figure 1. Frequency histogram showing the proportion of African ancestry in self-reported white, intermediate and black individuals living in Rio de Janeiro, Brazil.

Continuous distribution of African and European ancestries notwithstanding, self-reported white, intermediate or black Brazilians may differ significantly with respect to the distribution of polymorphisms in pharmacogenes, as we describe in Chapter 6.[7] In several cases, a significant trend for increasing (e.g., *GSTM3*B* and allele 825T in *GNB3)* or decreasing (e.g., *CYP3A5*3* and *GSTM1-null*) frequency from white, to intermediate to black individuals is detected. Nevertheless, an analysis based on non-linear logistic regression modeling[16] shows that the odds of having the *GSTM1-null* and the *GSTM3*B* polymorphisms is a continuous function of the individual ACA across all self-reported color categories. This pattern can hardly be reconciled with notions of "race-targeted" drug therapy (see below).

Among Hispanics, the proportions of African, Amerindian and European ancestry vary continuously, although significant east/west differences are discerned in the average proportions of the Amerindian (greater in the west) and the African (greater in the east) contributions.[13,14] The diversity of Hispanics with respect to individual ancestry, country of residence, cultural and socio-economic perspectives encompasses also PGx markers, as reviewed by Pedro Dorado, myself and Adrian Llerena (Chapter 5).[17] Most of the PGx data available derive from individuals of Mexican origin, including Mexican Mestizos and Mexican-Americans. These data display large (>12-fold) within-group variation in the frequency of relevant polymorphisms, such as allele *CYP2D6*10* (1–12.4%) and *CYP2D6* duplication (0.8–12.8%). Significant differences are also detected between the ranges reported for Mexican/Mexican Americans and other Hispanics in the frequency of functional *CYP* alleles, e.g., *CYP2D6*17* (0.2–2% in Mexicans vs. 6.4% in Cubans) and *CYP1A1*2C* (28 - 34% in Mexicans vs. 70% in Ecuadorians). The heterogeneity among Hispanics regarding PGx targets is not restricted to the CYP family of drug metabolizing enzymes, and may have important therapeutic consequences, as shown by the significant differences in β2-adrenergic receptor polymorphisms and bronchodilator responses to albuterol between Mexicans and Puerto Ricans asthma patients.[18] This observation highlights the hazards that are associated with ignoring stratification within "ethnic" groups, as is often done in the PGx literature.

The prevalence of polymorphisms in genes of pharmacological and toxicological relevance can vary over a large range (>10-fold) among the extant Amerindian groups living in Brazil (Chapter 6)[7] and in Spanish-speaking countries of the Americas (Chapter 5).[17] Some alleles are fixed in only one of the groups studied—e.g., *CYP1A1*2A* and **2C* in the Aché living in Paraguay—whereas other variants are absent in one group but present at relatively high frequency in others in the same geographical region—e.g., *GSTT1-null* in different Brazilian tribes. The very high levels of genetic drift characteristic of Amerindian populations is the most likely explanation for these dramatic differences.

Oceania

Admixture and founder effects determined the population structure of the peoples of Remote Oceania (aka Polynesia and Micronesia), of which the largest group are the Maori of New Zeland. Rod Lea and Geoff Chambers (Chapter 10)[19] revisit the process of the peopling of Oceania and estimate, on the basis of AIMs data, that between 40–50% of the Maori gene pool in the 21st Century is comprised of Caucasian genomic material of mainly United Kingdom origin. This is the result of intermarriage of indigenous Maori—whose ancestors are Austronesians from southeast Asia and Melanesians—with Europeans settlers in the last two hundred years. The impact of European admixture is evident in the frequency of the defective *ADH2*2* (called *ADH1B*47His*) allele in Maori (46%), midway between the average value in Europeans (4%) and the range reported for Austronesians (70–91%). Nevertheless, Lea and Chambers[19] emphasize that despite the extensive admixture in the Maori population, this indigenous group remains genetically distinguishable from the Caucasian population of New Zeland. Accordingly, the Maori differ markedly from Europeans in the haplotype signature at the *ADH* gene region as well as in the frequency distribution of polymorphisms in *CYP2A6, CYP2C9, CYP2C19* and *CYP2D6.* For some of the variant *CYP* alleles (e.g., *CYP2A6*4, CYP2C9*2* and *CYP2C19*2),* the frequency in Maori are similar to those reported for southeast Asians (see Chapters 7 and 9)[11,12] which is consistent with their Austronesian ancestry in southeast Asia.

Controlling the Impact of Admixture in PGx Studies

Genetic admixture creates population stratification, which if not controlled for, will affect gene mapping efforts and confound the association of other genetic and/or environmental factors with drug responses. This has been described as a "curse" for gene association studies requiring sifting through genetic noise, which might not be as necessary if non-admixed populations are used.[20] Eduardo Tarazona Santos, Sara Raimondi and Silvia Fuselli (Chapter 11)[21] thoroughly review the statistical approaches currently available to control for the confounding and modifying effects of population stratification and admixture on gene association studies. Three approaches are discussed: classical logistic regression, genomic controls and two recently developed structured association methods based on Bayesian inferences and computationally intensive use of Markov Chain Monte Carlo algorithms. These procedures rely on estimates of individual ancestry, and the proper choice of AIMs in order to avoid introducing bias in the results, is emphasized by Tarazona-Santos et al.[21] Their data demonstrate the inadequate use of general "ethnic" categories such as *admixed* (or, in the Americas, *Mestizo*) that have broad cultural and socio-economic basis, and do not necessarily reflect the genetic background of individuals or populations. Thus, in an urban sample from a shantytown in Lima, the capital of Peru, which was considered a priori as admixed or Mestizo, the genetic contribution of Amerindian ancestry (82%) proved to be higher than some populations traditionally categorized as Native American. This is reminiscent of the very high levels of African ancestry observed in some self-identified white Brazilians.

Global Perspectives

The challenges and opportunities associated with the implementation of PGx on a global scale are analyzed by Rashmi Shah (Chapter 12)[6] and by Béatrice Seguin and colleagues (Chapter 13).[22] A common concern is that therapeutic drugs are usually developed and investigated for their safety and efficacy in geographical and ethnical populations that may not comprise the ultimate targets of the drugs concerned. Indeed, most new drugs are predominantly tested in white males in industrialized countries and are subsequently licensed to other parts of the world, often with no bridging studies to verify their efficacy and toxicity in the concerned populations. Critical issues here are the clinical significance of inter-ethnic differences in drug response and the inherent difficulty in characterizing ethnicity, especially in admixed populations. Shah[6] summarizes the features of a medicinal product that may be indicative of greater or reduced susceptibility to ethnic factors and reviews the guidelines for collecting and reporting PGx data from clinical trials, issued by regulatory agencies from the United States (FDA), the European Union (CPMP) and Japan (PMDA), as well as by the International Conference on Harmonization (ICH). His analysis indicates that diversity in drug response across populations/ethnicities "(pharmacoanthropology" or "ethnopharmacogenetics") will continue to receive much greater regulatory attention than it has hitherto, with inevitable consequences on how drugs are developed, evaluated, approved, promoted and ultimately prescribed. Séguin et al[22] start from the premise that adoption of pharmacogenomics principles to improve drug therapy is indeed inevitable, at least in industrialized countries, and focus on the potential opportunities offered by PGx for developing countries. A major concern is to avoid a "pharmacogenomics divide", such that the advances in PGx would fail to benefit individuals living in the developing world. Drivers and barriers to the adoption of PGx in developing countries, and specific ways in which these countries could benefit from PGx-based drug therapy are discussed. Séguin et al[22] view the pharmaceutical companies in the developing world as major partners in this process, and emphasize the importance of regulatory frameworks and of intellectual property protection. The use of ethnicity or geographical ancestry as a tool to understand drug response variability and to capitalize on it for the purpose of improving public health in developing countries[22] although controversial[23,24] should gain novel insights from initiatives aimed at characterizing PGx profiles in different countries (Brazil, India, Mexico, Nepal, South Africa and Thailand).

Final Considerations: PGx in Admixed Populations

PGx studies in admixed populations face several challenges, such as the allocation of the enrolled subjects among impervious ethnic categories, the inherent confounding effect of population stratification, and the risk of epistasis (i.e., genetic variants that influence response to a drug in one group might not have the same effect in another group because of different gene-gene or gene-environmental interactions). By contrast, admixed populations display characteristics which might be advantageous for PGx research. I have recently discussed this issue[20] and will only mention three aspects here. Firstly, population admixture results in longer linkage disequilibrium blocks than in the previously isolated populations, allowing for a smaller number of markers for gene association studies. Second, it is possible to use admixed populations to gather information on peoples that are excluded or under-represented in clinical drug trials. For example, studies in admixed Brazilian and Hispanic populations could fill in PGx information gaps, pertinent to Africans and Native Americans. Third and most important for drug development programs, admixed populations, such as the tri-hybrid Brazilians, provide unique opportunities for gene association studies in individuals of different ancestry under identical, or very similar environmental and social-economic conditions.

Recognition of interethnic differences in drug response might be useful in the establishment of public health policies, the design and interpretation of clinical trials and possibly to help guiding clinicians to prospectively evaluate those patients with the greatest probability of expressing a variant genotype. A practical and highly controversial example of these possibilities is the approval by the US Food and Drug Administration of the drug BiDil® for use exclusively in African-Americans. The appropriateness of this decision—based according to one critic on "false promises, faulty statistics and reasoning"[25]—is discussed in different chapters of this book.[2,5,6,22] However, personalized drug therapy, the promise of pharmacogenomics, must be based on the recognition of the inherent genetic individuality. This notion is particularly relevant to admixed populations, in which substructure increases further the fluidity of racial/ethnic labels. Because interethnic admixture is either common or increasing at a fast pace in many, if not most populations, extrapolation on a global scale of PGx data from well-defined ethnic groups is plagued with uncertainty. To impact positively on global health, PGx must broaden its scope, with respect to both target and population diversity, and be inclusive of admixed populations with their perceived challenges and advantages. This goal is not likely to be achieved simply by mandates to include subjects from ethnic minorities in clinical drug trials, especially when these groups are represented in relatively small numbers and are labeled by phenotypes which do not accurately reflect genetic ancestry.

Acknowledgements

Grant support was provided by Conselho Nacional de Desenvolvimento Científico e Tecnológico (CNPq), Fundação de Amparo à Pesquisa do Estado do Rio de Janeiro, and Swiss Bridge Foundation.

References

1. Kalow W. Pharmacogenetics: Heredity and the response to drugs. Philadelphia: W.B. Saunders, 1962.
2. Pena SDJ. The evolution and structure of human genetic diversity. In: Suarez-Kurtz G, ed. Pharmacogenomics in Admixed Populations. Austin: Landes Bioscience, 2007.
3. Suarez-Kurtz G, Pena SDJ. Pharmacogenomics in the Americas: Impact of genetic admixture. Curr Drug Targets 2006; 7:1649-1658.
4. Patrinos A. "Race" and the human genome. Nat Genet 2004; 36(Suppl):S1-2.
5. McLeod HL. Pharmacogenetics in the African American population. In: Suarez-Kurtz G, ed. Pharmacogenomics in Admixed Populations. Austin: Landes Bioscience, 2007.
6. Shah RR. Pharmacogenetics, ethnic differences in drug response and drug regulation. In: Suarez-Kurtz G, ed. Pharmacogenomics in Admixed Populations. Austin: Landes Bioscience, 2007.
7. Suarez-Kurtz G, Pena SDJ. Pharmacogenetic studies in the Brazilian populations. In: Suarez-Kurtz G, ed. Pharmacogenomics in Admixed Populations. Austin: Landes Bioscience, 2007.

8. Akillu E, Dandara C, Bertilsson L et al. Pharmacogenetics of cytochrome P450s in African populations: Clinical and molecular evolutionary implications. In: Suarez-Kurtz G, ed. Pharmacogenomics in Admixed Populations. Austin: Landes Bioscience, 2007.
9. Kumar MR, Adithan C. Pharmacogenomics in the Indian population. In: Suarez-Kurtz G, ed. Pharmacogenomics in Admixed Populations. Austin: Landes Bioscience, 2007.
10. Ismail R, Teh LK. Genetic polymorphism of CYP2D6: Malaysian Indians have the highest frequency for CYP2D6*4 in Asia. Eur J Clin Pharmacol 2001; 57:617-618.
11. Choo SP, Lal S, Chowbay B. Pharmacogenetics and ethnicity: An Asian perspective. In: Suarez-Kurtz G, ed. Pharmacogenomics in Admixed Populations. Austin: Landes Bioscience, 2007.
12. Zhou HH, Zhang W. Pharmacogenetics in Chinese population. In: Suarez-Kurtz G, ed. Pharmacogenomics in Admixed Populations. Austin: Landes Bioscience, 2007.
13. Parra EJ. Admixture in North America. In: Suarez-Kurtz G, ed. Pharmacogenomics in Admixed Populations. Austin: Landes Bioscience, 2007.
14. Parra EJ, Kittles RA, Shriver MD. Implications of correlations between skin color and genetic ancestry for biomedical research. Nat Genet 2004; 36(Suppl):S54-60.
15. Sinha M, Larkin EK, Elston RC et al. Self-reported race and genetic admixture. N Engl J Med 2006; 354:421-422.
16. Suarez-Kurtz G, Vargens DD, Struchiner CJ et al. Skin color, genomic ancestry and the distribution of GST polymorphisms. Pharacogenetics/genomics 2007; (in press).
17. Dorado P, Suarez-Kurtz G, Llerena A. Pharmacogenetics of cytochrome P450 in Hispanic populations. In: Suarez-Kurtz G, ed. Pharmacogenomics in Admixed Populations. Austin: Landes Bioscience, 2007.
18. Choudhry S, Ung N, Avila PC et al. Pharmacogenetic differences in response to albuterol between Puerto Ricans and Mexicans with asthma. Am J Respir Crit Care Med 2005; 171:563-570.
19. Lea R, Chambers G. Pharmacogenetics in admixed Polynesian populations. In: Suarez-Kurtz G, ed. Pharmacogenomics in Admixed Populations. Austin: Landes Bioscience, 2007.
20. Suarez-Kurtz G. Pharmacogenetics in admixed populations. Trends Pharmacol Sci 2005; 26:196-201.
21. Tarazona-Santos E, Raimondi S, Fuselli S. Controlling the effects of population stratification by admixture in pharmacogenetics. In: Suarez-Kurtz G, ed. Pharmacogenomics in Admixed Populations. Austin: Landes Bioscience, 2007.
22. Séguin B, Essajee S, Jimenez-Sanchez G et al. Human genomic variation studies and pharmacogenomics are critical for global health. In: Suarez-Kurtz G, ed. Pharmacogenomics in Admixed Populations. Austin: Landes Bioscience, 2007.
23. Suarez-Kurtz G. Letter to the editor re. Pharmacogenetics, pharmacogenomics and population admixture: Implications for drug development and prescription. Rev Genet 2005, (doi:10.1038/nrg1559-c1).
24. Daar AS, Singer P. Reply re. Pharmacogenetics, pharmacogenomics and population admixture: Implications for drug development and prescription. Rev Genet 2005, (online: 1; doi:10.1038/nrg1559-c1).
25. Kahn J. BiDil: False promises: Faulty statistics and reasoning have lead to the first "racial medicine". Genewatch 2005; 18:6-9.

Index

D

E

F

G

H

I

M

N

O

P

Q

R

S

T

U

V

W

Y

For Product Safety Concerns and Information please contact our EU representative GPSR@taylorandfrancis.com Taylor & Francis Verlag GmbH, Kaufingerstraße 24, 80331 München, Germany

Printed and bound by CPI Group (UK) Ltd, Croydon, CR0 4YY
14/09/2026
02216223-0012